ENVIRONMENTAL CHEMISTRY:
ASIAN LESSONS

Environmental Chemistry: Asian Lessons

by

Vladimir N. Bashkin

Moscow State University

KLUWER ACADEMIC PUBLISHERS

DORDRECHT / BOSTON / LONDON

A C.I.P. Catalogue record for this book is available from the Library of Congress.

ISBN 1-4020-1003-6 (HB)
ISBN 1-4020-1004-4 (PB)

Published by Kluwer Academic Publishers,
P.O. Box 17, 3300 AA Dordrecht, The Netherlands.

Sold and distributed in North, Central and South America
by Kluwer Academic Publishers,
101 Philip Drive, Norwell, MA 02061, U.S.A.

In all other countries, sold and distributed
by Kluwer Academic Publishers,
P.O. Box 322, 3300 AH Dordrecht, The Netherlands.

Printed on acid-free paper

Printed in the Netherlands

TABLE OF CONTENTS

PREFACE

Environmental chemistry is becoming an increasingly popular subject in both undergraduate and graduate education in all Asian countries. Courses in ecology, geography, biology, chemistry, environmental science, public health, and environmental engineering all have to include environmental chemistry in their syllabuses to a greater or lesser extent. Humanity's ever growing impact on the Environment, and the consequent local, regional, and global effects demand a profound understanding of the mechanisms underlying the impacts on human health and sustainability of the biosphere.

Many textbooks have appeared in recent years aiming to fulfill these requirements; however, most of these deal mainly with examples from the developed countries of North America and Europe. Taking into account the geographical boundaries of environmental pollution, which is especially pronounced in Asia, and the specific peculiarities of pollution in developing countries, this course aims to close this gap by providing regionally oriented knowledge in basic and applied environmental chemistry.

The rapid growth in energy production, industrialization and urbanization and a significant increase in agricultural production in Asia have caused a remarkable increase in pollutant (SO2, NOx, heavy metals, persistent organic compounds, etc) emissions during the past several decades. The consequences of these growing emissions have been closely connected with enforced pollution loading on human and ecosystems health, which results in actual and potential risk for many sensitive individuals and receptors.

Many environmental problems in Asia are closely connected with increasing energy demand and production. Air pollution owing to energy production is an emerging important environmental problem in Asia as a whole and especially in East and Southeast Asia. Ever growing emissions of sulfur dioxide, nitrogen oxides, ammonia, and ground-level ozone concentrations in many cities in this region, and the potential synergetic effects of this pollution, have caused considerable concern to the local populations. The haze pollution resulting from forest fires enhances these synergetic interactions.

Another large set of environmental problems is related to waste production and utilization. All Asian countries are touched by these consequences of urbanization, industrialization and agricultural development, and there are many specific regional peculiarities in how this problem is solved regarding the air, soil, fresh and marine water pollution, human and ecosystem health.

Thus, the requirement for Asian textbook in environmental chemistry is obvious. I have spent more than three years in Asian Universities (Seoul National University and King Mongkut's University of Technology Thonburi, Bangkok) teaching environmental classes for upper undergraduate and graduate students. Moreover, I have delivered similar lectures in various US (Cornell) and European (Lund, Kiel, Moscow) Universities and this gave me the basis for making comparisons between both regional problems in environmental pollution and chemistry and students' abilities to understand these problems. My experience showed me that students are much more interested in local examples, which are very numerous in the Asian region and which are of much more personal concern. For instance, accumulation of PCB and HM in many soils of Korea and Thailand, terrestrial and marine water pollution in many countries, waste problems, urban air quality, greenhouse gases, etc.

The above-mentioned peculiarities were been taken into account as I thought about and selected topic for this textbook. Not all required topics have been elucidated to their full extent; preference has been given to the first priority pollutants. For example, heavy metal pollution, chloro-organics, nitrogen, acid rain are of the first concern, and some other topics are of the second or even third priority, let us say, radioactive pollution. Some urgent problems, like soil pollution, are of special concern, and this important topic is directly considered in Chapter 7 and indirectly in many other chapters devoted to specific pollutants (Chapters 1, 12, 13, 14, 15, and in some others where it is related). A similar approach is used for the description of air and water pollution in the Asian region.

In general, this textbook progresses towards a critical analysis of the data and how they can be used to solve environmental problems. For example, in Chapter 2 sufficient attention is paid to both sources of GHG pollution, present trends, basic chemistry and physics, and implementation of the international conventions and protocols for abatement strategies. Less attention has been paid to technological approaches, which in general lie outside the scope of this book.

Finally, I would like to thank all my students in various universities where I have delivered classes for their great interest in environmental chemistry. My special thanks are due to my colleagues and academics in various Asian (and other) universities who have made numerous valuable comments and remarks about the draft of this book.

Vladimir Bashkin,
Professor,
Moscow State University, Russia
and
Joint Graduate School of Energy & Environment,
King Mongkut's University of Technology Thonbury,
Bangkok, Thailand

PART I

ENERGY AND ATMOSPHERIC POLLUTION

CHAPTER 1

SOURCES OF ENVIRONMENTAL POLLUTION IN ASIA

1. INTRODUCTION

In recent years three factors have contributed most directly to the excessive pressure on the environment and natural resources in Asia: a doubling of the region's population in the last 35 years; a tripling of the regional economic output in the last 20 years; and the persistence of poverty. The environmental impact of ever increasing population has been exacerbated by the growing inequity between rich and poor. The comparatively wealthy because of their affluence have access to lifestyles which demand large sources of energy, raw materials and manufactured goods. The poor are deprived of basic resources and forced to make unsustainable use of natural resources available to them immediately to meet their survival needs.

The rapid industrialization, remarkable economic achievements and population growth of today have continued into the 21st century.

Asia is a region of sharp contrasts. It has the world's most populous countries, the People's Republic of China and India, and some of the world's smallest countries. Similarly the region includes some of the least developed countries such as Bangladesh, but also Japan, which is one of the world's richest nations in term of Gross Domestic Product (GDP) per capita. The physical diversity of the region is evident: it includes the highest mountain and the deepest sea, the rainiest areas and driest deserts with their associated variations in ecosystem diversity, culture and living standards.

This diversity in demography, economic conditions, physical size, resources endowment, and ecology presents numerous challenges for accounting anthropogenic and natural sources of environmental pollution in Asia.

2. STATIONARY EMISSION SOURCES

In Asia, stationary sources of pollution include industrial sector, mining and construction. All these sources result from the rapid development in many Asian countries and they present an abundant amount of numerous pollutants.

The industrial sector is one of the most dynamic sectors of the economy and plays an essential role in economic development. If environmental considerations are not effectively integrated into the design of industrial processes, severe problems can result.

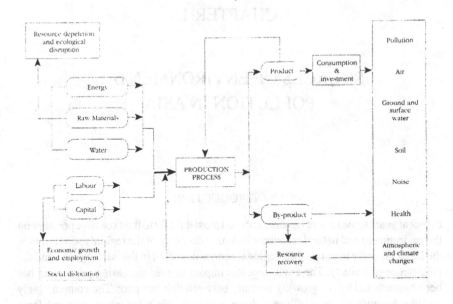

Figure 1. Industrial production flow chart in Asian countries (ESCAP, 1995).

Industry consumes 37% of the energy and emits 50% of the carbon dioxide, 90% of the sulfur dioxide and nearly all the toxic chemicals. Concentrations of heavy industry have caused severe local environmental damage and global environmental threats such as Global Warming (see Chapter 2) and Stratospheric Ozone Depletion (see Chapter 6). Whilst during recent years industrial expansion has always been regarded as essential to economic development in Asia, resource conservation and pollution control have been regarded as a secondary consideration, or even as an actual constraint on wealth creation. Only recently the severity of some local impacts of industry and the high cost of soil remediation (see Chapter 7) once environmental damage has occurred has led to dramatic changes in these attitudes.

2.1. Manufacturing

An industrial process basically involves extraction and exploitation of materials from the natural resource base and creation of useful products. Figure 1 illustrates activities extending over the entire chain of events, from raw material extraction or pre-processing, through the manufacturing processes themselves, right up to disposal of the discarded products. At each stage in the process there will be wastes, which will enter into the environment as pollution.

The environmental effects of the manufacturing industry fall into two broad categories:

- The depletion of non-renewable natural resources such as fossil fuel (industry is the major consumer of energy), minerals and construction materials; and

- The emission of wastes and polluting substances to air, water and land.

The extent of the potential and actual damage by an industrial installation depends on its size and nature and, to some degree, on the technology of production. A coincident review of the impacts of manufacturing processes on the various components of the physical and human environment is presented in Table 1.

2.2. Mining

Major users of non-renewable resources in the Asian region are the energy sector (see below) and the mining sector, which covers an enormous variety of activities and potential environmental effects. Mineral extraction and, to a lesser degree, mineral processing are almost always environmentally harmful, the problems varying greatly with geographic and climate conditions, as well as with the type of mining operation and the environmental protection measures in place. Box 1 describes some of the adverse effects of lignite mining in Northern Thailand.

Box 1: Lignite mining activities in Mae Moh valley, North Thailand (Bashkin et al, 2000)

The Mae Moh basin is located in the Lampang Province, approximately 500 km NNE of Bangkok and 125 km SE of Chang Mai. In the Mae Moh basin there is an estimated 1.4×10^9 tons of Thai lignite. The Mae Moh mine is the largest lignite mine in Thailand, with reserves of 1.46 million tons. It started operation in 1955 and at present can supply a maximum 100,000 tons of lignite daily. The mining operation is an open pit and divided into two activities: overburden removal and lignite mining. Overburden removal activities include excavation of topsoil and hard materials in the mining area. The uncovered lignite is drilled, blasted, and then excavated and loaded into rear dump trucks, which carry it to the crushers. The crushed lignite is then transferred to the lignite stockyard by three sets of conveyor belts.

The major sources of particulate matter in the Mae Moh area are reported to be the power plant, coal storage, coal mine (Table 2), and road dust.

The Mae Moh area is now facing serious and worsening problems of air pollution from TSP and PM_{10}. At many monitoring sites the measured concentrations of both greatly exceed health-based norms for a 24 hour average and the annual average. The most exceedances occur between January and March, during the dry season.

In all cases, considerable care is needed in controlling waste water from ore processing and rain water runoff contaminated with toxic heavy metals. Unregulated discharge of contaminated water, particularly into sluggish rivers, can be disastrous. Non-ferrous metals, in spite of their relatively low volume of production,

Table 1. Environmental effects of selected industrial sectors.

Impacts	Industry				
	Textiles	Iron & Steel	Chemicals	Electronics	Biotechnology
Raw material used	Wool synthetic fibers, chemicals for treating	Iron ore, limestone, recycled scrap	Inorganic and organic chemicals	Chemical substances, e.g., solvents and acids	
Air	Particulate, odors, SO_2, HC	Major polluter: SO_2, HC, CO, hydrogen sulfide, acid mist	Major polluter: organic chemicals (benzene, toluene), odors, CFCs	Toxic gases	Biochemical pollution
Water resources quantity	Process water	Process water	Process water		Process water
Water resource quality/ effluents	BOD, suspended solids, salts, toxic metals, sulfates	BOD, suspended solids, oil, metals, acids, phenol, sulfides, sulfates, ammonia, cyanides, effluents from wet gas scrubbers	Organic chemicals, heavy metals, suspended solids, COD, cyanide	Contamination of water by toxic chemicals, accidental spillage of toxic material	Used effluent
Solid wastes and soil	Sludges from effluent treatment	Slag, wastes from finishing operations, sludges from effluent treatment	Major polluter: sludges from air and water pollution treatment, chemical process wastes	Sludge, contamination of soil	Positive use for cleanup of contaminated land
Risk of accidents		Risk of explosions and fires	Risk of explosions, fires and spills		
Others: noises, workers health and safety	Noise from machines, inhalations of dust	Accidents, exposure to toxic substances and dust, noise	Exposure to toxic substances, potentially hazardous products	Risk of exposure to toxic substances due to spills/ leaks	Fears of hazard from the release of modified genetic species

Table 2. Emission of PM from coal mining.

Sources	Row matter, ton/day	Rate of PM kg/ton	PM, ton/day
Soil	74,000	0.075	55.5
STB	740,000	0.25	185
Soil pile	740,000	0.075	117.2
CD	100,000	0.075	7.5
CC	100,000	0.65	154.0
CTB	100,000	0.025	2.5
Coal pile	100,000	0.075	15.9
Total			556.6

STB—soil transmitting belt; CD—coal digging; CC—coal crushing;
CTB—coal transmitting belt.

pose a significant threat because they require hazardous processes and create toxic byproducts during mining and refining (see Chapter 12 for more detailed description).

Additionally, the mining of low-grade ores, especially in aquatic environments, requires that extremely large quantities of rock and overburden are removed and processed. Furthermore, potentially toxic chemicals, as well as inorganic reagents (zinc and copper sulfate, sodium cyanide and sodium dichromate), are used in processing certain ores. The tailings left by the metal extraction contain residues of both the minerals and flotation agents.

Some ores contain considerable quantities of sulfides, which transform in the environment by oxidizing processes into sulfuric acid. Other problems associated with mining operations, besides physical change, are noise, dust and high radon levels. Workers' health is a serious consideration in all mining operations.

2.3. Construction

The construction industry produces key infrastructure components essential for economic development. In recent years the Asian region has experienced high pressure on its existing infrastructure leading to rapid growth of construction of transport systems, utilities and office developments.

Although the extent of adverse environmental impacts from the construction sector is not so high as in the case of the other industrial sectors, major impacts do occur including:

- Water, air, and land pollution, and ecological destruction from industries producing construction materials (e.g., bricks, cement, asbestos);

- The extraction of construction materials (building stone, marble, lime) from geological deposits, disrupting the physical forms of landscapes and destroying land cover while also poisoning water and emitting dust into the atmosphere;

- Dust emission during construction activities, land vibration and production of construction waste and rubble;

- Large-scale construction industry often creates risk of accidents and health hazards.

2.4. Energy

Energy is one of the crucial determinants of development. The amount of energy used by a country is, in general, closely linked to its output of goods and services. At the same time, of all human activities, the production and use of energy has perhaps the largest impact on the environment. Thus the need to use energy, whilst minimizing the adverse environmental effects associated with such use, is a major challenge that faces countries at all levels of development.

During the mid-1970s to mid-1980s, concerns over energy were related to its availability, resulting in a significant growth in prices. Currently, with energy in abundant supply, concerns have shifted to consideration of the environmental impacts of energy production and use in many Asian countries.

Rapid economic growth in the Asian region during the past few decades has been accompanied by a steep rise in energy use, leading to increased environmental impacts. The close links between the 3 'E's (economic growth, energy, environment) are well recognized. For example, the use of energy has several adverse environmental impacts such as deterioration of air quality (see Chapter 3) and acid deposition (see Chapter 4). At the same time, environmental change, such as replacement of forest by agricultural land or urban areas, effects the availability of energy, in this case firewood, which is the principal source of energy supply in rural areas for most developing Asian countries.

Each society's choice of energy sources depends on a range of factors such as availability, convenience, price, nature of the risk, and more recently, potential environmental impact. No energy source, not even solar or wind, is benign from an environmental perspective, as each step in the energy life cycle—exploration, production, transportation, conversion, and end use—has environmental effects associated with it. Therefore, in order to estimate the environmental impacts of a particular energy source (fossil fuel, nuclear, renewables, etc.), it is essential to analyze the impacts from all parts of the full life cycle.

A range of potential environmental impacts is associated with the energy system in general, including:

- *Air Pollution*: Energy conversion is the largest anthropogenic source of air pollution (fossil fuel and biomass burning). Altogether it is estimated that energy activities contribute about 90% of the total sulfur dioxide and lead, and 85% of nitrogen oxides, 55–80% of the carbon dioxide, and 30–50% of carbon

Table 3. Importance of energy activities in the generation of air pollutants (ESCAP, 1995).

Pollutants	Man-made as per cent of total	Energy activities		Contributions as per cent of energy-related releases
		As per cent of total	As per cent of man-made	
SO_2	45^3	40^3	90^3	Coal combustion: 80; Oil combustion: 20
CO_2	4^3	$2.2–3.3^3$	$55–80^3$	Natural gas: 15; Oil: 45; Solid fuel: 40
CO	50^3	$15–25^3$	$30–50^3$	Transport: 75; Stationary sources: 25
VOC	5^1	2.8^3	55^1	Oil industry:15; Gas industry:10; Mobile sources:75
SPM	11.4^3	4.5^3	40^1	Transport: 17; Electric utilities: 5; Wood combustion: 12
NO_x	75^3	64^3	85^3	Transport:51; Stationary sources:49
N_2O	$37–58^3$	$24–43^3$	$65–75^3$	Fossil fuel combustion: 60–75; Biomass burning: 25–40
CH_4	60^3	$9–24^3$	$15–40^3$	Natural gas losses:20–40; Biomass burning:30–50
Pb	100^3	90^3	90^3	Transport:80; Stationary combustion:20
Radionuclides	10^3	2.5^3	25^3	Mining, milling of uranium:25; Nuclear power stations and coal combustion:75

Notes: [1] Estimates for OECD countries only;
[2] Estimates for USA only;
[3] Global estimates.

monoxide emitted into the atmosphere from man-made sources (ESCAP, 1995). The relative importance of energy activities in the generation of air pollutants is illustrated in Table 3. See also Chapter 12 for estimation of HM emission from lignite-burning power plants.

- Water pollution: The energy system also contributes to water pollution (e.g., oil and coal runoff, heat from power plants and heavy metals from disposal of combustion ashes).

- Solid Waste: Large quantities of solid waste are generated from the operation of fossil fuel plants. These wastes, predominantly combustion ashes, contain heavy metals, which can leach out of the ash and cause pollution if not disposed of in proper landfill facilities. In the case of nuclear power, the disposal of waste poses significant health risks.

- Noise: The noise from the operation of fossil fuel and nuclear power stations can be considerable.

- Radiation: Energy conversion is the largest anthropogenic source of radiation through the disposal of wastes from nuclear stations and the mining of uranium (see Table 3).

Of all the regions of the world, the Asian region is the one in which energy use has been increasing the most rapidly. This increase reflects the region's dynamic economic growth, given the recognized close positive correlation between energy use and measures of economic activity such as the Gross Domestic Product (GDP). Accordingly, the Asian region now consumes about a quarter of the world's energy, from traditional fuels such as firewood to advanced sources such as nuclear fission reactors, liquefied natural gas, and photovoltaic conversion of solar energy. All the major energy sources such as coal, oil, natural gas, hydropower, nuclear power, firewood, and other biomass play an important role in at least some of the countries of the region. In addition, renewable energy sources, such as solar, wind, and geothermal energy, are becoming increasingly important in the more developed countries of the region. Geothermal power, for example, plays an important role in energy supply in the Philippines, although its overall contribution to energy supply in the Asian region is small.

3. EMISSION OF POLLUTANTS IN VARIOUS ASIAN COUNTRIES

The rapid growth of industrialization and urbanization and a significant increase in agricultural production in East Asia have caused a remarkable increase in SO_2 and NO_x emissions during the past several decades. The consequences of these growing emissions have been closely connected with enforced acidification loading on ecosystems and that results in actual and potential risk to many sensitive ecosystems.

Air pollution is an emerging important environmental problem in Asia as a whole, and especially in East Asia. Ever growing emissions of sulfur dioxide, nitrogen oxides, ammonia and ground level ozone concentration in many big cities in this region make the people worry about the synergetic effects of various pollutants.

Emission projections indicate that potentially large increases in emissions may occur during the next 10–25 years in Asia, especially in China, Thailand, India, etc, in accordance with the planned development programs. If this occurs (the probability of which is very high), the negative impacts which have been experienced in Europe and North America (USA and Canada) during the last centuries will become increasingly apparent in large parts of Asia within the next 20–30 years. According to the

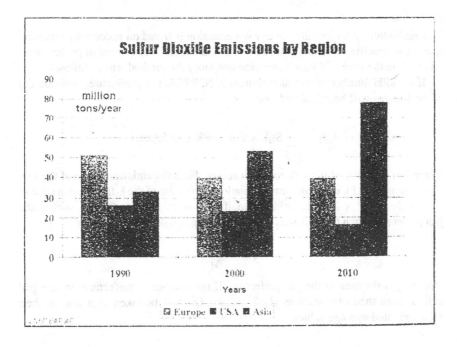

Figure 2. World sulfur dioxide emission by regions (ESCAP, 2000).

present estimates SO_2 emission in Asia will surpass the combined emissions of North America and Europe by the year 2010 if the current growth rate of energy consumption continues (Figure 2).

The primary man-made source of sulfur and nitrogen acidifying compounds in Asia is low quality fossil fuel with high content of sulfur (up to 7% in Thai lignite, Chinese and Indian brown coal, etc.) and nitrogen (heavy oil and gas).

3.1. SO_2 Emission Inventory in China

The eastern coast of China is the highest emission region in the Asian domain and possibly, with sharply decreasing sulfur dioxide emissions in Central Europe and East USA, in the whole world. This emission coincides with both energy production and population growth in this part of China. The other parts of China, such as the western part, are not so abundant in population and industrial sources as the eastern coast and these differences point out the spatial heterogeneity in emission of sulfur dioxide. As an example we can cite estimates of spatial distribution of SO_2 emission in China from 1990 until 1995 with $1° \times 1°$ Longitude-Latitude (LoLa) resolution (Bai Naibin, 1997).

Methodology

The methodology for calculating any gas emission is based on accounting emission sources in specified regions, types of fuel consumed and the content of pollutants in this fuel. In the case of China's emission inventory the method was as follows.

If the total number of emission sources is N, for the j-th prefecture, emission Qij of the I-th gas will be calculated based on

$$Qij = \sum_{k=1}^{N} Sijk \times Fijk \times PK \ (Tg \ S/year),$$

where k is intensity of the k-th emission source, Fk is the emission factor of the k-th emission source, Pk is the unit transformation coefficient of the k-th emission source.

For the 1st $1° \times 1°$ LoLa grid square, if there is only one prefecture, such as the j-th prefecture, then the emission Qijl of the I-th gas will be

$$Qil = \frac{Qijl}{Aj},$$

where Aj is the area of the j-th prefecture. If the number of prefectures in one grid cell is more than one, such as $j1, j2, \ldots, jm$, Qil will be taken as a sum of their area-weighted average values

$$Qij = \sum_{m=1}^{M} \left[(Wijm \times Qijl)/Aj \right],$$

where Wijl is a ratio of the area occupied by the jm-th prefecture in the 1st $1° \times 1°$ LoLa grid square/total area of the grid.

Emission sources

The major anthropogenic sources of sulfur dioxide emissions are fossil fuel and biomass burning, iron and non-ferrous metal smelting and sulfur acid production. The natural emissions from volcano eruptions and massive forest fires should be also taken into account if any occur in the considered period.

Statistical data of fossil fuel (coal, oil, natural gas, etc) consumption, biomass (corn, rice, wheat, etc) burning, metal smelting, and chemical industry production are useful for emission sources inventory.

Emission factors

Emission factors represent the amount of any gas emitted into the atmosphere upon burning or any other technological operations that proceed the oxidation of organic and inorganic sulfur to sulfur dioxide, per unit of product.

Coal burning is the main source of SO_2 in China as well as in many other Asian countries and it represents about 90% of total emission. The emission factor for coal burning operations depends on the content of S in the coal and desulfurization

Table 4. Relative growth of SO_2 emissions in 1990–1995 in China (Bai Naibin, 1997).

Region	1990	1995	Increase, % to 1990
	SO_2, kg/km^2		
North China	2477	3274	32.17
The Northeast	2526	3006	19.47
East China	6779	9274	36.80
The Middle-South	3877	5188	33.81
The Southwest	1849	2782	50.46
The northwest	629	828	24.98
The whole country	2153	2691	24.98

equipment facilities of coal-burning power plants. The average content of sulfur in Chinese coal varies from 0.92% in the North China deposits to 1.42% in the Northwest deposits, being on average 1.08% for the whole country. The effectiveness of desulfurization processes depends on the equipment installed and can achieve 90–95% from produced sulfur dioxide mass.

The sulfur contents of oil products used in China are as follows: aviation fuel, 0.02–0.1%; gasoline, 0.15%; light diesel oil, 0.2%; heavy diesel oil, 0.5–1.5%; heavy oil, 1.0–3.0%. Sulfur content in natural gas ranges from 270 mg/m^3 (domestic consumption) to 400 mg/m^3 (industrial consumption).

The content of sulfur in straw ranges from 0.10% (sorghum and rice straw) to 0.18% (wheat straw). The relative degree of sulfur oxidation fraction is also necessary to take into account when the calculations of emission factors are made up. These values are 0.71–0.91% and 0.84% is a reasonable average.

Emission factors for sulfuric acid production are 27.4 kg/ton acid for traditional technology and only 3.3–5.3 kg/ton acid for the advanced technologies. During smelting of sulfide ores, a simple stoichiometric relationship is assumed to give SO_2 released per metal production in kg/ton: 2000 for copper ($CuFeS_2$); 1000 for zinc (ZnS) and 320 per lead (PbS). The effectiveness of desulfurization in this industry was about 50% in China during 1990–1995.

Similar emission factors for SO_2 inventory have been used in other Asian countries (Akimoto and Narita, 1994).

Trends in SO_2 emissions in 1990–1995

Table 4 presents the increments in SO_2 emissions in 1990–1995 for the separate regions and the whole China area. We can see that during a 5-year period the increase in emission was 25% for the whole country and ranged from 19.5% in the Northwest to 50% in the most growing industrial areas of the Southwest. The total amount of

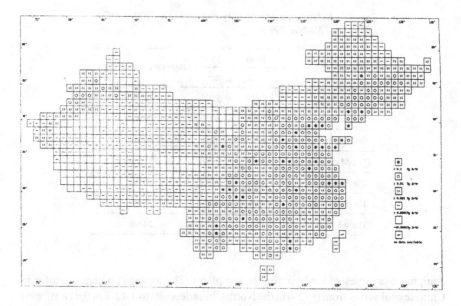

Figure 3. Spatial distribution of SO_2 *emissions in China in 1995 (Bai Naibin, 1997).*

emitted sulfur dioxide was in 1995 about 25.8 million tons for the whole country area of 9.6 million km^2. The spatial distribution of emissions in 1995 is shown in Figure 3.

3.2. Emission Inventory in Japan

The Environmental Agency of Japan has conducted questionnaire surveys on the emission of air pollutants from stationary sources every three years, in order to obtain information for promotion of air pollution control policy. Based on the results of these surveys, together with the data on emission factors for mobile and other sources, the emission trends for SO_2 and NO_x are shown in Table 5.

Figure 4 shows the trend of SO_2 and NO_2 concentrations observed in air pollution monitoring stations. We can see that during the 1990s the concentration of SO_2 has decreased significantly, whereas NO_2 remains about the same level. The sources of nitrogen oxide pollution are connected mainly with transport.

3.3. Sulfur Dioxide Emission Inventory in Thailand

Total fossil fuel consumption in Thailand increased from 1,058 PJ in 1990 to 2,064 PJ in 1997. It was accounted for at an average annual increase of 10%. Although the energy demand in the country was decreasing sharply in 1997–1998, due to the economic crisis, in 2000 the Thai economy recovered very rapidly, achieving almost the same level of development as had been recorded in 1997.

Table 5. Emission of SO_2 *and* NO_x *in Japan in 1990–1995, ton/year (After Toda, 1999).*

Pollution sources	1990	1991	1992	1993	1994	1995
Nitrogen oxides						
Energy industries	265	268	265	253	264	251
Manufacturing and construction	452	439	434	451	468	510
Transport	916	959	951	928	932	963
Other sources	103	85	94	110	107	137
Forest and grassland conversion	1	1	1	1	1	1
Waste incineration	52	55	55	57	59	64
All Japanese sources	1789	1807	1800	1800	1831	1926
Transboundary input	390	410	417	421	440	498
Aviation	54	57	58	58	61	69
Marine	336	353	359	362	379	429
TOTAL	2569	2627	2634	2641	2711	2922
Sulfur dioxide						
Energy industries	239	236	249	230	251	226
Manufacturing and construction	380	359	357	368	382	362
Transport	186	202	150	88	91	95
Other sources	61	67	67	69	77	85
Waste incineration	33	32	38	40	42	48
TOTAL	899	896	861	795	843	817

The statistical data of SO_2 emissions from different sectors in Thailand is shown in Table 6. The total SO_2 emission accounted for 1.2–1.7 millions tons (1Tg = 1 million tons) in 1991–1997. The power generation sector, as the largest point source, takes up to 60–70% of the total emissions from all sectors.

The lignite-fired Mae Moh power plant is one of largest power sources in Thailand, providing about 25% of the country's total energy supply. The history of the power plant began as a small lignite-fired power plant with two 6.25 MW generating units that started operation in 1960 and lasted until 1978. When the Electricity Generating Authority of Thailand (EGAT) took over the operation of the Lignite Authority in 1969, lignite exploration was expanded to locate and exploit potential lignite reserves for power generation.

The first three 75 MW generation units were installed at Mae Moh power plant in 1978–1981. In 1980, with advanced technology and high efficiency equipment, further surveys by EGAT revealed the lignite reserves to be over 500 million tons.

CHAPTER 1

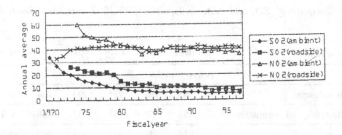

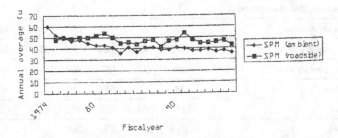

Figure 4. Annual trends for So_2 *and* NO_2 *(top) and* **SPM** *(bottom) concentrations in ambient and roadside air in Japan (Toda, 1999).*

Table 6. Amount of SO_2 *emissions categorized by Thai economic sector in 1997–1999 (Khummongkol, 1999).*

Sector	1991	1992	1993	1994	1995	1996	1997
	SO_2 emission, ton/year × 1000						
Industry	180	218	227	231	263	258	237
Power generation	825	897	895	935	990	839	1294
Residential	2	2	4	4	10	5	2
Transport	187	195	149	165	189	185	197
Others	45	46	276	23	22	28	26
TOTAL	1239	1358	1550	1359	1474	1316	1756

Therefore, the Mae Moh lignite power plant had planned 10 generating units, with a total capacity of 1,724 MW. Further exploration in 1978 resulted in lignite reserves of over 1,000 million tons and led to expansion of the power plant to 13 units, with a total combined capacity of 2,625 MW. The daily rated lignite consumption at the plant's 100% load is about 70,000 ton, which ensures the rated output capacity of the units (Table 7).

Table 7. Characteristics of units of the Mae Moh power plant, Thailand (Khummongkol, 1999).

Units	Rated Output (electr.), MW	Date of FGD installed
Units 1–3	75	None
Units 4–7	150	1999
Units 8–11	300	1998
Units 12–13	300	1995

The installation of desulfurization equipment in 1998–1999 in Units 4–11 allowed Thailand to decrease the SO_2 emission significantly.

4. TRANSBOUNDARY TRANSPORT OF SULFUR AND NITROGEN COMPOUNDS IN THE EURASIAN CONTINENT

There is a growing concern about the long-range transboundary air pollution problem which is not limited to geographical domains. Consequently the assessment of acidification loading on ecosystems, monitoring of acid rains, experimental assessment of terrestrial ecosystem response to acid input, calculation and mapping of critical loads as the indicators of ecosystem sensitivity to acid deposition in East Asia, are all of great scientific and political interest (Bashkin and Park, 1998).

It is well known that during the winter the major weather patterns in East Asia facilitate the transboundary transport of air pollutants from west to east, from land to sea and the reverse in summer. Pollutants can thus be transported from country to country in the whole region of East Asia.

The results of the preliminary calculations of sulfur and nitrogen transboundary transport in the Eurasian continent consist of two main parts—deposition maps and mass budget tables (Sofiev, 1998). Some examples of them are presented below.

Deposition maps

Deposition maps of oxidized sulfur, oxidized nitrogen and ammonia are shown in Figures 5–7. It is clearly seen that there are three main sources of pollution in the Eurasian continent—Europe, the Urals industrial region and East Asia. The main direction of atmospheric transport is from West to East. As a result the influence of Europe is detectable practically up to the Lake Baikal area (Middle Northeast Asia), especially for nitrogen compounds. The highest deposition of sulfur compounds is in the Black triangle region in the center of Europe and on the East Coast of China— up to $7 \, g/m^2$ of sulfur per annum. Nitrogen deposition is more homogeneous with less pronounced peaks. According to current results, in 1995 the highest deposition was on the East Coast of China and accounted for $2 \, g/m^2$ of nitrogen per square meter per annum. In central Europe the corresponding value is as much as half of

Figure 5. Deposition of oxidized sulfur in the Eurasian continent, mg/m² per year (Sofiev, 1998).

this amount. Ammonia deposition is driven by spatial emission distribution, which is almost artificial in current calculations.

East Asia is practically a self-polluted region. The bulk of emission sources are located close to the coast, whilst the prevailing wind direction leads to further transport of these masses to the ocean and relevant islands. This effect is clearly seen in all deposition maps. the quite short transportation distance of all compounds over the marine surface is probably explained by a fast increase of the dry deposition velocity onto the water surface with growing of wind speed and the creation and crashing of waves.

Source-receptor mass budget matrices

As was mentioned above, the most important characteristic of acidification is deposition of sulfur and nitrogen oxides. Deposition of bound nitrogen is crucial for eutrophication. Because of the atmospheric redistribution of pollutants, a considerable part of the deposition onto some region can originate from remote sources. This pollution exchange should be taken into account during the development of emission reduction measures. Necessary information can be presented in the form of mass budget matrices. Corresponding tables for Eurasia are presented in Tables 9–11. Notations of emitters and receptors are shown in Table 8. Data in tables are averaged over 4 years—1991–1994—in order to reduce the influence of meteorological variability.

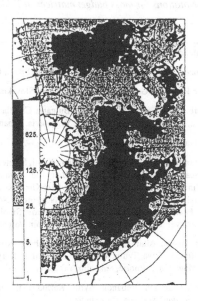

Figure 6. Deposition of oxidized nitrogen in the Eurasian continent, mg/m^2 per year (Sofiev, 1998).

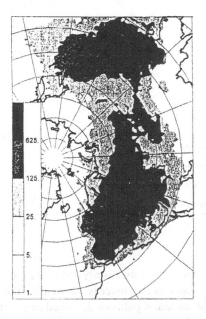

Figure 7. Deposition of reduced nitrogen in the Eurasian continent, mg/m^2 per year (Sofiev, 1998).

Table 8. Notations for mass budget matrices in Tables 9–11.

Notation	Description
Ner	Denmark, Finland, Norway, Sweden, Baltic Sea
Wer	Belgium, France, Luxembourg, Netherlands, Portugal, Spain, Ireland, Great Britain
Cer	Austria, Czech Republic, Germany, Hungary, Poland, Slovakia, Switzerland
Ser	Armenia, Azerbaijan, Bosnia & Hercegovina, Bulgaria, Croatia, Cyprus, Georgia, Greece, Italy, Moldavia, Romania, Slovenia, Turkey, Yugoslavia, Albania, Macedonia
Eer	Belarus, Estonia, Latvia, Lithuania, Ukraine
Rue	European part of Russia: Arkhangelsk, Karelian, Komi, S.Peterburg, Murmansk, Ladoga lake, White Sea, Astrakhan, Bashkortostan, Belgorod, Bryansk, Chuvashiya, Ivanovo, Jaroslavl, Kaliningrad, Kalmykia, Kaluga, Kavkaz, Kirov, Kostroma, Krasnodar, Kursk, Lipetsk, Marial, Mordovia, Moscow, Nihznii Novgorod, Novgorod, Orel, Orenburg, Penza, Perm, Pskov, Rostov, Ryasan, Samara, Saratov, Smolensk, Stavropol, Tambov, Tatarstan, Tula, Tver, Udmurtiya, Ulyanovsk, Vladimir, Volgograd, Vologda, Voronehz
rua	Asian part of Russia: Altai, Chelyabinsk, Ekaterinburg, Kemerovo, Kurgan, Novosibirsk, Omsk, Tomsk, Tuva, Tyumen, Krasnoyarsk, Amur, Buryatiya, Chita, Kamchatka, Khabarovsk, Magadan, Primorie, Sakhalin, Yakutiya, Yrkutsk, Baikal lake
cas	Iran, Iraq, Israel, Jordan, Lebanon, Syria, Kazakhstan, Kirghizia, Tajikistan, Turkmenistan, Uzbekistan, Aral Sea, Caspian Sea
chn	China
jpn	Japan
nkr	North Korea
skr	South Korea
mgl	Mongolia
asi	Asia other
atl	Atlantic
arc	Arctic, oceanic part
pas	Pacific ocean
med	Mediterranean sea
sum	Sum of column/row
bsm	Transport outside the grid
qa	Annual emission

Columns in these tables reflect the distribution of the emission from a particular source. Total annual emission of the source is shown in the bottom cell of the column (row qa). Rows show the amount of pollution deposited onto the territory of a particular receptor. Hence each cell contains the amount of the deposition onto the specific receptor originating from the specific emitter.

Table 9. Mass budget matrix for oxidized sulfur. Mean over 1991–1994. Unit = ktonne S per annum (Sofiev, 1998).

Rc/Em	ner	wer	cer	ser	eer	rue	rua	cas	chn	Jpn	nkr	skr	mgl	sum
ner	136	144	323	19	51	65	2	0	0	0	0	0	0	739
wer	29	2991	509	72	20	12	5	10	0	0	0	0	0	3648
cer	21	282	2163	149	28	8	0	1	0	0	0	0	0	2650
ser	3	79	421	1377	78	16	1	25	0	0	0	0	0	1999
eer	16	42	492	128	597	47	1	2	0	0	0	0	0	1324
rue	34	61	401	108	426	1057	59	21	0	0	0	0	0	2165
rua	9	24	124	25	103	318	1516	249	106	1	4	5	5	2486
cas	4	96	112.	142	108	198	91	864.	16	0	0	0	0	1630
chn	0	1	8	2	6	13	52	35	7152	0	11	9	10	7299
jpn	0	0	1	0	1	1	9	1	79	196	8	43	0	338
nkr	0	0	0	0	0	1	2	1	111	0	42	56	0	213
skr	0	0	0	0	0	0	1	0	59	1	9	139	0	208
mgl	0	0	4	1	3	7	33	20	44	0	0	0	30	141
asi	0	0	0	0	0	0	0	10	46	0	0	0	0	56
arc	12	46	89	7	26	135	216	.7	1	0	0	0	0	539
pas	2	15	13	2	8	18	143	14	1235	175	83	314	4	2024
med	4	354	340	963	127	21	1	109	0	0	0	0	0	1919
sum	269	4134	4998	2992	1581	1914	2131	1368	8848	372	155	565	48	29375
bsm	1	115	63	177	15	7	10	221	671	100	12	84	1	1829
qa	270	4250	5069	3172	1598	1928	2161	1590	9531	472	166	647	50	30904

Accumulation of artificial errors during the long chain of computation results in some small differences between actual annual emission (row qa) and the sum of depositions originating from the source (row sum plus transport outside the calculation domain bsm). Such a deviation never exceeds 1% of the source emission, which confirms the stability of the applied scheme.

Brief analysis of the tables confirms that three above-mentioned main regions of pollution (Europe, the Urals industrial region, and East Asia) have a definite influence on each other. There is a considerable transport from Europe to the East. In particular, up to 40% of the oxidized nitrogen deposition onto the European territory of Russia comes from Europe. For the Asian part of Russia this fraction is about 10%. Corresponding figures for sulfur and ammonia are not so large, which is explained by comparably short transport distance of these species.

CHAPTER 1

Table 10. Mass budget matrix for oxidised nitrogen. Mean over 1991–1994. Unit = 100t N per annum (Sofiev, 1998).

Rc/Em	ner	wer	cer	ser	eer	rue	rua	cas	chn	Jpn	nkr	skr	mgl	sum
ner	152	107	122	10	14	10	0	0	0	0	0	0	0	415
wer	44	1403	263	35	5	3	0	1	0	0	0	0	0	1756
cer	24	220	643	68	8	2	0	0	0	0	0	0	0	965
ser	5	63	143	468	18	11	0	4	0	0	0	0	0	713
eer	25	44	161	52	125	24	0	0	0	0	0	0	0	433
rue	63	57	123	46	111	383	19	6	0	0	0	0	0	806
rua	14	14	24	6	18	119	335	49	43	1	4	2	2	633
cas	6	47	27	59	25	88	30	165	5	0	0	0	0	454
chn	0	0	1	0	1	3	18	7	1470	1	9	3	5	1519
jpn	0	0	0	0	0	0	2	0	20	182	7	13	0	225
nkr	0	0	0	0	0	0	1	0	38	0	29	11	0	80
skr	0	0	0	0	0	0	0	0	17	1	9	28	0	55
mgl	0	0	0	0	0	2	12	3	11	0	0	0	11	40
asi	0	0	0	0	0	0	0	3	7	0	0	0	0	10
arc	28	35	19	1	5	19	13	1	0	0	0	0	0	121
pas	2	5	1	0	1	4	50	5	373	212	77	90	2	821
med	7	204	125	397	31	18	0	16	0	0	0	0	0	797
sum	370	2200	1653	1143	362	686	483	263	1984	398	134	146	22	9844
bsm	1	54	14	61	3	3	3	52	132	171	9	23	0	911
qa	372	2255	1667	1205	365	690	487	315	2119	569	143	169	22	10378

As noted earlier, East Asia is mainly a self-polluted region. Only a few countries are affected by remote sources (first of all, by emitters located in the Asian part of Russia and in Central Asia). The most affected territory belongs to Mongolia, where the fraction of sulfur deposition from remote sources amounts to 30%. The sources in Asia affect Eastern Russia, the shelf seas, and the Pacific Ocean. The main source of all types of pollutants is China. Its annual emission exceeds the values of other countries taken together by a factor of several fold (for sulfur and oxidized nitrogen) or by one order of tar (for ammonia). Consequently this country is the most important donor of the acidifying and euthrophic substances in Asia. The relevant recipients are South Korea and Japan.

Thus, it is impossible for an individual country alone to solve the problem of air pollution and its impact on the ecosystems. It requires regional inter-governmental

Table 11. Mass budget matrix for reduced nitrogen. Mean over 1991–1994. Unit = 100t N per annum (Sofiev, 1998).

Rc/Em	ner	wer	cer	ser	eer	rue	rua	cas	chn	Jpn	nkr	skr	mgl	sum
ner	129	30	92	8	38	14	0	0	0	0	0	0	0	312
wer	37	1120	140	20	14	5	0	2	0	0	0	0	0	1336
cer	21	131	678	54	30	5	0	0	0	0	0	0	0	919
ser	2	26	101	793	59	27	0	15	0	0	0	0	0	1022
eer	11	10	117	84	564	56	0	1	0	0	0	0	0	843
rue	20	10	57	69	219	835	8	26	0	0	0	0	0	1245
rua	3	2	7	8	26	142	298	70	194	2	1	1	10	762
cas	2	23	10	91	31	151	26	433	48	0	0	0	1	816
chn	0	0	0	0	1	4	21	20	6403	0	4	2	33	6489
jpn	0	0	0	0	0	0	2	0	56	139	1	5	1	204
nkr	0	0	0	0	0	0	1	0	80	0	12	5	1	99
skr	0	0	0	0	0	0	0	0	28	1	2	21	0	52
mgl	0	0	0	0	0	2	13	5	35	0	0	0	46	101
asi	0	0	0	0	0	0	0	10	82	0	0	0	0	93
arc	7	6	6	1	6	14	3	1	1	0	0	0	0	44
pas	2	1	0	1	1	7	43	7	762	132	20	43	9	1027
med	2	94	58	332	85	29	0	21	0	0	0	0	0	621
sum	236	1451	1268	1461	1073	1289	415	613	7687	273	40	76	102	15983
bsm	0	36	8	95	8	5	2	107	309	57	1	7	2	696
qa	236	1486	1275	1557	1079	1295	417	720	8007	330	41	82	104	16629

cooperation. Currently, regional/sub-regional agreements on a pollutant emission abatement strategy do not exist at all in this region. However, some urgent concerns are emerging. In order to reduce the actual and potential risk of environmental degradation the general idea should be to negotiate, for example, the possible expansion of the UN/ECE Convention on Long-Range Transboundary Air Pollution to cover the whole Eurasian continent.

FURTHER READING

1. ESCAP, 1995. *State of the Environment in Asia and Pacific.* United Nations, 638 pp.

2. Bashkin V. N. and Park S.-U., 1998. *Acid Deposition and Ecosystem Sensitivity in East Asia.* Nova Science Publisher, USA, 427 pp.

3. PDC, 1999. *Proceedings of East Asian Workshop on Acid Deposition.* 6–8 October 1999, Bangkok, 276 pp.

4. ESCAP, 2000. *State of the Environment in Asia and Pacific.* United Nations, 905 pp.

WEBSITE OF INTEREST

1. http://www.unescap.org

QUESTIONS AND PROBLEMS

1. Describe the types of pollution sources in Asia. Present briefly the peculiarities of environmental pollution in the Asian region in comparison with the World's other regions.

2. Consider the role of industrial production in environmental pollution. What types of industry are of the most environmental concern?

3. Compare the influence of energy and construction in producing pollutants in various Asian countries.

4. Describe the methodology for an emission inventory of air pollutants. Note the peculiarities of individual sources of sulfur dioxide emission in various countries.

5. Estimate the content of sulfur and nitrogen in local fossil fuels and biomass burning in the area where you live.

6. Apply the methodology of SO_2 emission inventory for calculating this pollutant emission in your region.

7. What are the reasons of transboundary pollution in Eurasian continent? Present your own speculations and explanations.

8. If you are living in a country listed in Tables 9–11, try to calculate the role of transboundary sources in air pollution by sulfur and nitrogen compounds in your country.

9. Calculate the contribution of marine salts to sulfur deposition in any island or coastal area of your country.

CHAPTER 2

GREENHOUSE GASES

1. INTRODUCTION

Climate change, the increase in average temperatures or Global Warming, and the associated sea level rise pose distinct and serious threats to the Asian countries. Anthropogenic climate change takes place because of the Greenhouse Effect, a process by which gases in the Earth's atmosphere alter the radiation balance of the Earth. These gases are generally called greenhouse gases (or GHG), and include notably carbon dioxide, CO_2, methane, CH_4, nitrous oxide, N_2O, water vapor, H_2O, chlorofluorocarbons, CFCs, and ozone, O_3. Most of them are natural gases and the only exception is the CFCs. Greenhouse gases allow the bulk of the energy in visible light from the Sun to pass through the atmosphere to the ground, but prevent some of the heat generated when sunlight strikes the ground from radiating back to space.

This notion is not new. The famous European chemist and physist Arrhenius, in 1896, was one of the first to warn that human activities could be effectively "thickening" the layer of CO_2 by the emission of carbon into the atmosphere through the burning of fossil fuels. Indeed, many scientists believe that such global warming has already been under way for some time, and it is largely responsible for the average global temperature increase of about two thirds of a degree Celsius from the second part of the 19th century.

At present, no one is sure of the extent or timing of the future temperature increase, nor is it likely that reliable predictions for individual regions will ever be available much in advance of the events in question. However, if current models of the atmosphere are correct, significant global warming will occur in the coming decades or even years. We should remember here that the average annual temperatures, for example, for North America, have permanently increased during the last 20 years.

In this chapter the mechanism by which global warming could arise is explained and the nature and sources of the chemicals which are responsible for the effect are analyzed. The principal attention will be paid to the Asian inputs to global warming though emissions of various GHG.

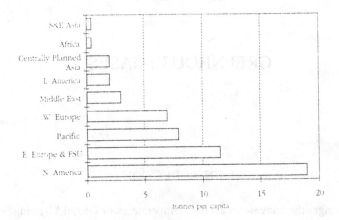

Figure 1. Annual regional carbon dioxide emissions per capita (ESCAP, 1995).

2. SOURCES OF GREENHOUSE EFFECTS IN ASIA

In Asia, greenhouse gases are emitted through a number of human activities, which include: energy production and use, land use changes (most notably deforestation), livestock raising and disposal of livestock wastes, CFCs production, cement production, disposal of human sewage and municipal solid wastes, fertilizer use, and rice cultivation. We will consider the emissions of various greenhouse gases in global and regional Asian scales, paying the most possible attention to the Asian domain.

2.1. Carbon Dioxide

Each inhabited region of the globe is responsible for emissions, but the extent of these emissions (and their impacts) has varied and continues to vary substantially between regions. Differences are particularly evident when presented on a per capita basis, as Figure 1 illustrates for the principal GHG, carbon dioxide.

In accordance with these data the Asian per capita emissions are by far behind those in Europe and North America. However, if we consider the total per region CO_2 emission results (Figure 2) on the example of fossil fuel combustion, the data are rather opposite.

The shown data are related to the beginning of the 1990s and during the last decade emissions were arising, in spite of economic crisis in 1997–1999. Moreover, the projected emissions of carbon dioxide in the 11 Asian countries are still very dramatic (Table 1).

The data show that the predicted emissions will exceed the 1995 levels in various countries from 2.7 (China) to 7.6 (Indonesia) times, being in total for the estimated Asian countries as much as 3.5 times.

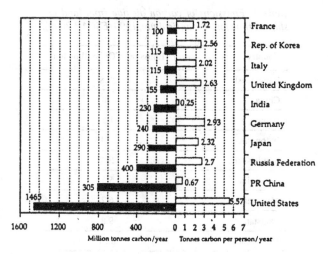

Figure 2. Carbon dioxide emissions from fossil fuels by World regions (ESCAP, 1995).

Table 1. Recent and predicted emissions of CO_2 from the energy sector in 11 Asian countries, CO_2, million tons (ESCAP, 2000).

Country	1995	2020
Bangladesh	21	89
China	2,325	6,221
India	565	2,308
Indonesia	157	1,191
South Korea	248	932
Mongolia	14	74
Myanmar	6	19
Pakistan	70	482
Philippines	44	238
Thailand	80	777
Vietnam	28	328
Total	3,558	12,269

In addition to energy use, changes in land use are a major source of carbon dioxide emissions. In some countries the forest might actually be a net sink for carbon dioxide, i.e., the removal of carbon from the atmosphere might be larger than the emissions. According to ESCAP study (ESCAP, 2000), two of 11 Asian countries indicated that the land use changes represented a net sink for carbon. These data are shown in Table 2.

Table 2. Emissions of carbon dioxide from land use shifting from forest to agriculture, 1990 (ESCAP, 2000).

Country	Emission, CO_2, million tons
Bangladesh	29.8
China	−281
India	1.5
Indonesia	−334.9
South Korea	26.2
Mongolia	5.5
Myanmar	6.7
Pakistan	9.8
Philippines	82.1
Thailand	78.1
Vietnam	31.2

2.2. Methane

The concentration of atmospheric methane is increasing, probably as a result of increasing cattle population, rice production, and biomass burning. Increasing methane concentrations are important because of the role it plays in stratospheric and tropospheric chemistry. Methane is also important to the radiation budget of our planet.

Agriculture and livestock are the largest sources of anthropogenic methane in most Asian countries. Coal mines, the production of oil and gas and transmission of natural gas are also large contributors to methane emissions in some countries. The aggregate emissions of methane from each country are shown in Table 3, which also provides the data on nitrous oxide emissions.

2.3. Nitrous Oxide

Emissions of nitrous oxide, both natural and anthropogenic, in the main Asian countries are shown in Table 3. According to the total global emissions of 17 million per year in the 1990s, the share of Asia is about 45–50% (see details in Chapter 6).

2.4. Chlorofluorocarbons

Anthropogenic emissions of chlorofluorocarbons, CFCs, both on global and regional scales are discussed in more detail in Chapter 6, devoted to the stratospheric ozone depletion. The global response to this environmental threat is generally considered to be one of great successful international cooperation. The rapid decline in the emissions of ozone depleting substances on a global basis is shown in Figure 3.

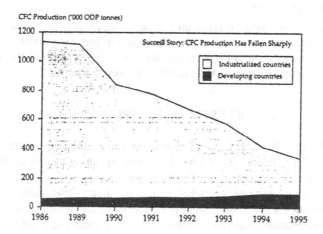

Figure 3. Decline in the global production of ozone depleting CFCs (ESCAP, 2000).

*Table 3. Emissions of methane and nitrogen oxide in
11 Asian countries, 1990, million tons (ESCAP, 2000).*

Country	Methane	Nitrous oxide
Bangladesh	1.7	0.05
China	29.1	3.6
India	18.5	2.6
Indonesia	4.9	0.2
South Korea	1.4	0.1
Mongolia	0.3	0.001
Myanmar	2.1	0.08
Pakistan	2.7	0.002
Philippines	1.5	0.30
Thailand	2.7	0.11
Vietnam	2.6	0.15
Total	67.50	7.2

In the 1980s, most emissions of CFCs originated in the industrialized countries. In the period since 1995 emissions of CFCs from these countries have declined further, and are now less than those from the developing countries, mainly the countries of the Asian region. In contrast, the emissions of CFCs in many Asian countries such as China, India, and Pakistan, were increasing at the end of the 1990's.

2.5. *Tropospheric Ozone*

Tropospheric ozone levels are increasing in many Asian cities as the principal products of photochemical smog formation. The main cause of this phenomenon is the rapid development of the transport system and the tremendous growth of the number of vehicles in all Asian countries. More details are given in Chapter 3.

3. CHEMISTRY OF THE TROPOSPHERE

The greenhouse effect takes place in the lower part of the atmosphere, called the troposphere. This phenomenon is an enhancement of natural processes which manages the radiation balance at the Earth's surface owing to a permanent increase of the GHG concentrations. For a better understanding of this problem we should consider the chemical composition of the troposphere and residence times of some species of interest, the temperature profile in the whole atmosphere and the increasing concentrations of greenhouse gases.

3.1. *Chemical Composition of the Atmosphere*

The biogeochemical history of the atmosphere composition buildup provides a good example of the impact of living organisms on the environment (Bashkin and Howarth, 2002). Recent monitoring indicates that the actual composition of the Earth's gas shell has been the ultimate stage of a long lasting process, in which a major role was assigned to the biogeochemical activity of living matter.

The mass of the atmosphere is about 5.14–5.27×10^{15} tons (Walker, 1977; Voitkevich, 1986). The major part of gaseous matter (about 80%) occurs in the troposphere whose upper equatorial boundary reaches as high as about 17 km, lowering to 8–10 km at the poles. The troposphere is restricted by the tropopause, which is characterized by a sharp temperature drop and the absence of water vapors (Figure 4).

The active physical, chemical, and biogeochemical interactions of the atmosphere with ocean and land occur mainly in the troposphere. It also contains the bulk of water vapors and airborne particulate matter. It is also well known that many important biosphere photochemical reactions take place in the troposphere.

Above the troposphere, in the stratosphere and mesosphere, the gas density becomes increasingly more pronounced, and the thermal conditions become subject to complex variations, both temporal and spatial. At a height of 25–35 km the oxygen molecules are split by the influence of solar UV radiation to form ozone in accordance with the Chapman mechanism, which is responsible for the absorption of 97% of harmful ultraviolet solar radiation. Devoid of this shield, life on the terrestrial surface would have been doomed to extinction (see Chapter 6).

Up to 80–100 km from the Earth's surface is the ionosphere, a region of highly rarefied and ionized gas molecules. The exosphere, the outermost belt of the gas shield, might extend as far as 1800 km. The losses of the lightest gases, hydrogen and helium, into space from the Earth's atmosphere occur in the exosphere.

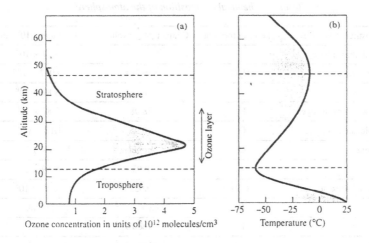

Figure 4. Temperature profile of the atmosphere (Baird, 1999).

Table 4 shows the globally and seasonally averaged composition of the atmosphere. The Earth's atmosphere composition is to a significant degree determined by the activity of living organisms and is maintained owing to a system of biogeochemical cycles. At present 99.8% of the gaseous matter of atmosphere is nitrogen, oxygen, and argon. The content of water is greatly variable and no average values can be presented. Whilst the concentration of nitrogen and oxygen are nearly invariant, the concentration of other constituents can vary both spatially and temporally.

Atmospheric components, which are present in small amounts, are water vapor, inert gases, and species produced by photochemical reactions and biological processes. The geochemistry of inert gases is of particular interest for the historical re-constructions of the atmosphere. The relative high percentage of argon is associated with the ^{40}Ar isotope, which is derived from the decay of the widespread ^{40}K. In contrast, the amount of atmospheric helium is smaller by a factor of 1000, as might be expected. This is owing to the continuous extra-atmospheric dissipation of this element in the exosphere. The rest of the inert gases are present in the amounts in which they have evolved during the entire period of the Earth's existence. An analysis of the isotopic xenon production supports an idea that the gas shell emerged during a very short period of time, which was roughly coincident with the period of the Earth's accretion (Dobrovolsky, 1994).

Environmental concentration units for gases are shown in Box 1.

Box 1. Environmental concentration units for gases (After Baird, 1999)

Two types of concentration scales are commonly used for gases present in air. For *absolute* concentrations the most common scale is **molecules per cubic centimeter** of air.

Table 4. Chemical composition of the atmosphere.

Constituents/species	Chemical formula	Content (% by volume)	Total mass, 10^9 tons
Total atmosphere			5.27×10^6
Water vapor	H_2O	variable	0.017×10^6
Dry air		100	3.87×10^6
Molecule of nitrogen	N_2	78.08	3.87×10^6
Molecule of oxygen	O_2	20.95	$1.18 \times x10^6$
Ozone	O_3	Variable	~ 3.3
Argon	Ar	0.93	6.59×10^4
Carbon dioxide	CO_2	0.032	2.45×10^3
Neon	Ne	1.82×10^{-3}	6.48
Helium	He	5.24×10^{-4}	2.02
Krypton	Kr	1.14×10^{-4}	1.69
Xenon	Xe	1.87×10^{-6}	2.02
Methane	CH_4	1.5×10^-4	~ 4.3
Molecule of hydrogen	H_2	$\sim 5.0 \times 10^{-5}$	~ 1.8
Nitrous oxide	N_2O	$\sim 3 \times 10^{-5}$	~ 2.3
Carbon monoxide	CO	$\sim 1.2 \times 10^{-5}$	~ 0.59
Ammonia	NH_3/NH_4^+	$\sim 1.0 \times 10^{-6}$	~ 0.03
Nitrogen dioxide	NO_2	1.0×10^{-7}	~ 0.0081
Sulfur dioxide	SO_2	$\sim 2 \times 10^{-8}$	0.0023
Hydrogen sulfide	H_2S	$\sim 2 \times 10^{-8}$	0.0012

Relative concentrations are usually based on the chemists' familiar **mole fraction** scale (called mixing rates by physicists), which is also a molecule fraction scale. Because the concentrations for many constituents are so small, atmospheric and environmental scientists often re-express the mole fraction as "part per ____". Thus, a concentration of 100 molecules of air gas such as carbon dioxide dispersed in one million (10^6) molecules of air would be expressed as "100 parts per million", or "100 ppm" rather than as a molecule or mole fraction of 0.0001. Similarly, "ppb" and "ppt" stand for "parts per billion" (one in 10^9) and "part per trillion" (one in 10^{12}), respectively.

It is important to emphasize for gases, these relative concentration units express the number of molecules of a pollutant (i.e., the "solute" in chemists' language) that are present in one million or billion or trillion molecules of air. Since according to the ideal gas law (PV = nRT), the volume of a gas is proportional to the number of molecules it contains, the "parts per" scales represent the volume a pollutant's gas

Table 5. Composition of gases of the Great Tolbachin-sky fissure eruption, Kamchatka, Asian Russia (after Dobrovolsky, 1994).

Species	Content (% by volume)	
	Water vapor included	Dehydrated
H_2O	78.56	—
N_2	11.87	55.36
CO_2	4.87	22.71
H_2	3.01	14.04
HCl	0.57	2.66
CO	0.39	1.86
CH_4	0.44	2.05
H_2S	0.16	0.75
NH_3	0.11	0.51
HF	0.06	0.26
Ar	0.06	0.30
SO_2	0.03	0.14
O_2	0.01	0.05
He	0.001	0.005

would occupy, compared to that of the stated volume of air, if the pollutant were to be isolated and compressed until its pressure equaled that of the air. In order to emphasize that the concentration scale is based upon molecules or volumes rather than upon mass, a "v" (for volume) is sometimes shown as part of the unit, e.g., 100 ppmv. Concentrations are sometimes expressed in terms of the partial pressure of the gas, which is directly proportional to its mole fraction but expressed in units of atmospheres or Pascals or bars.

3.2. Residence Times of Atmospheric Gases

The concept of mean residence time (MRT) is useful in consideration of average atmospheric composition shown in Table 5. For any tropospheric reservoir, the MRT value can be calculated as

$$MTR = Mass/flux,$$

where "Mass" is related to the quantity of any species in reservoir, and "Flux" may be either the input or loss from the reservoir. One example of this calculation can be presented for N_2O (Schlesinger, 1991). The average nitrous oxide content in the atmosphere is about 300 ppb. Multiplied by the mass of the atmosphere,

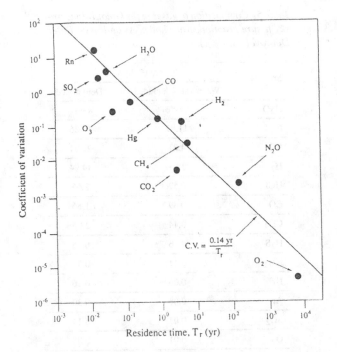

Figure 5. Variability in the concentration of atmospheric gases (expressed as the coefficients of variation in measurements) as a function of their estimated mean residence times in the atmosphere (Schlesinger, 1991).

we obtain 2.3×10^9 tons for the content of this species in the entire atmosphere. Our best estimates of the sources of N_2O suggest an annual production of at least 20×10^6 ton/yr, giving the mean residence time of over 100 yr for nitrous oxide in the atmosphere. During this period, this gas will be relatively evenly distributed within the global atmosphere. The higher concentrations will be near strong point sources of pollution. On the other hand, the average volume of water in the atmosphere is equivalent to $\sim 13,000 \, \text{km}^3$ at any time, or 25 mm above any point on the Earth's surface. The global and seasonal average daily precipitation would be about 2.7 mm. Thus, the MRT for H_2O in the atmosphere is

$$\text{MTR} = 25 \, \text{mm}/2.7 \, \text{mm day}^{-1} = 9.3 \, \text{days}.$$

In comparison with circulation of global and regional atmosphere mass, this is a short time. It means that water vapor content is very variable in time and space.

The MTR values for other atmospheric gases are shown in Figure 5.

The biogeochemical processes which control the atmospheric levels of oxygen and carbon dioxide play the most important role in maintaining normal environmental conditions on the Earth. Free oxygen is a prerequisite for existence of major forms of life, and carbon dioxide, being the basic building material for photosynthesis, is also the leading factor for the greenhouse effect, which, in turn, determines the thermal and climatic conditions on the surface of our planet.

The major atmospheric species, N_2, O_2, CO_2, and Ar, are relatively unreactive. They show nearly uniform concentrations and relatively long mean residence times. Biogeochemically, molecule nitrogen is almost inert; with the exception of symbiotic microbes, all living systems can only assimilate nitrogen from bound or fixed forms, such as ammonium and nitrate. The atmosphere contains only a small portion of the total O_2 released by photosynthesis through geological time. This value does not depend on the present activity of the photosynthetic process. If we can imagine the instantaneous combustion of all the organic matter that is stored now on land, this would reduce the atmospheric oxygen content by only 0.035% (Schlesinger, 1991). It has been suggested that atmospheric O_2 controls the storage of reduced carbon, not vice versa. However, despite potential reactivity of molecule oxygen (see Chapter 6), these rates of reactions with reduced compounds are sufficiently slow and O_2 is a relatively stable atmospheric species with mean residence time of about 10,000 years (Figure 4).

In the atmosphere, CO_2 is affected by processes that operate on different time scales, including interaction with the silicate cycle (for more details, see Bashkin and Howarth, 2002), dissolution in the oceans, and annual cycles of photosynthesis and respiration (see also Section 4). The relative effect of these processes is described below taking into account the interrelationship between sources and sinks of carbon dioxide in the Asian region. Here, it is important to note that carbon dioxide is not reactive with other atmospheric species; its MRT is three years (Figure 5). This value is largely determined by exchange with seawater (see below).

3.3. Chemical Composition of Volcanic Gases

As regards other elements, indicated in Table 4, their original natural sources are mainly related to volcanoes (H_2S, NH_3, H_2 and many others). The typical example of these gases' composition is given in Table 5.

4. INCREASING GHG CONCENTRATIONS IN THE TROPOSPHERE

At present in the troposphere the contents of many minor gases are monitored at being in concentrations well in excess of what is predicted under equilibrium geochemistry (Table 4). In most cases, the monitored tropospheric concentrations are maintained by the action of living matter, mainly microbes. Natural biogeochemical cycles of nitrogen and sulfur are driven by biota; however at present these cycles are complicated by the modern anthropogenic influence (see Chapters 4 and 13). Unlike major atmospheric constituents, these gases are highly reactive and correspondingly they show short residence times and variable contents, both spatially and temporally (Figure 5).

The anthropogenic emissions of different gases increase their concentrations in the troposphere and enhance the greenhouse effects. Here we will discuss the relationship between sources (see Section 2) and sinks of the main GHG.

4.1. Carbon Dioxide

Cyclic processes of carbon mass exchange are of particular importance for the global biosphere both in terrestrial and oceanic ecosystems.

This element is distributed in the atmosphere, water, and land as follows. According to existing data, there are 6160×10^9 tons or 1.4×10^{16} mol of CO_2 in the atmosphere (1680×10^9 tons of C). A major source of atmospheric carbon dioxide is respiration, combustion, and decay, which is different from oxygen, the main source of which is photosynthesis. In its turn, an important sink of CO_2 is photosynthesis (about 66×10^9 tons/yr or 1.5×10^{15} mol/yr). Since carbon dioxide is somewhat soluble in water ($K_H = 3.4 \times 10^{-2}$ mol/L/atm), exchange with the global ocean must also be considered. The approximate global balance of atmosphere-ocean water exchange is 7×10^{15} mol/yr (308×10^9 tons/yr) being taken up and 6×10^{15} mol/yr (264×10^9 tons/yr) being released in different parts of the oceanic ecosystem (Bunce, 1994). The residence time of CO_2 in the atmosphere is about two years, which makes the atmospheric air quite well mixed with respect to this gas. However, a more recent analysis shows that the terrestrial ecosystems have much stronger sinks of carbon dioxide uptake. The details of major ecosystem-level CO_2 experiments have been shown recently (Koch and Mooney, 1996).

In the global ocean, along with the occurrence in living organisms, carbon is present in two major forms: as a constituent of organic matter (in solution and partly in suspension) and as a constituent of exchangeable inorganic ions HCO_3^-, CO_3^{2-}, and CO_2

$$CO_2 \text{ (g)} \longleftrightarrow H_2CO_3 \text{ (aq)} \longleftrightarrow -H^+/+H^+ \longleftrightarrow HCO_3^- \text{ (aq)} \longleftrightarrow$$
$$\longleftrightarrow -H^+/+H^+ \longleftrightarrow CO_3^{2-} \longleftrightarrow +Ca^{2+} \longleftrightarrow CaCO_3 \text{ (s)}.$$

The amount of CO_2(aq) in the oceans is sixty times that of CO_2(g) in the Earth's air, suggesting that the oceans might absorb most of the additional carbon dioxide injected at present into the atmosphere. However, there are some drawbacks restricting this process. First of all, CO_2 uptake into surface oceanic waters (0–100 m) is relatively slow ($t_{1/2} = 1.3$ yr). Secondly, these surface waters mix with deeper waters very slowly ($t_{1/2} = 35$ yrs). Consequently, the surface oceanic waters have the capacity to remove only a fraction of any increase in the anthropogenic CO_2 loading (Figure 6).

The known analytical monitoring data obtained over many years at the Mauna Loa Observatory in Hawaii, a location far from any anthropogenic sources of carbon dioxide pollution, show a pronounced one year cycle of CO_2 content (Figure 7).

One can see the peak about April and the drop through around October each year. These data indicate that the content of carbon dioxide in the Earth's atmosphere is not perfectly homogeneous. Some explanations would be of interest for understanding this figure.

Hawaii is in the Northern Hemisphere where the photosynthetic activity of vegetation is maximal in summertime (May–September). In this period CO_2 is removed from the air a little bit faster than it is added. The reverse situation is exhibited during the winter (see http://mlo.hawaii.gov).

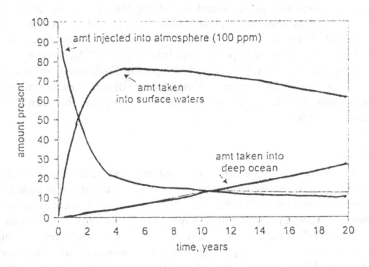

Figure 6. Calculated uptake of CO_2 from atmosphere to the surface and deep oceanic waters (Bunce, 1994).

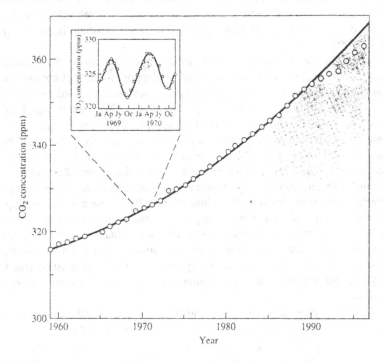

Figure 7. Observations of CO_2 concentration at the Mauna Loa Observatory for the period of 1958–1999 (Baird, 1999).

This is a reasonable explanation and accordingly the monitoring stations in the Southern Hemisphere show the highest concentration of CO_2 in October, and the lowest in April. A gradual increase in the partial pressure of carbon dioxide over the last few decades is clearly pointed out from Figure 7. The value of $p(CO_2)$ was ca. 315 ppmv in 1958; it had reached 350 ppmv in 1988, and 365 pmv in 1999. Accordingly, this trend can give the doubling of carbon dioxide content in the Earth's atmosphere sometime during the end of the 21st century and this seems to be a reasonable prediction.

4.2. Methane

After carbon dioxide, methane is the next most important GHG, the concentration of which has been increasing during the 20th century. As illustrated in Figure 8, the atmospheric concentration of methane has more than doubled as compared with its pre-industrial values; almost all of this increase has occurred in the 20th century.

Historically, before the mid-19th century, the methane concentration was approximately constant at about 0.75 ppm, but it has risen continually since then and had reached almost 1.75 ppm by 1994. In the 1970s, the rate of increase was twice as much as that in the 1980s, and in the early 1990s it temporarily became almost zero. It is not known with certainty why the growth rate decreased and has fluctuated recently, though some scientists have speculated that it is related to the air temperature decreases associated with the explosion of Mt. Pinatubo in the Philippines (Baird, 1999).

The rise in the anthropogenic CH_4 level is presumed to be the consequence of such human activities as increased food production, fossil fuel combustion and forest cutting. The annual rate of anthropogenic methane emissions increased from 100×10^6 tons to 400×10^6 tons at the end of the 1990s (Stern and Kaufman, 1996). The estimates show that in Asia about 75% of current methane emissions are anthropogenic in origin. Methane is produced biologically in the anaerobic decomposition of plant materials. Such processes are common in rice paddies and waste landfills, which are greatly represented in the Asian region. Ruminant animals, including cattle and sheep, produce huge amounts of methane as a byproduct in their stomachs when they digest the cellulose in their food. The animals subsequently belch the methane into air. The burning of biomass, such as forest and grasslands in tropical and semitropical areas of Asia, releases methane to the extent of about 1% of the carbon consumed, along with large amounts of carbon monoxide; both compounds are products of incomplete (poorly ventilated) combustion (see Chapter 3). Coal mining and natural gas flaring and supply are also sources of methane emissions.

In summary, there are six different significant sources of anthropogenic methane emissions and their relative importance may be indicated as follows:

Livestock > rice ≫ landfills = coal mining > biomass burning = natural gas transport and supply.

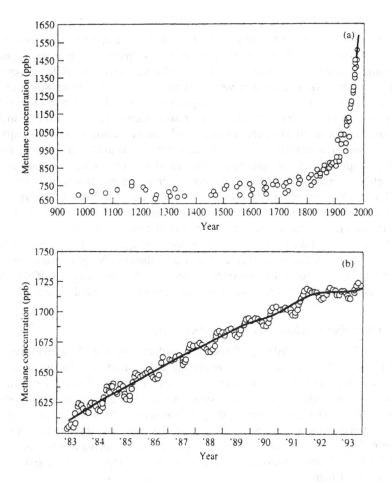

Figure 8. Growth of atmospheric methane concentration: (a) over the past 1,000 years, using data from various ice cores, and (b) in recent times. The smooth curve in the latter averages out the seasonal fluctuations shown by circles (Baird, 1999).

The dominant sinks for atmospheric methane, accounting for about 90% of its loss from air, are the reaction with hydroxyl radical, OH, (reaction (1))

$$CH_4 + OH \longrightarrow CH_3 + H_2O. \tag{1}$$

As discussed further in Chapter 3, this reaction is only the first step of a sequence which transforms methane ultimately to CO and then CO_2.

The other sinks for methane gas are reaction with soil and loss to the stratosphere (see more details on stratospheric chemistry and methane sink in Chapter 6).

4.3. Nitrous Oxide

The absorption band of IR light for nitrous oxide is 7.8 and 8.6 μm. Per molecule, N_2O is 206 times as effective as CO_2 in causing an immediate increase in global warming. Terrestrial and aquatic nitrogen in the form of NH_4^+ and NO_3^- are cycled through the biosphere to make proteins and nucleic acids. The processes of decay return the nitrogen to the atmosphere as N_2 and as N_2O by the action of denitrifying bacteria. Almost all the nitrogen fixed by the Haber process is used as fertilizer, so that increased use of fertilizers increases the rate of return of nitrogen to the atmosphere through biological denitrification. Under rice cultivation in the Asian countries, about 1.25% of applied fertilizer nitrogen is released to the atmosphere as nitrous oxide. This effect is reflected in the gradual increase in the global atmospheric levels of N_2O, which have risen from 0.29 to 0.31 ppmv over the past 25 years. Nitrous oxide, N_2O, is rather unreactive in the troposphere and has a residence time of $\sim$ 20 yr. This species contributes also to ozone depletion potential (see Chapter 6).

There are no sinks for nitrous oxide in the troposphere. Instead, all of it rises eventually in the stratosphere where each molecule absorbs UV light and decomposes, usually to N_2 and atomic oxygen (90%) or reacts with atomic oxygen (10%). More details about the role of N_2O as ozone-depleting species are contained in Chapter 6.

4.4. Chlorofluorocarbons and their Replacements

Gaseous compounds consisting of molecules with carbon atoms bonded exclusively to fluorine and/or chlorine atoms have perhaps the greatest potential amongst trace gases to induce global warming. They are both very persistent and absorb strongly in the 8–13 μm range. Owing to their persistence and to their efficiency in absorbing thermal IR, each CFC molecule has the potential to cause the same amount of global warming as do tens of thousands of CO_2 molecules. However, the net effect of CFCs on global warming is small owing to their concentrations, being much smaller than that for CO_2. These compounds have great ozone-depleting potential and will be discussed in Chapter 6.

4.5. Tropospheric Ozone

Like methane and nitrous oxide, tropospheric ozone is a natural greenhouse gas, but one which has a short tropospheric residence time. Ozone's bending vibration occurs at 14.2 μm, near that for CO_2, and thus it does not contribute much to the enhancement of the greenhouse effect since atmospheric carbon dioxide already removes much of the outgoing light in this wavelength.

As instanced further in Chapters 3 and 5, ozone is formed in the troposphere as a result of urban air pollution and forest and grass fires. As a result of these anthropogenic activities, the levels of ozone in the troposphere have probably increased since pre-industrial times. The best guess is that approximately 10% of the increased global warming potential of the atmosphere results from increases in tropospheric ozone, although this value is very uncertain.

Table 6. Estimated relative global warming potentials
of greenhouse gases, weight basis (Bunce, 1994).

Greenhouse gases	Global Warming Potential
CO_2	1
CH_4	11
N_2O	270
CH_3Cl	3400
CF_2Cl_2	7100
CHF_2Cl	1600

4.6. Global Warming Potential of GNG

Although CO_2 has gained most of the attention of the news media in terms of its increased atmospheric concentration and the implications for global warming, recent information suggests that the other trace GHGs (methane, nitrous oxide, tropospheric ozone, CFCs) will probably have a combined effect comparable with that of CO_2. Estimated relative warming potentials of these gases are CO_2, 50%; CH_4, 20%; CFCs, 20%; N_2O, 5%, based on the absolute levels of emission, efficiency of radiation trapping, and atmospheric lifetime of each of these greenhouse gases (Bunce, 1994). For example, CFCs contribute almost half as much to greenhouse warming as CO_2, even though they are many orders of magnitude less abundant, because each ton of CFC emitted is over 6,000 times as effective at trapping infrared radiation (IR) as a ton of CO_2 (Table 6). The global warming potential (GWP) of carbon dioxide is equal to 1 by definition.

However, the absolute effects of various GHG depend on their GWPs, and on their concentrations and residence times in the troposphere.

5. REGIONAL ASIAN IMPACTS OF GLOBAL WARMING

Although climate change and its associated sea level rise may already be affecting the region, the major impacts of these changes will be felt in the future. The uncertainty surrounding the future encourages policy makers to use specific scenarios that refer to options for future behavior of humans, and the related impacts on climate. Projections of the degree and timing of climate change are derived from scenarios of future emissions. These scenarios, in turn, can be used as the basis for estimating the biological/physical (biophysical) and social/economic impacts associated with each scenario.

A set of impact scenarios specific to the countries of the Asian region was prepared under the sponsorship of the Asian Development Bank (ADB). This study has given country-specific scenarios for economic and population growth, and other parameters, which may make the impact analyses more useful for country-scale strategies.

5.1. Regional Impacts of Climate Change and Sea Level Rise

Assessment studies have shown how climate changes and sea level rise may generate a vast array of biological and physical impacts. Particular examples of estimated regional Asian impacts include:

- changes in temperatures—one study provides estimates of annual average temperature increases for inland South Asia and the Indo-China peninsula of 0.3 to 0.7 °C by 2010 and 1.2–4.5 °C by 2070 (ESCAP, 2000);

- the Goddard Institute for Space Studies (GISS) model, based on a doubling of atmospheric CO_2 concentration, provides predictions for the East Asian Seas region with an increase in precipitation in most places, but with a slight decrease for the Philippines (ESCAP, 2000);

- changes in the timing of precipitation, such as the shifting of monsoon patterns and changes in the severity of monsoons;

- changes in plant growth, both positive and negative, are predicted. Increases could occur, induced by higher concentration of CO_2 and increased activity of photosynthesis. The decreases could occur as a result of intolerance of high temperatures or from changes in the amount and timing of precipitation;

- changes in biodiversity and species distribution, including both domesticated crops and livestock, and native flora and fauna;

- changes in oceanic temperatures and their effects on ocean productivity, including the productivity of and growth rates of reef ecosystems, since the organisms that build coral reefs are especially sensitive to temperature;

- the most obvious and the most severe impact of global warming in the Asian region (as well globally) is the rise of the sea level. We should consider this especial impact in more detail.

Physical and biological impacts of sea level rise

The impact of sea level rise (SLR) brought on by climate change is inundation of coastal lands by the higher water level of the oceans. Estimates of the extent of inundation by one meter SLR in the region includes 23,000 km^2 of land in Bangladesh, 126,000 km^2 in China, and significant losses of land in many other nations. Similar estimates from other scenarios are shown in Table 7.

According to the best recent estimates (ESCAP, 2000), a doubling of carbon dioxide concentration in the atmosphere will probably result in a sea level rise of about 30–60 cm, with a most likely rise of 45 cm. If present trends continue, this may be the situation by 2070. The threat that this presents to the low plain regions of the Asian nations is very substantial. Here we would like to illustrate the impact that sea level rise could have on Bangladesh.

Table 7. Land loss and population displacement in various sea level rise scenarios (ADB, 1992).

Country	SLR scenario, cm	Land loss		People displaced	
		km^2	per cent	Millions	per cent
Bangladesh	100	29,846	20.7	14.8	5.0
India	100	5.763	0.4	7.1	0.8
Indonesia	60	34,000	1.9	2.0	1.1
Malaysia	100	7,000	2.1	0.05	0.3
Pakistan	100	1.700	2.1	—	—
Thailand	60			—	—
Vietnam	90	20,000	6.1	—	—

Bangladesh is one of the most densely populated countries of the World, with a large population subjected to frequent flooding and storms. A rise of about 45 cm in sea level could inundate about 11% of the total land area of the country, displacing about 5% of the present population, i.e., about 7 million people. If the sea level reached one meter, about 21% of the land would be inundated, affecting about 20 million people. The area likely to be affected is shown in Figure 9.

In addition to the physical hardship brought upon the population by the loss of land, agricultural output would also suffer considerably. The loss in rice production alone is estimated to be in the range of 0.8–2.9 million tons per year by 2030, and would exceed 2.6 million tons per year by 2070.

6. INTERNATIONAL AGREEMENTS ON GLOBAL CLIMATE CHANGE

An important outcome of the United Nations Conference on Environment and Development, held at Rio de Janeiro in June 1992, was the signing by World leaders of the United Nations Framework Convention on Climate Change (UNFCCC). That agreement has already been signed by more than 165 countries. By definition, UNFCCC provides a general framework for steps which individual countries might take to address the problem of global climate change. It does not set specific targets and timetables for reducing emissions. The Conference of the Parties has been meeting each year to work out the details of implementing the Framework.

At the meeting in Kyoto at the end of 1997 the countries agreed to a Protocol that commits the industrialized countries to reduce their combined emissions of greenhouse gases from their 1990 levels by the period 2008–2012. Specific targets were agreed to by each of the industrialized countries. Of the Asian region, Japan agreed to a reduction of 6% from its 1990 level.

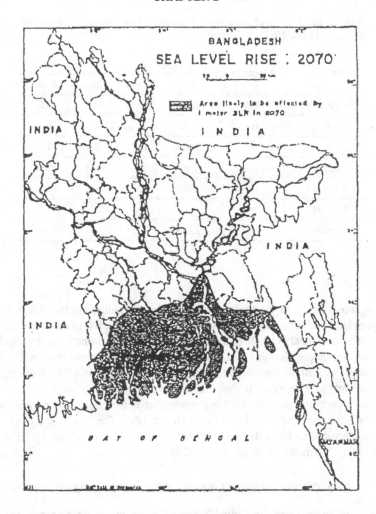

Figure 9. The area of Bangladesh likely to be affected by sea level rise of 1 meter (ESCAP, 2000).

Although the Kyoto Protocol represented a major step forward, a great deal still needs to be worked out in future sessions of the Conference of Parties. At present there are no binding requirements for the developing countries to reduce even the rate of growth of their CO_2 emissions. Many groups in the industrialized countries are citing this as a reason for their countries not yet having ratified the Kyoto Protocol. The developing countries of the Asian region believe, in general, that the industrialized countries are responsible for most of the emissions of GHG, and should reduce them substantially before asking the developing countries to do likewise.

The last problem is more applicable on a per capita basis, but not a true per country basis, as is shown in Section 2. Thus the Asian countries are still the most responsible for regional and global GHG production both at present levels and future projections.

FURTHER READING

1. Bunce N., 1994. *Environmental Chemistry*, second edition, Wuerz Publishing Ltd, Winnipeg, Chapter 1.

2. ESCAP, 1995. *State of the Environment in Asia and Pacific*. United Nations, NY, Chapter 6.

3. Baird C., 1999. *Environmental Chemistry*, second edition, W. H. Freeman and Company, NY, 173–221.

4. Daniels E. J., Jody B. J., Brockmeier N. F. and Wolsky A. M., 1999. Carbon dioxide recovery from fossil-fueled power plants. In: Meyers R. A. (Ed.), *Encyclopedia of Environmental Pollution and Cleanup,* John Wiley, NY, 271–285.

5. ESCAP, 2000. *State of the Environment in Asia and Pacific*. United Nations, NY, Chapter 21.

WEBSITES OF INTEREST

1. Carbon Dioxide Information Analysis Center: Data on CO_2 levels, emissions, etc., http://cdiac.esd.ornl.gov/pns/top10.html

2. Goddard Institute: up-to-date data on average global surface temperatures, http://giss.nasa.gov/research/observe/surftemp.html

3. United Nations: Framework on Climate Change Secretariat, http://www.unfccc.de

4. Global Climate Change Digest, Center for Environmental Information, http://www.globalchange.org/dgsample/sample.htm

5. Observations of CO_2 concentration at the Mauna Loa Observatory, http://mlo.hawaii.gov

QUESTIONS AND PROBLEMS

1. When did the problem of global warming arise and who was the first scientist to put this question?

2. What are the greenhouse gases? Discuss their principal similarities in terms of chemical and physical properties.

3. Discuss the radiation balance of the Earth's surface. Present quantitative data on the trapping of IR radiation in the troposphere.

4. Characterize the main sources of carbon dioxide emissions in the Asian region. Present the data on a per capita and per country basis.

5. Describe the regional sources of methane emissions in the Asian countries. Discuss the role of rice production in the methane emission pattern.

6. What chemical properties of a nitrous oxide molecule allow a researcher to consider this as both a global warming and an ozone-depleting compound?

7. Characterize the emission pattern of CFCs in the Asian countries and make a prognosis of their future influence in global warming.

8. Characterize the chemical composition of the atmosphere. Discuss the major and minor constituents.

9. Present the definition of mean residence time and give characteristic examples for different GHGs.

10. Discuss the role of mean residence time of a chemical species in developing their global warming potentials.

11. Describe the relationship between sources and sinks of carbon dioxide on a regional scale.

12. Discuss the actual and potential role of ocean waters as a sink of carbon dioxide. Indicate the limitations of CO_2 dissolving in sea waters.

13. Describe the relative influence of various sources of methane production in the Asian region and point out the characteristic peculiarities of methane emissions in this area.

14. Compare the global warming potentials of different GHGs and discuss the relative role of various chemical compounds in greenhouse effects.

15. What are the biological and physical consequences of global warming in the Asian region?

16. Present a discussion of the possible threat from sea level rise in various Asian countries and make a conclusion about the most dangerous areas in the region.

17. Discuss the political aspects of emissions of greenhouse gases. Present the role of developed and developing countries in the mitigation of global warming.

CHAPTER 3

URBAN AIR POLLUTION

1. INTRODUCTION

In Asia the concept of urban air pollution has changed significantly during the past several decades. Thirty or fifty years ago air pollution was only associated with smoke, soot, and odor. At present we would suggest the following definition that describes the concentration of many chemical species in urban air. *Air pollution is the presence of any substance in the atmosphere at a concentration high enough to produce an objectionable effect on humans, animals, vegetation, or materials, or to alter the natural biogeochemical cycling of various elements and their mass balance.* These substances can be solids, liquids, or gases, and can be produced by anthropogenic activities or natural sources. In this chapter, however, only non-biological materials will be considered. Airborne pathogens and pollens, molds, and spores will not be discussed. Airborne radioactive contaminants will not be discussed either. The natural urban air pollution due to forest fires and corresponding haze problems will be considered in Chapter 5.

Air pollution in cities can be considered to have three components: sources, transport, and transformations in the troposphere, and receptors. The sources are process, device, or activity, which emits airborne substances. When the substances are released, they are transported through the atmosphere and are transformed into different substances. Air pollutants which are emitted directly to the atmosphere are called *primary pollutants.* Pollutants which are formed in the atmosphere as a result of transformations are called *secondary pollutants.* The reactants that undergo the transformation are referred to as precursors. An example of a secondary pollutant is troposphere ozone, O_3, and its precursors are nitrogen oxides ($NO_x = NO + NO_2$) and non-methane hydrocarbons, NMHC. The receptors are the person, animal, plant, material, or urban ecosystems affected by the emissions (Wolff, 1999).

2. MODERN STATE OF URBAN AIR POLLUTION IN ASIA

The rapid growth of cities has, together with associated industry and transport systems, resulted in an equally rapid increase in urban air pollution in the Asian region. Air pollution is principally generated by fossil fuel combustion in the energy, industrial and transportation systems. Use of poor quality fuel (e.g., coal with high sulfur content

Table 1. Predominant outdoor pollutants and their sources.

Pollutants	Sources
Sulfur oxides	Coal and oil combustion, smelters
Ozone	Photochemical reactions
Lead, manganese	Automobiles, smelters
Calcium, chlorine, silicon, cadmium	Soil particulate and industrial emissions
Organic substances	Petrochemical solvents, unburned fuel

and leaded gasoline), inefficient methods of energy production and use, poor condition of automobiles and roads, traffic congestion and inappropriate mining methods in developing countries are major causes of increasing airborne emissions of sulfur dioxide (SO_2), nitrogen oxides (NO_x), suspended particulate matter (SPM), lead (Pb), carbon monoxide (CO) and ozone (O_3). Predominant outdoor pollutants are shown in Table 1.

Air quality is worsening in virtually all Asian cities, except perhaps in Singapore, South Korea and Japan. Air pollutants, mainly in the form of suspended particulate and sulfur dioxide is most common in the cities of the developing countries. Among mega-cities in the region, and in the world for that matter, Beijing and Bangkok are the two most polluted cities. In general, cities in high-income countries like Tokyo, Osaka, and Seoul, have relatively lower levels of SPM and SO_2 in air than the cities in the developing countries, for instance, Shenyang, New Delhi, Tehran and Jakarta, where WHO Guidelines for these species are invariably exceeded. Air pollution by nitrogen oxides is one of the major problems in the cities of developed countries like Japan (see Chapter 1). In China the annual average concentration of SO_2 is 66 $\mu g/m^3$, nitrogen oxide, 45 $\mu g/m^3$, and total SPM, 291 $\mu g/m^3$. In New Delhi air pollution is so heavy that one day of breathing it is comparable to smoking 10 to 20 cigarettes (ESCAP, 2000). You can see these data in comparison with WHO Guidelines in Figures 1–3.

The deterioration of air quality of the urban areas is mainly the result of increases in industrial and manufacturing activities and in the number of motor vehicles. Motor vehicles normally concentrate in the urban areas and contribute significantly to the production of various types of air pollutants, including carbon monoxide, hydrocarbons, nitrogen oxides and particulates. For example, it is estimated that around 56 tons of CO, 18 tons of hydrocarbons, 7 tons of NO_x, and less than one ton each of SO_2 and particulate matter are discharged daily through the tile pipes of vehicles in Kathmandu alone. In Shanghai the contribution of CO, hydrocarbon, and NO_x emission by automobiles to the air was over 75, 93 and 44%, respectively. These figures are estimated to increase further to 94% for NO_x, 98% for hydrocarbons, and 75% for NO_x by 2010. In Delhi vehicles already account for 70% of the total emissions of nitrogen oxides, not to mention the amount of lead pollution from using leaded gas.

Total Suspended Particulates

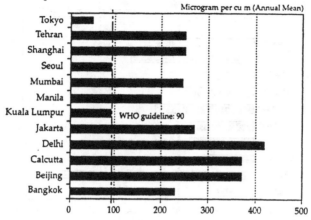

Figure 1. Ambient levels of TSP in Asian cities (ESCAP, 2000).

Sulphur Dioxide

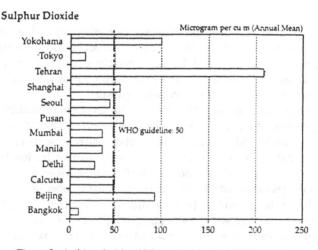

Figure 2. Ambient levels of SO₂ in Asian cities (ESCAP, 2000).

In the wake of a growing number of motor vehicles, the problem is likely to become more acute in the future. Many Asian cities with more prosperous economics have already tripled or quadrupled in the number of passenger cars over the last 10–15 years. In Bangkok, for example, the number of road vehicles grew more than sevenfold between 1970 and 1990 and more than 300,000 new vehicles are added to the streets of this city every year. In China it is projected that by 2015 there will be 30 million trucks and 100 million cars, and that the scope for future growth will still be huge. The forces driving this level of growth in the vehicle number in the region

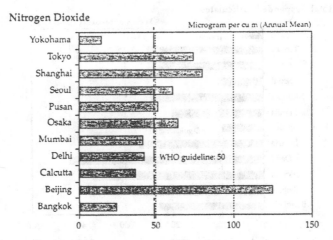

Figure 3. Ambient levels of NO$_x$ in Asian cities (ESCAP, 2000).

range from demographic factors (urbanization, increasing population, and smaller households), to economic factors (higher incomes and declining car prices), to social factors (increased leisure time and the status associated with vehicle ownership), to political factors (powerful lobbies and governments which view the automobile industry as an important generator of economic growth).

Most of the growth in motor vehicle fleets in the developing countries is concentrated in large urban areas. Primary cities draw the largest concentration of vehicles. For instance, in Iran, South Korea, and Thailand, about half of these countries' automobiles are in the capital cities. In Shanghai, the number of automobiles doubled between 1985 and 1990, and at present is more than half a million. However, the growth in the vehicle fleet results primarily from increases in the number of motorized two- and three-wheel vehicles, which are more affordable than cars for large segments of the population and often serve as a stepping-stone to car ownership. In Thailand, Malaysia and Indonesia, for instance, two- and three-wheelers make up over half of motor vehicles. The number of two- and three-wheel vehicles is expected to grow most rapidly in China, India and in other densely populated low income countries. In China, it is projected that there will be 70 million motorcycles by 2015. Production of motorcycles and cars in India is also increasing 20% annually, outstripping that for buses, which grow at 3% per year. In Nepal the registered motor vehicles as of 1998 totaled over 200,000, with more than half comprised of two-wheelers. Over half of these were concentrated in Kathmandu.

Owing to the tremendous rise in the number of vehicles in several countries of the region, the increase in per capita energy consumption has also been quite dramatic. It is projected that energy use in the region will double between 1990 and 2010. In kilograms of oil equivalent, it has increased from 91 to 219 in Indonesia, 80 to 343 in

Thailand, 312 to 826 in Malaysia, and 670 to 2,165 in Singapore. In urban areas high energy use contributes to local air pollution. Cars consume about five times more energy and produce six times more pollutants than buses. Another environmental impact of this development besides the related air pollution is the depletion of non-renewable natural resources. Like the air pollution problem, the depletion of non-renewable sources of energy also has global implications.

The mounting cost of pollution in the cities of the developing Asian countries is a waste of human and physical resources. In Bangkok, Jakarta, and Kuala Lampur the annual cost from dust and lead pollution is estimated at US$5 billion, or about 10% of combined city income. Air pollution also pushes up the incidence and severity of respiratory-related diseases. Mortality from cardiovascular disease, particularly of the elderly (over 65 years) population, increases with air pollution because labored breathing strains the heart. Studies in China revealed that air pollution, along with smoking, also greatly increases the risk of lung cancer.

The more developed nations in the region have exhibited improvement in air quality in recent years due to a number of measures taken to mitigate air pollution problems. For instance, in South Korea levels of sulfur dioxide and total suspended particulates have been declining in Seoul and Pusan since 1990. However, slight increases in concentration of other pollutants such as nitrogen oxides, ozone, and carbon dioxide in major cities of South Korea have been recorded. In Hong Kong, SO_2, NO_2 and TSP levels averaged $80\,\mu g/m^3$ in 1998. Air quality in Singapore has also significantly improved with the adoption of various strategies to prevent air pollution at the source. Several countries of the region are now promoting the use of unleaded gas. China was planning to convert fully to unleaded gas in 2000 (ESCAP, 2000).

3. DEVELOPMENT OF PHOTOCHEMICAL SMOG IN ASIAN CITIES

The above mentioned urban air pollution in Asian cities drives the tropospheric chemical reactions. This tropospheric chemistry is dominated by the oxidation of trace atmospheric components, as a result of which organic compounds such as methane and other hydrocarbons are converted into carbon dioxide and water. The consequences of these chemical transformations are known as photochemical smog (photosmog) and the associated problem of ground level ozone. Here we should consider also the effects of particulate matter, one of the major pollutants of urban air in Asia.

3.1. Formation of Hydroxyl Radical

The hydroxyl radical is central in tropospheric chemistry and photochemical smog formation. The hydroxyl radical OH is continually being formed and consumed in the troposphere and it has very short half-life due to its high reactivity, especially in urban polluted air. This species carries no charge, and is therefore chemically distinct from hydroxyl ion, OH^-, which has an additional electron. The major route for the formation of hydroxyl radical in the troposphere occurs by a complicated mechanism, which is driven by sunlight

$$NO_2 \xrightarrow{h\nu,\ \lambda < 400\,nm} NO + O, \tag{1}$$

$$O + O_2 \xrightarrow{M} O_3, \tag{2}$$

$$O_3 \xrightarrow{h\nu,\ \lambda < 320\,nm} O_2{}^* + O{}^*, \tag{3}$$

$$O{}^* + H_2O \longrightarrow 2\,OH. \tag{4}$$

Other sources of OH are shown in equations (5)–(7), and reaction (6) can give a peak in OH shortly after sunrise under some conditions of local air pollution

$$O{}^* + CH_4 \longrightarrow OH + CH_3OH, \tag{5}$$

$$HNO_3\ (or\ HNO_2) \xrightarrow{h\nu} NO + OH, \tag{6}$$

$$H_2O_2 \xrightarrow{h\nu} 2\,OH. \tag{7}$$

3.2. Tropospheric Concentration of OH in Urban Areas

The globally, seasonally, and diurnally averaged value for concentration of OH is 3×10^{-5} ppbv (Bunce, 1994). Local concentrations of OH are not well monitored experimentally in the Asian domain. The values $> 1 \times 10^7$ radicals cm^{-3} are registered under highly polluted air conditions (Bangkok, Metro Manila, New Delhi). However, the background rural values are 2.5–25×10^5 radicals cm^{-3}, e.g., of two orders less. In rural areas, the concentration of OH parallels closely the intensity of sunlight (see reaction (3) as a main source of OH). In a polluted urban troposphere, the dependence between sunlight and OH formation is complicated by many chemical reactions.

3.3. Chemistry of Hydroxyl Radical

The main two reactions of OH radical are connected with addition and abstraction of an H^+ atom.

1. Addition to an unsaturated center:

$$OH(g) + NO_2(g) \xrightarrow{M} HNO_3(g), \tag{8}$$

where M is a catalyst.

2. Abstraction from a suitable substrate:

$$OH(g) + CH_4(g) \longrightarrow CH_3(g) + H_2O(g). \tag{9}$$

The whole spectrum of hydroxyl radical reactions is shown in Figure 4. These reactions will be partly considered in the text of this chapter.

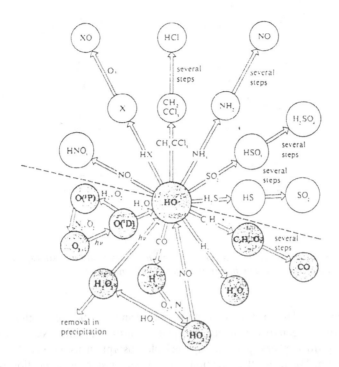

Figure 4. Control of trace gas reactions by OH *in polluted urban troposphere. Processes below the dashed line are those largely involved in controlling the concentration of* OH *and those above the line control the concentrations of the associated reactants and products. Reservoirs of atmospheric species are shown in circles, reactions denoting conversion of one species to another are shown by arrows, and the reactants and products needed to bring about a particular conversion are shown along the arrows. Hydrogen halides are denoted by* HX *and hydrocarbon by* C_xH_y *(Davis and Chameides, 1982).*

3.4. Physical Description of Photosmog

Physical characteristics of photosmog include a yellow-brown haze, which reduces visibility, and the presence of substances which irritate the respiratory tract and cause eye-watering. The yellowish color is owed to NO_2, whilst the irritant substances include ozone, aliphatic aldehydes, and organic nitrates. The four conditions necessary before photosmog can develop are:

- sunlight

- hydrocarbons

- nitrogen oxides

- temperatures above 18 °C

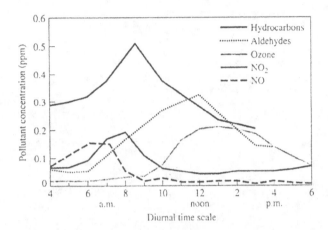

Figure 5. Sketch of the diurnal variation in the concentrations of nitrogen oxides, hydrocarbons, ozone and aldehydes under conditions of photosmog (Manahan, 1994).

Sunlight is needed for the formation of OH radical and initiation of the photochemical reactions of nitrogen oxides in the troposphere. Nitrogen dioxide, NO_2, is important as the only tropospheric gas with appreciable absorption in the visible region of the spectrum. We recall that reaction (1) is the first step in production of ozone and ultimately the OH radical. The chemical reactions of photosmog involve the attack of a hydroxyl radical on organic substances. The temperature being above 18 °C gives an idea of the temperature-dependent reactions that increase production of obnoxious by-products to build up to the levels associated with urban air pollution.

The reader can easily estimate whether or not the local conditions in his/her region are suitable for photochemical smog formation. We recall Section 2 that describes the modern state of urban air pollution and the known climatic conditions of subtropical and tropical zone where most Asian mega-cities are located.

Photochemical smog was recognized as an urban air pollution problem in Los Angeles, California, USA, in 1949. From that time this phenomenon has been documented in many other sunny locations in the United States and elsewhere in the world like San Paulo, Brazil; Mexico City, Mexico; Metro Manila, Philippines; Bangkok, Thailand; New Delhi, India; Shanghai, China and in many other Asian cities with urban pollution from automobile transport. As long ago as the 1950s the automobile was identified as the leading contributor to photochemical smog. Los Angeles was the first major American city to build an extensive freeway system and to rely principally on private automobiles rather than public facilities for transportation. At present this is common in many Asian cities (see Section 3).

The evidence against the automobile is illustrated in Figure 5, which can be interpreted as follows.

Early in the morning pollution levels are low. Nitrogen oxide and unburned hydro-carbon concentrations rise as people drive to work. As the sun'rises higher in the sky, NO is converted to NO_2, and subsequently levels of ozone and aldehydes increase. The latter maximize towards midday, when the solar intensity is highest. Notice that the concentration of NO_x falls after about 10 a.m. and does not rise again during the evening rush hours. There is no second peak at the evening rush hour, because by then the free radical chain reactions are already fully under way.

The previous paragraph can be restated in the context of reactions (1)–(4). Automobile emissions cause elevated concentrations of NO, which is oxidized to NO_2. Nitrogen dioxide is photolyzed in sunlight, and this reaction proceeds faster the higher the photon intensity. Tropospheric ground level (as opposed to stratospheric ozone that will be considered in Chapter 6) ozone is formed, and its photolysis leads to the formation of OH. Automobile emissions also provide the organic substances (substrates) for reaction with OH; intermediates and by-products—such as alde-hydes and organic nitrates—of the oxidation of these substrates to CO_2 and H_2O are the irritating compounds of the smog. Many of these reactions are temperature-dependent, and so photochemical smog becomes increasingly noticeable the hotter the weather.

An overview of the processes responsible for the formation of photosmog is summarized in Figure 6. The chemical reactions for the processes illustrated in this figure are explained in the following section.

All four conditions for photochemical smog must be met simultaneously; conse-quently, the location and the seasons where this phenomenon is likely to be observed may be predicted. Since automobiles provide the NO_x and HC, photosmog is a big city phenomenon. Sunlight and high temperatures are needed. Other factors contributing to photosmog include orografic features, which may hinder the dispersal of the pollutant plume; this is a factor in the Los Angeles district, where mountains to the east tend to trap the air close to the city. Temperature inversions and lack of wind both serve to localize the pollutant plume and hinder its dispersal.

Thus this type of urban air pollution is called photochemical smog of Los Angeles type. This air pollution phenomenon is very frequent in many Asian cities, especially those in subtropical and tropical zone.

3.5. Chemical Aspects of Photosmog

In chemical terms the simplest way of considering the interplay between NO_x and tropospheric ground level ozone in urban air is in terms of the reactions below

$$NO_2 \xrightarrow{h\nu,\ \lambda < 400\,nm} NO + O,$$

$$O + O_2 \xrightarrow{M} O_3,$$

$$NO + O_3 \longrightarrow NO_2 + O_2. \tag{10}$$

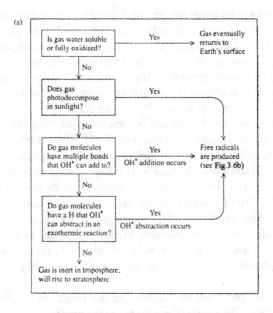

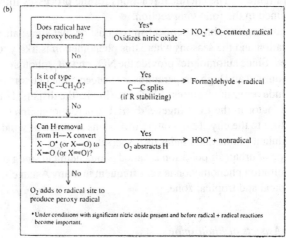

Figure 6. Generalized scheme for the formation of photochemical smog (Baird, 1999).

Tropospheric ozone and NO_x exist in a pseudo-equilibrium (equation (11)), which is driven by sunlight ($\lambda < 400\,nm$) and reverts to the left thermally

$$NO_2 + O_2 \xleftrightarrow[\text{thermal}]{h\nu} NO + O_3. \qquad (11)$$

In the absence of other reactions between these gases, the relative concentrations of NO, NO_2 and O_3 should depend on the solar flux and hence upon the season, geographical location, and time of day. For example, $[O_3] \approx 20 \cdot [NO_2]/[NO]$ at 40 °N *in the absence of urban air pollution.*

The peak levels of ground ozone under conditions of photosmog cannot be predicted just from Equilibrium (11). The concentrations of oxidizable substrates as well as NO_x must be considered. Maximum $[O_3]$ will be produced when the ratio $[NMHC] : [NO_x]$ is near 10. Here NHHC is non-methane hydrocarbons. This ratio depends both on the total concentrations of NO_x and oxidizable substrates, and also on the aging of the pollutant plume. The ratio of substances in the plume changes as the plume ages, because of photochemically induced reactions, and therefore the composition of plume may be very different in rural areas downstream of the original pollution source (Bunce, 1994). The oxidation chemical mechanisms of photosmog formation in polluted urban air can be best seen by writing it in terms of a cycle in which the stoichiometric reaction contains no reactive intermediates and also leaves the NO : NO_2 ratio undisturbed. We will discuss these cycles using the simplest example of oxidation of carbon monoxide as the fuel. You should understand that the oxidation of any other compound is much more complicated.

Cycle 1:

$$CO + OH \longrightarrow CO_2 + H, \tag{12}$$

$$H + O_2 \longrightarrow HO_2, \tag{13}$$

$$HO_2 + NO \longrightarrow NO_2 + OH, \tag{14}$$

$$NO_2 + O_2 + h\nu \longrightarrow NO + O_3.$$

Net reaction:

$$CO + 2O_2 + h\nu \longrightarrow CO_2 + O_3.$$

Under polluted urban air conditions, where sufficient NO is present to reduce HO_2 rapidly, oxidation of CO produces ozone as a by-product, additional to that formed by pseudo-equilibrium reaction (11) alone.

Cycle 2:

$$CO + OH \longrightarrow CO_2 + H,$$

$$H + O_2 \longrightarrow HO_2,$$

$$HO_2 + O_3 \longrightarrow 2O_2 + OH. \tag{15}$$

Net reaction:

$$CO + O_3 \longrightarrow CO_2 + O_2.$$

As we have mentioned already, similar, but more complicated cycles can be written for the oxidation of other substrates. For example, cycle 3 is drawn for methane.

Cycle 3a, methane to formaldehyde:

$$CH_4 + OH \longrightarrow CH_3 + H_2O,$$

$$CH_3 + O_2 \longrightarrow CH_3OO,$$

$$CH_3OO + NO \longrightarrow NO_2 + CH_3O, \tag{16}$$

$$CH_3O + O_2 \longrightarrow CH_2O + HO_2, \tag{17}$$

$$NO + HO_2 \longrightarrow OH + NO_2, \tag{18}$$

$$2NO_2 + 2O_2 + 2h\nu \longrightarrow 2NO + 2O_3 \quad 2 \times (11).$$

Net reaction:

$$CH_4 + 4O_2 + 2h\nu \longrightarrow CH_2O + H_2O + 2O_3.$$

Cycle 3b, formaldehyde to carbon monoxide:

$$CH_2O + OH \longrightarrow HCO + H_2O, \tag{19}$$

$$HCO + O_2 \longrightarrow H_2O + CO, \tag{20}$$

$$NO + HO_2 \longrightarrow OH + NO_2,$$

$$NO_2 + O_2 + h\nu \longrightarrow NO + O_3.$$

Net reaction:

$$CH_2O + 2O_2 + h\nu \longrightarrow CO + H_2O + O_3.$$

The combination of Cycles 1, 3a, and 3b indicates that as much as 4 moles of ozone can be produced per mole of methane oxidized (reaction (21))

$$CH_4 + 11O_2 \longrightarrow CO_2 + 2H_2O + 4O_3. \tag{21}$$

Cycles such as (1), (3a) and (3b) explain why ground level ozone is frequently present at higher concentration down wind of an urban area than in the city itself. The concentration of OH in the urban polluted atmosphere is reduced when NO_x levels are high, because NO_2 reacts with OH (reaction (22))

$$HO + NO_2 \xrightarrow{M} HNO_3. \tag{22}$$

Reaction (22) is important under conditions of urban air pollution when both $[NO_2]$ and $[OH]$ tend to be high. The influence of reaction (22) is seen in the variation of the concentrations of nitrogen oxides and ozone with time under conditions of photosmog. In Figure 5 the NO_x content peaks about 8 a.m., and the oxidant concentration rises as NO_x falls. This observation is considered with reaction (22) acting as a sink for NO_x, in addition to the cycling back and forth between NO and NO_x implied by Equilibrium (11). Reaction (22) represents the most important sink for NO_x in the polluted air of Asian cities. The HNO_3 is either broken down by photolysis, releasing NO_2 and OH again, or is removed permanently from the atmosphere by deposition in rain or adsorbed in particles. This process has been monitored in Seoul (Yoon and Won, 1998).

Reaction (22) links also photosmog and acid deposition (see Chapter 4), especially in Japanese cities and Bangkok, where nitric acid is the major contributor to acidic deposition.

We can summarize the major chemical pathways involving NO and NO_2 as follows:

$$N_2 + O_2 \xrightarrow{\text{heat}} 2\,NO \underset{h\upsilon}{\overset{\substack{O_3, HO_2 \\ \text{other oxidants}}}{\rightleftharpoons}} NO_2 \underset{h\upsilon}{\overset{OH}{\rightleftharpoons}} HNO_3 \longrightarrow \text{acid deposition.}$$

Peroxyacetyl nitrate (PAN)

The irritation of air enriched solely with NO or NO_2 did not produce photo-oxidants; this only occurred when hydrocarbons were also present in the polluted urban atmosphere. This leads to a build-up of tropospheric ozone, and hence to faster rates of photoinitiation through photodissociation of ozone, and then to a further build-up of ozone, and so on. Besides ozone, which is toxic at low concentrations (0.1–1 ppmv), other intermediates responsible for adverse effects include aldehydes and organic nitrates, such as peroxyacetyl nitrate (PAN).

The difference between the oxidation chemistry occurring in unpolluted background areas and that involved in photosmog formation in Asian cities, is that the latter condition occurs when the concentrations of NO_x and HC are much higher than in rural areas (Table 2).

Aldehydes may be formed as intermediates in the oxidation of alkanes and alkenes, initiated by a hydroxyl radical, and by direct attack of ozone upon alkenes. Attack by OH is the dominant sink as shown in reaction (23)

$$CH_3CH = O + OH \longrightarrow CH_3C = O + H_2O. \tag{23}$$

Organic nitrates are formed by radical coupling reactions between NO_2 and either peroxyalkyl or peroxyacyl free radicals (see list of air free radicals in Box 1.). These nitrates redissociate both thermally and photochemically, and thus represent temporary reservoirs for NO_x, as follows from reaction (24)

$$ROO + NO_2 \longrightarrow ROONO_2, \tag{24}$$

where R = alkyl oracyl group.

Table 2. *Typical concentrations of chemical species in unpolluted background and polluted urban areas of the Asian region, ppbv (ESCAP, 1995; 2000, with additions).*

Air species	Unpolluted areas	Heavily polluted areas
Nitrogen oxides	< 1	1,200–3,000
Carbon monoxide	< 250	15,000–60,000
Ozone	< 60	120–500
*Hydrocarbons	< 65	> 1500
PAN	< 0.05	20–70
Aldehydes	< 2	20–80

*Note: Excluding methane (ca. 1700 ppbv), most of which is of natural origin.

Box 1. *Important free radicals in atmospheric chemistry and their mean concentrations (Barnes, 1999).*

Radical	Formula	Main source(s)	Concentration, radicals cm^{-3}
Oxygen atom	O	Photolysis of ozone and NO_2	2.5×10^4
Hydroxyl	OH	Reaction of O with water vapor	8×10^5
Hydroperoxy	HO_2	Reaction of O_2 with alkoxy radicals	6.5×10^8
Methoxy	CH_3O	Photodissociation of CH_4	1.3×10^6
Methylperoxy	CH_3OO	Photodissociation of CH_4	1.0×10^8
Nitrate	NO_3	Reaction of NO_2 with O_3	3.0×10^8

It is known that PAN is thermally labile, and reverts to NO_2 and acetylperoxy radical with the activation energy of $112 \, kJ \, mol^{-1}$, thereby re-initiating free radical oxidation of hydrocarbons (Grojean et al, 1994). If air containing PAN cools quickly, the PAN may be transported to remote locations. This is believed to be the manner in which PAN has come to be found in the Arctic and mid-Atlantic, far from any industrial pollution sources (Bunce, 1994).

Higher homologues such as peroxypropinyl nitrate are also present. Like ozone, peroxyacyl nitrates are toxic and irritant at very low concentrations (< 0.1 ppbv).

The production of PAN from ethane is shown in Figure 7. The principal scheme for methane oxidation is similar to that for methane and PAN is not the main product. The other reactions lead eventually to carbon dioxide as a chief product.

Nitrate radical, NO_3

The nitrate free radical is uncharged, and is chemically distinct from the nitrate anion, NO_3^-. It is formed in atmosphere by oxidation of NO_2

$$NO_2 + O_3 \longrightarrow NO_3 + O_2. \tag{25}$$

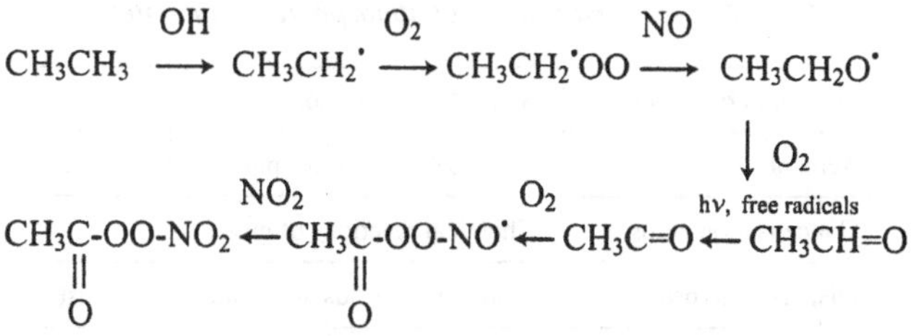

$$\begin{array}{ccccccc}
& \text{OH} & & \text{O}_2 & & \text{NO} & \\
\text{CH}_3\text{CH}_3 & \longrightarrow & \text{CH}_3\text{CH}_2^{\textbf{·}} & \longrightarrow & \text{CH}_3\text{CH}_2^{\textbf{·}}\text{OO} & \longrightarrow & \text{CH}_3\text{CH}_2\text{O}^{\textbf{·}}
\end{array}$$

$$\downarrow \text{O}_2$$

$$\begin{array}{ccccccc}
& \text{NO}_2 & & \text{O}_2 & & h\nu, \text{ free radicals} & \\
\text{CH}_3\text{C-OO-NO}_2 & \longleftarrow & \text{CH}_3\text{C-OO-NO}^{\textbf{·}} & \longleftarrow & \text{CH}_3\text{C=O} & \longleftarrow & \text{CH}_3\text{CH=O} \\
\| & & \| & & & & \\
\text{O} & & \text{O} & & & &
\end{array}$$

Figure 7. Production of PAN from ethane.

Of the various tropospheric intermediates we have discussed, NO_3 is unique in that its reactions are more important at night than during daytime. Its daytime concentration is extremely low due to rapid photolysis by visible light

$$NO_3 \xrightarrow{h\nu, \ \lambda < 670 \, nm} NO_2 + O. \tag{26}$$

Some of the NO_3 escape destruction by conversion to N_2O_5, which acts as a temporary reservoir

$$NO_3 + NO_2 \longleftrightarrow N_2O_5, \tag{27}$$

N_2O_5 dissociates thermally (i.e., during both day and night), and releases NO_3 again. At night, in the absence of photodissociation, the major sinks for NO_3 are hydrogen abstraction and addition to the unsaturated center. The chemistry of NO_3 is thus similar to that of the hydroxyl radical, although NO_3 is intrinsically less reactive and its average concentration is higher (see Box 1)

$$RH + NO_3 \xrightarrow{O_2} HNO_3 + R \quad \{\rightarrow ROO, \text{ etc.}\}. \tag{28}$$

Tropospheric HNO_3 is formed by several routes; reaction (25) is the major source, with hydrogen abstraction by HNO_3 (reaction (28)) and hydrolysis of (reaction (29)) as the minor sources

$$N_2O_5 + H_2O \longrightarrow 2\,HNO_3. \tag{29}$$

4. ROLE OF PARTICLES IN URBAN AIR POLLUTION IN ASIAN CITIES

Particles or particulates in the urban atmosphere, which range in size from about one-half millimeter (the size of sand or drizzle) down to molecular dimensions, are made up of an amazing variety of materials and discrete objects that may consist of either solids or liquid droplets. A number of terms are commonly used for different particulates depending on size and sources (Table 3).

We can see that particulates originate from a wide variety of sources and processes, ranging from simple grinding of bulk matter to complicated chemical and biochemical syntheses. For the most part, aerosols consist of carbonaceous material, metal oxides

Table 3. Types and sources of atmosphere particulates.

Type of particulates	Sources of formation
Aerosol	Colloidal-sized atmosphere particle
Condensation aerosol	Chemical reactions of gases
Dispersed aerosols	Grinding of solids and dispersion of dust
Fog	Fine dispersed water droplets
Haze	Forest fires
Mists	Liquid particles
Smoke	Incomplete combustion of fuel

and glasses, dissolved ionic species (electrolytes), and ionic solids. The predominant constituents are carbonaceous material, water, sulfate, nitrate, ammonium, and silicon. The composition of aerosol particles varies significantly with size. The very small particles tend to be acidic and often originate from gases, such as from conversion of SO_2 to H_2SO_4. Larger particles tend to consist of materials generated mechanically, such as by grinding of limestone, and have a greater tendency to be basic.

In Asia, the sources of particalates in urban air are both natural (Yellow sand phenomenon) and of anthropogenic origin.

4.1. Physical Behavior of Particulates in the Atmosphere

The most important physical characteristic of particulates is their size. This size is usually expressed as the diameter of a particle. The rate at which a particle settles is a function of particle diameter and density. The settling rate is very important in determining the fate and effects of particulates in the urban air. For spherical particulates greater than ≈ 1 μm in diameter, Stokes law can be applied:

$$v = \frac{gd^2(\rho_1 - \rho_2)}{18\eta},$$

where v is the settling velocity in cm sec^{-1}, g is the acceleration of gravity in cm sec^{-2}, ρ_1 is the density of particle in g cm^{-3}, ρ_2 is the density of air in g cm^{-3}, and η is the viscosity of air.

We can not apply this equation to particles with diameter less than one μm since the behavior of these particles is under stochastic motion in accordance with Brown's law.

Yellow sand phenomenon in East Asia

The vast area of Arid and semi-Arid ecosystems of Central and East Asia is subject to wind erosion. The major natural sources of dust emission are the Gobi desert (Xinjiang Province, Northwestern China), and the Karakum and Kazylkum deserts

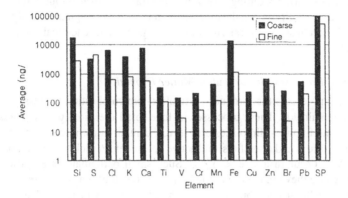

Figure 8. Average concentration of various elements in particulates during Yellow sand case in Seoul (Choi et al, 1998)

(Central Asian areas of Kazakhstan, Uzbekistan and Turkmenistan). The Gobi desert is the second largest desert of the world. The climate is extremely dry and windy. The strongest winds occur during the winter and spring periods. These factors influence the deflation of soil surface layers, which with scarce vegetation are not protected, especially during the springtime, after thawing of the frozen upper soil.

In accordance with the joint effects of aridity and soil texture, dust emission rates increase from the east of the Gobi desert to west by as much as five orders. The maximum emission rate is 1.5 ton/ha/year. The total dust emission rate of the Gobi desert is estimated as 25×10^6 tons per year and that in spring is 15×10^6 tons per year. The seasonal dust emission rates in summer, autumn and winter are 1.4×10^6, 5.7×10^6 and 2.9×10^6 tons, correspondingly.

Chemical composition of particulate matter from Yellow sand phenomenon was studied in Seoul in spring 1998. The mean total suspended particle (TSP) mass concentration in urban air was 98.9 μg m^{-3} with the maximum 264.8 μg m^{-3} and the minimum 23.9 μg m^{-3} (Cho et al, 1998). The ratio of mean SPS mass concentration for Yellow sand cases to that for the non-Yellow sand was 1.50, showing that the SPM tends to be greater in Yellow sand periods. The mean ratio of particles with diameter less than 10 μm (PM10) to the TSP was 0.69 during Yellow sand cases.

The content of various elements, including heavy metal components in coarse particles (PM8) and fine particles (PM0.4) of TSP were analyzed and compared (Figure 8).

Analysis of these data show that concentration of each chemical element was higher in TSP during Yellow sand cases than during non-Yellow sand cases by 2–3 orders. Most elements were indicated to have higher concentrations in coarse particles than in finer ones. This could be related both to the chemical composition of soil from the Gobi desert as the main source of coarse TSP and to the adsorption of heavy metals on particle surfaces during coarse particle transport over polluted regions of East China and South Korea. On the other hand, fine particles originate mainly due to anthropogenic combustion of fossil fuels and correspond to the chemical composition of these fuels.

4.2. *Chemical Processes for Particulate Matter Formation*

Most chemical processes that produce particulates are combustion processes, as we have mentioned already. These processes include fossil fuel fired power plants, incinerators, home furnaces, fireplaces and stoves, cement kilns, internal combustion engines, forest, brush and grass fires, and active volcanoes. Particles from combustion sources tend to occur in a size $< 1\,\mu$m. Such small particulates are particularly important because they are readily carried into the alveoli of lungs and are relatively enriched in more hazardous constituents, such as toxic heavy metals and arsenic.

Inorganic particulates

Metal oxides constitute a major class of inorganic particles in the urban air. These are formed whenever fuels containing metals are burned. For example, particulate iron oxide is formed during combustion of pyrite-containing lignite

$$3\,FeS_2 + 8\,O_2 \longrightarrow Fe_3O_4 + 6\,SO_2. \tag{30}$$

Organic vanadium in residual fuel oil is converted to particulate vanadium oxide. Part of the calcium carbonate in the ash fraction of coal is converted to calcium oxide and is emitted to the atmosphere through the stack

$$CaCO_3 + heat \longrightarrow CaO + CO_2. \tag{31}$$

A common process for the formation of aerosol mists involves the oxidation of SO_2, which content is typically high in the urban air of most Asian cities (see Section 2) to H_2SO_4. This acid is a hydroscopic substance that attracts water vapor in the atmosphere to form small liquid droplets, as follows

$$2\,SO_2 + O_2 + H_2O \longrightarrow 2\,H_2SO_4. \tag{32}$$

We should remember that urban air in Asian cities like Bangkok, Beijing or Seoul contains basic compounds, such as ammonia from automobile emissions or calcium oxide from cement kiln and dust. In these conditions, sulfuric acid will react to form salts

$$H_2SO_4\ (droplets) + CaO(s) \longrightarrow CaSO_4\ (droplet) + H_2O. \tag{33}$$

or

$$H_2SO_4\ (droplets) + 2\,NH_3(g) \longrightarrow (NH_4)_2SO_4\ (droplet). \tag{34}$$

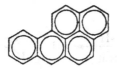

Figure 9. Chemical structure of benz(a)pyrene (Manahan, 1994).

These reactions are common during rainy seasons. Under low air humidity during dry seasons, water is lost from these droplets and solid particulates are formed.

The preceding examples show various ways in which solid or liquid inorganic particulates are formed in polluted urban air by chemical reactions. Such reactions constitute the most important general processes for the formation of aerosols, particularly the smaller particles in air of Asian cities.

Organic particulates

A significant amount of organic particulate matter is produced by automobile engines in combustion processes that involve pyrosynthesis and nitrogenous compounds. These products may include N-containing species and oxidized HC polymers. Lubricating oil and its additives contribute also to organic particulates. Analyses of particulate matter emitted by gasoline auto engines (with and without catalysts) and diesel truck engines determine more than 100 different compounds. Among the prominent classes of compounds measured were *n*-alkanes, *n*-alkanoic acids, benzaldehydes, benzoic acids, azanaphthalenes, polycyclic aromatic hydrocarbons (PAH), pentacyclic triterpanes, and sterans (Manahan, 2000).

Synthesis of polycyclic aromatic hydrocarbons (PAH)

The organic particulates of greatest concern are PAN hydrocarbons, which consist of condensed ring aromatic molecules. The most often cited example of a PAH compound is benz(a)pyrene, a compound that can be metabolized in human organisms to a carcinogenic form (Figure 9).

PAHs may be synthesized from saturated hydrocarbons under oxygen-deficient conditions. Hydrocarbons with very low molecular masses, including even methane, may act as precursors for the polycyclic aromatic compounds. The process of PAH formation from low molar mass HCs is called pyrosynthesis. This happens at temperatures exceeding $\approx 500\,^{\circ}C$ at which $C - H$ and $C - C$ bonds are broken to form free radicals. These radicals undergo dehydrogenation and combine chemically to form aromatic ring structures (Figure 10).

The ability of HCs to form PAHs is changing in order: aromatics > cyclo-olefins > olefins > paraffins.

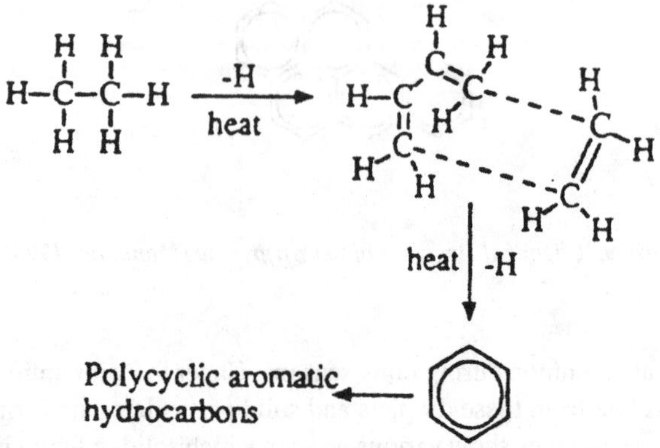

Figure 10. Formation of PAH from methane (Manahan, 1994).

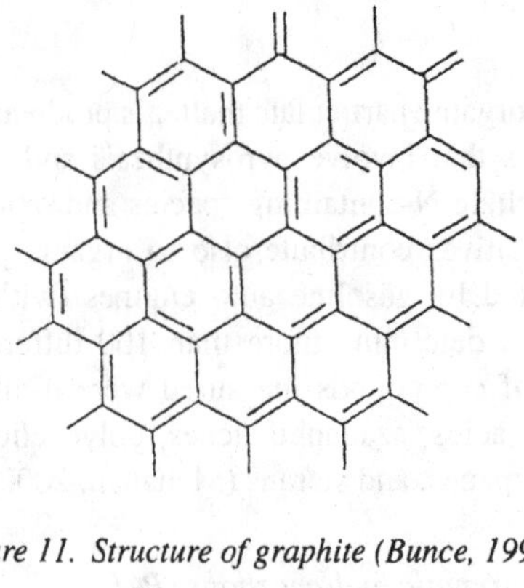

Figure 11. Structure of graphite (Bunce, 1994).

PAH compounds may be also formed from higher alkanes present in fuels and plant materials by pyrolysis, the process of cracking of organic compounds to form smaller and less stable molecules and radicals.

We can consider incomplete combustion of organic products that produces soot. Soot is an impure form of elemental carbon (graphite). Soot particles are roughly spherical, whereas pure graphite has a layered structure (Figure 11).

Soot forms through accretion of graphite-like precursors, but precursors contain many structural defects. Soot is black and it is especially effective in reducing visibility (see below).

Important environmental sources of soot particles in Asian cities are burning coal, petroleum products, and wood. Coal burning produces large amounts of soot particles and still is an important source of reducing visibility in many cities of China and India.

In the case of petroleum products, diesel engines are overall much dirtier than gasoline engines (3 g of carbon per kg of diesel fuel v/v 0.1 g per kg of gasoline). Particulates from wood burning include those from both forest fires (see more details in Chapter 5) and residential wood stoves. The latter will be considered in the following section on indoor air quality). Here we only mention that hardwood (oak, dipterocarp) typically produces 0.4 g soot per kg whereas softwood (pine, beach, palm matter), which is used more often in Asia, produces 1.3 g kg^{-1}.

Pyrolysis of soot produces PAHs. Car and truck exhausts also contain volatile PAHs, such as naphthalene, antracene, phenathrene, etc., which are products of incomplete combustion.

4.3. Role of Particulates in Visibility Degradation in Urban Air

Visibility impairment in urban air is greatly affected by ambient particulate substances: organics, soot (elemental carbon), soil dust, sulfates and nitrates. One of the primary particles, soot is emitted directly to urban air by diverse biomass burning, diesel engines and agricultural activities. All these process are greatly active in Asian cities.

Visibility degradation, one of the most obvious manifestations of air pollution, results from light extinction by atmospheric particles and gases. The effects of pollution on light absorption are particularly strong in urban and industrial sites. Recent studies have indicated that among all the pollutants present in a typical urban environment, airborne particulate matter, especially fine particles, are the main contributors to visibility degradation. The characteristic example is shown for one of the Korean cities, Kwangju (Figure 12).

Here we should compare the visibility decrease due to particalates concentration in urban air, so called London smog, with a similar atmospheric phenomenon, Los Angeles photochemical smog.

London type of smog was common in all industrial countries of the 19th and early 20th centuries. The causative agent in this kind of smog is coal, upon which the Industrial Revolution was founded, and indeed still is in emerging economics of most Asian countries, such as India, China, and Thailand.

Air pollution owing to burning coal has been recorded in London since the Middle Ages. The incidence of the nuisance probably peaked in Victorian times, but it was only in the 1950s that concerns about health finally forced remedial action in the form of clean air legislation. On one particularly smoggy day in January 1955, the light extinction was for a time 99.9% or, in other words, light intensity fell to 0.1%, of what it would have been under sunny conditions, leading to a period of almost total darkness. Under such conditions, solid particulate loading, higher than 4000 μg m^{-3}, was recorded. During the smog of December 1952, which lasted for nearly a week, the death rate in London was over 4000 more than normal, especially among elderly and those with respiratory ailments. Cleaner air in London was only achieved with a ban on burning coal in domestic fireplaces.

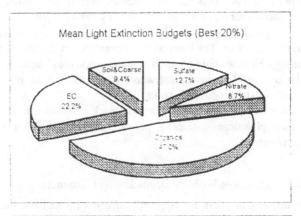

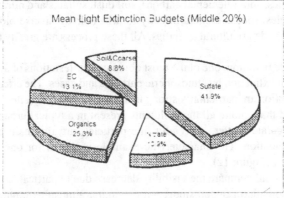

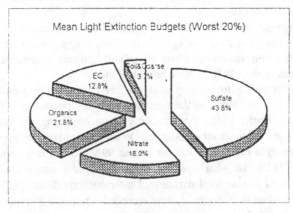

Figure 12. Mean light extinction budgets in Kwangju, South Korea (Kim et al, 1999(b)).

Table 4. Comparison of London and Los Angeles types of smog (Bunce, 1994).

Parameters	Smog types	
	London's type	Los Angeles' type
Pollution source	Coal	Oil
Cause of reduced visibility	Smoke, soot	Liquid aerosols
Chemistry of the air	Reducing (SO_2)	Oxidizing (O_3, PAN)

London type of smog is quite different from photochemical, Los Angeles type of smog. This is shown in the following comparisons (Table 4).

The reader can determine the sources of air pollution in his/her own city and estimate the type of smog, which is predominant during a year or separate seasons. However, we should note that in many Asian cities like Bangkok, New Delhi, Bombay, Jakarta, and Beijing, both types of smog are developing, and very often, simultaneously (ESCAP, 2000).

5. INDOOR AIR QUALITY

Indoor air pollution in urban centers occurs both in the home and in the workplace. It can often pose a greater threat to human health than outdoor air pollution, both in developed and developing countries of the Asian region. In particular, women and young children from low-income households are often at significant risk from exposure to high concentrations of pollutants from cooking in poorly ventilated houses.

In Ahmedabad, India, mean values of SPM during cooking were as high as $25,000 \, \mu g \, m^{-3}$ in coal-burning households, and 15,000 to $20,000 \, \mu g \, m^{-3}$ where wood and dung were used. This is 130 times higher than the threshold set by US safety standards, not to be exceeded more than once a year. In addition, mean levels for the carcinogen, benz(a)pyrene, BaP, were as high as $9,000 \, ng \, m^{-3}$ during cooking: some 45 times higher than US standards for occupational (8 hours) exposure (ESCAP, 1995).

Another example is the high concentrations of SPM, SO_2, CO and BaP, which have been recorded in coal-burning households in many Chinese cities. In Shenyang, lung cancer risk is thought to be 50–70 per cent higher among those who spend most of their lives indoors. For further discussion see Section 6 below.

A study on indoor air pollution abatement through household fuel switching in three cities—Poona, India, Beijing, China, and Bangkok, Thailand—revealed that moving up the "energy ladder", from biomass to kerosene, leads to substantial reductions in health damaging emissions (Tables 5 and 6). Similar effects were noticed when shifting from kerosene to LPG, and from coal (vented) to gas (Smith et al, 1994).

Table 5. Estimated daily exposures from cooking fuel along the energy ladder in Poona, India (Smith et al, 1994).

Fuel	Estimated daily exposure, mg-h/m³	
	PM_{10}	NO_2
Biomass	17–26	0.22–0.66
Kerosene	2.4–3.6	0.08–0.11
LPG	0.4	0.05
Indian standards	0.98	0.28
WHO recommendation	0.56	0.70

Table 6. Estimated daily exposure from household fuel use along the energy ladder in Beijing (Smith et al, 1994).

Fuel	Estimated daily exposure, mg-h/m³		
	PM_{10}	NO_2	CO
Coal (vented)	2.3–3.5	0.31–0.51	310–430
Gas	1.4	0.15	60
Chinese standards	0.7	0.7	56
WHO recommendation	0.56	0.70	140

6. URBAN AIR POLLUTION AND HEALTH EFFECTS

High exposure to both indoor and outdoor ambient air pollution has been associated with a number of illnesses, including:

- acute respiratory infections (particularly in children);

- chronic lung disease, such as asthma, tuberculosis, and associated heart diseases;

- pregnancy stillbirths; and

- cancer.

Since urban air is generally more polluted than rural air, the incidence of related diseases is more common in cities. For example, lung cancer mortality is higher in Chinese cities than in the nation as a whole; and 60% of Calcutta residents, suffer from respiratory diseases, compared to the national average of 3%. Furthermore, it has been estimated in Bangkok that SPM could cause up to 1,400 deaths in the city per year, and that lead pollution could cause 200,000 to 500,000 cases of hypertension, 300 to 900 cases of heart attack and stroke, and 200 to 400 deaths per year (ESCAP, 1995). In spite of significant reduction of lead concentration in the Bangkok urban air due to leaded gasoline being phased out during the second part of the 1990s, some studies indicate that the long-time effects of lead poisoning on children in Bangkok could cause an average loss of 3.5 IQ points per child before the age of seven, i.e., an estimated total loss of 400,000 to 700,000 IQ points per year.

6.1. Acute Respiratory Infections

Acute Respiratory Infection (ARI), such as pneumonia, is one of the biggest causes of death for young children in the Asian region. ARI is also responsible for more episodes of illness than any other disease, with the exception of diarrhoea, and it is well known that ARI is aggravated by exposure to pollutants and indoor environmental tobacco smoke (ETS).

According to WHO estimates, Bangladesh, India, Indonesia and Nepal together account for about 40 per cent of global mortality in young children caused by pneumonia, with infant mortality rates above 40.0 per 1,000 live births. The case-fatality rates due to pneumonia among hospitalized children are between 4.2 and 18.3%. Furthermore, a study in Nepal involving a weekly examination of 240 children under two years of age over a six month period, for ARI incidence and severity, demonstrated a strong relationship between the number of hours per day the children spent indoors by the fire, and the incidence of moderate and severe ARI cases.

6.2. Chronic Obstructive Lung Disease (COLD) and Cor Pulmonale

COLD is known to be an outcome of chronic air pollution exposure. Although tobacco smoke is known to be the major risk factor, studies in India and Nepal have found that non-smoking women who regularly cook on biomass stoves exhibit a higher prevalence of COLD than would be expected, or than appears in women who use them less frequently. Indeed, owing to indoor exposure, nearly 15% of non-smoking women in Nepal (20 years and older) had chronic bronchitis; a very high rate for nonsmokers (ESCAP, 1995).

In China, COLD was associated with long term exposure to indoor coal smoke (Chen et al, 1990), and in India, Cor pulmonale (heart disease secondary to chronic lung disease) has been found to be more prevalent, and on average to develop earlier, in non-smoking women who cook with biomass fuels than those who do not.

6.3. Tuberculosis

Tuberculosis, which continues to be a serious public health problem in the region, is also aggravated by air pollution. In 1991, more than 2,000,000 cases were reported in India and Indonesia alone (Table 7), and just 10 countries in the Asian region accounted for over one million deaths.

Table 7. Tuberculosis notification in selected countries of Asia, notification rate per 100,000 population (ESCAP, 1995).

Country	All forms		Smear positive, of total reported	
	Number	Rate	Per cent	Rate
Bangladesh	56052	47.2	35	16.5
Bhutan	996	169.6	30	50.9
India	1555353	182.5	22	40.6
Indonesia	469832	245.4	16	39.6
Maldives	380	172.4	31	53.4
Mongolia	1611	71.6	7	5.0
Myanmar	16440	38.3	72	27.9
Nepal	8993	45.8	47	21.5
Sri Lanka	6174	35.4	54	19.1
Thailand	50185	88.7	76	66.5

Around 40% of the population in South and Southeast Asia were infected with HIV and M. Tuberculosis is extremely high in the region and the number of active cases may rise by a factor of seven during the coming decade. Furthermore, the age group which is economically most productive (15–59 years) is particularly vulnerable to the disease. As shown in Table 7, notification rates per 100,000 population were particularly high for all forms of the disease in Indonesia, India, Maldives, and for smear-positive cases were above 50 per 100,000 in Thailand, Maldives and Bhutan. Also of note is the extremely low smear-positive rate in Mongolia (5 per 100,000) where the level of urban air pollution is minimal among Asian countries.

6.4. Pregnancy Stillbirths

A study in India found that pregnant women who cook over open biomass stoves have almost 50% greater chance of stillbirth, although there was no measured increase in neonatal death rates (Mavalankar et al, 1991). The main threat to pregnancy appears to come from carbon monoxide, which enters the blood in substantial amount during cooking.

6.5. Cancer

Air pollution has also resulted in an increased incidence of lung and other cancers in the Asian region. One study in Japan (Subue, 1990) found that women who reported cooking with straw or wood fuel when they were 30 years of age subsequently had an

Table 8. Probability of dying from cancer and cardiovascular diseases between ages 15 and 60 years for males and females in China and India, in per cent (Murray and Lopez, 1994).

Disease	China		India	
	Male	Female	Male	Female
Stomach cancer	1.06	0.58	0.37	0.21
Colorectal cancer	0.25	0.23	0.12	0.1
Liver cancer	1.81	0.59	0.14	0.1
Lung cancer	0.65	0.34	0.38	0.1
Diabetes mellitus	0.14	0.17	0.49	0.65
Rheumatic heart disease	0.39	0.64	0.35	0.86
Ischaemic heart disease	0.87	0.45	2.61	1.09
Cerebrovascular disease	2.06	1.75	1.08	1.32
Inflammatory cardiac disease	0.20	0.17	2.31	1.48

80% increased chance of developing lung cancer in later life. We know that cancer, like chronic lung disease, takes many years to develop after exposure. Furthermore, there is mounting evidence that lung cancer can also be associated with exposure to coal smoke. The corresponding statistical assessments have been shown in China (Chen et al, 1990).

Current knowledge suggests that social habits and diet are as much a cause of cancer and cardiovascular diseases as are exposure to environmental hazards such as carcinogens and viruses. Apart from the obvious activities of smoking and hard drinking, other examples of ways in which diet can increase the risk of cancer include preserving food through salting, smoking and pickling, which have been shown to increase the risk of oral and stomach cancer.

Indoor and outdoor exposure to many chemical substances (formaldehyde, asbestos, PVC, many metals, like Cr, As, Be, Ti, V, pesticides and nitrosoamines) can also spur the development of cancer (see Chapter 11 for more detailed discussion of ecotoxicochemistry and ecotoxicology of pollutants). Here we can only state that the International Agency for Research on Cancer has identified 60 environmental agents that can aggravate cancer for humans during exposure to polluted urban air (Misch, 1994). The probabilities of cancer development in urban environments of two of the most polluted Asian countries are shown in Table 8.

The development of all above-mentioned diseases is related to the exceeding of air quality standards in many cities of the Asian regions. These standards for the most frequent pollutants are shown in Table 9. The readers can compare the air quality monitoring results in his/her cities with environmental standards.

Table 9. Ambient air quality standards in some Asian countries, all concentrations in μg m^{-3}, except CO in mg m^{-3} (Radojevic and Bashkin, 1999).

Pollutant	Averaging time	Malaysia	Thailand	Indonesia	Philippines	Singapore	Japan
SO_2	10 min	500	—	—	—	—	—
	1 h	350	—	900	850	—	285
	24 h	105	300	300	370	365	114
	Annual	—	100	60	—	80	—
NO_2	30 min	—	—	—	300	—	—
	1 h	320	320	—	—	—	205
	24 h	—	—	—	—	—	102
	Annual	—	—	100	—	100	—
CO	1 h	35	50	30	35	40	25
	8 h	10	20	10	10	10	—
	24 h	—	—	—	—	—	12
O_3	30 min	—	—	—	200	—	—
	1 h	200	200	160	—	235	128
	8 h	120	—	—	—	—	—
SPM	24 h	260	330	230	180	260	200
	annual	90	100	90	—	75	100
Lead	24 h	—	10	2	—	—	—
3 months		1.5	—	20	1.5	—	

FURTHER READING

1. Manahan S. E., 2000. *Environmental Chemistry*, seventh edition, Lewis Publishers, NY, 265–398.

2. Baird C., 1999. *Environmental Chemistry*, second edition, W. H. Freeman and Company, NY, 112–169.

3. Bunce N., 1994. *Environmental Chemistry*, second edition, Wuerz Publishing Ltd, Winnipeg, Chapter 3.

4. ESCAP, 1995. *State of the Environment in Asia and Pacific*. United Nations, NY, Chapters 8 and 16.

5. ESCAP, 2000. *State of the Environment in Asia and Pacific*. United Nations, NY, Chapter 18.

WEBSITES OF INTEREST

1. Air pollution information network,
http://www.york.ac.uk/inst/sei/APIN/welcome.html
http://www.york.ac.uk/inst/sei/SEI/welcome.html

2. http//:www.unescap.org

QUESTIONS AND PROBLEMS

1. Give the definition of air pollution. What are the primary pollutants, secondary pollutants, and precursors? Present examples.

2. Characterize the general sources of air pollution in the Asian cities. Compare the average data with those for your own city.

3. Describe urban air pollution trends in the developing and developed countries of the Asian region.

4. Characterize the role of transport in urban air pollution in various Asian cities. Explain why two-wheeler engines emit more pollutants than car engines.

5. Explain the chemistry of hydroxyl radical in photochemical smog. What are the main sources of OH in rural and urban areas?

6. What are the tropospheric concentrations of OH in urban areas? Compare these values with those shown in Box 1 for other radicals.

7. Describe the control of trace gas reactions by OH in a polluted urban troposphere and present the scheme of these reaction sets.

8. Characterize the role of automobiles in formation of photochemical smog. Present a generalized scheme for the formation of photochemical smog.

9. Characterize the chemical aspects of photosmog. Write the cycles of chemical reactions in polluted urban air using carbon monoxide and methane as the example species.

10. Explain the role of nitrogen dioxide and nitrate radicals in photochemical smog reaction during daytime and nighttime.

11. Characterize four main parameters of photochemical smog and discuss the probability of photosmog formation in your city.

12. Why is photochemical smog called Los Angeles type of smog? What was the reason for appearance of this phenomenon?

13. Characterize the sources of particulate matter formation in polluted urban air. Present examples.

14. Discuss physical characteristics of various particles. Present Stocks law and describe what type of particles may be considered using this law.

15. Describe the chemical reactions that lead to the formation of inorganic particulates in urban air.

16. Discuss the chemistry of organic particle formation and explain the role of fuel combustion in these processes.

17. Characterize the role of different particles in visibility degradation. Explain the example given for the Korean city Kwangju.

18. What is the characteristic feature of the London type of smog? Discuss the history of this type of smog.

19. Compare two different types of smog and discuss the chemical reactions which are involved in both smog formation processes.

20. Discuss the role of biomass fuel in indoor air quality in Asian cities. Give examples.

21. Discuss the connection between urban air pollution and human health in Asian cities. Present examples of airborne diseases.

CHAPTER 4

ACID DEPOSITION

1. INTRODUCTION

Acid rain is not a new phenomenon either in Asia or in other parts of the world. Acid rain was first reported in Europe in the 19th century (Ducros, 1854). Acid rain and its harmful effects have been extensively studied in the industrialized countries of Europe and North America over the last thirty years (Radojevic and Harrison, 1992).

Normal, unpolluted rainwater has a pH close to 5.6, in consequence of the rain-drops being in equilibrium with atmospheric concentration of CO_2 (reaction (1) and Section 3)

$$CO_2(g) + H_2O(l) \longleftrightarrow H_2CO_3(aq) \longleftrightarrow H^+(aq) + HCO_3^-(aq). \tag{1}$$

We should stress here that even completely unpolluted rainwater does not have a pH of 7.0, because it is not pure distilled deionizing water and it contains equilibrium amounts of atmospheric gases, namely, carbon dioxide at an average present concentration of 365 ppbv. Acid rain is generally defined as having pH lower than 5.6. Low acidic rainwater has a pH between 5.6–5.0, acidic rainwater, 5.0–4.5, strong acidic rainwater, 4.5–3.5, and extremely acidic, < 3.5.

Over the last ten years, the presence of acid rain has been identified at numerous sites in Asia. Recent industrial development in Asia has raised concerns about actual and potential acidification in the region (Rodhe et al, 1992; ESCAP, 1995). Current and projected manifestation of acid forming compound emissions and deposition with indication of main sensitive ecosystems is shown in Table 1.

Accordingly, in this chapter, we consider acid rain monitoring patterns and relative chemistry of acid rain formation in the Asian region.

2. ACID RAIN MONITORING IN ASIA

2.1. Japan

Of all the Asian countries, Japan has the longest tradition of environmental research and environmental awareness. Studies of rainwater composition were reported as long ago as 1894 (Kellner et al, 1894). Annual mean rainwater pH values were reported to be 4.1 in Tokyo, 5.2 in Kobe and 5.6 in Hamamatsu in the period 1936–1937

Table 1. Acid rain in Asia.

Region	High emission		High deposition		Ecological
	Current	Future	Current	Future	Effects
Asian Russia		x	x	x	Forest
NE China	x	x		x	Forest
South China	x	x	x	x	Soil, forest
Japan			x	x	Soil, forest
South Korea			x	x	Soil, forest
Thailand	x	x .	x	x	Soil, forest
India	x	x	x	x	Soil, forest
Iran		x		x	Mountains
Shi Lanka			x	x	Soil, forest
Indonesia	x	x	x	x	Soil, forest

(Miyake, 1939). Several instances of rainwater pH values less than 3 were observed in the 1970s (Hashimoto, 1989; Hara, 1993). The incidence of acid rain seems to have declined in Japan due to the control of SO_2 and NO_x pollution (Hashimoto, 1989).

The Japan Environment Agency has been investigating acid rain since the early 1970s, and in 1983 it established the National Acid Deposition Monitoring network. The aim of this network is to accurately assess the state of acid deposition in Japan, its spatial distribution, long-term trends, and long range transport (Iijima, 1997). The network comprises 48 urban, rural and remote stations throughout Japan. Wet-only deposition is collected using automatic samplers on a daily, weekly and forthnightly basis. Samples are refrigerated on-site and sent to the laboratory every two weeks for analysis of pH, conductivity and major ions (SO_4^{2-}, NO_3^-, Cl^-, NH_4^+, Na^+, K^+, Ca^{2+}, and Mg^{2+}). Three five year surveys have been conducted so far; 1983–1988, 1988–1993, and 1993–1998. The range of pH values for all stations was 4.5 to 5.8 for the period 1988–1993, and this is similar to pH values observed in Europe and North America. The modern results of SO_2 content in air as the main precursor of pH in rain water are shown in Figure 1.

2.2. *Asian Russia*

Siberia is the vast region situated between the Ural Mountains and the Pacific Ocean. There are different large local sources of pollution inside this area such as nonferrous metal smelters, oil and gas wells, oil chemistry plants and electric power stations

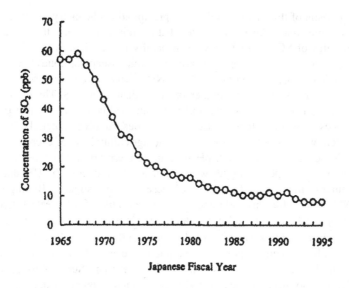

Figure 1. Trends in annual concentration of SO$_2$ *at 15 monitoring stations in Japan (Bashkin and Park, 1998).*

located mostly within industrial zones of big cities. At the present, significant inputs of acid forming pollutants are related to trans-regional and trans-boundary pollution from the European part of Russia, Eastern, Central and Western Europe (Sofiev, 1998) because of circumpolar wind directions, which are predominant in the Northern hemisphere. In the future, ecosystem damage due to the atmospheric deposition in southern Siberian regions caused by trans-boundary pollutant transport from rapidly industrializing China is also possible. Owing to neutralizing effects of base cations, the pH values are close to weak acid, 5.6–5.0. The acid and even strong acid rains have been monitored in the Far East and in industrial agglomerations of South Siberia (Ryaboshapko et al, 1994)

2.3. China

After Japan, the most extensive chemistry data set in the Asian region is for China. The Institute of Environmental Chemistry initiated a survey of precipitation chemistry in some cities in the late 1970s, and nationwide surveys have been reported since 1982. The results of these studies show that acid rain occurs in many parts of China, especially in the southwest (Zhao and Sun, 1986; Galloway et al, 1987; Zhao and Xiong, 1988, Narita et al, 1997).

The main cause of acid rain in China is the industrial and domestic combustion of coal. Many of the furnaces have short chimney stacks and have minimal controls. The content of sulfur varies between different regions. In Beijing, the sulfur content of coal is circa 1%, whereas in Guizhou province in the south it is between 3 and 5%.

A comparison of the survey results with precipitation chemistry results from the US and a remote area in Australia showed that the rainwater concentrations and wet deposition rates of SO_4^{2-} in China were generally higher than in the US and this was attributed to the high sulfur content of coal and absence of controls (Galloway et al, 1987). The same applied to Ca^{2+} and NH_4^+ concentrations in rainwater and their deposition rates; they were higher in China than in the US. The high content of Ca^{2+} was ascribed to calcareous soils, uncontrolled emissions of dust particles from furnaces, and the extensive use of calcareous building materials. The high NH_4^+ content was attributed to the widespread agricultural use of excretory wastes and the release of NH_3 from high pH soils in northern areas. However, for NO_3^-, concentrations and deposition rates were lower in China than in the US, due to the lower number of motor vehicles. Of the Chinese cities investigated, Beijing had the highest NO_3^- concentrations and deposition rates because of the greatest density of motor vehicles. Cities in the south had higher SO_4^{2-} concentrations and deposition rates than Beijing and this was explained in terms of the lower sulfur content of coal in Beijing, although neutralization by alkaline particles and NH_3 emitted from the alkaline soils in the north may also play a role (Zhao and Sun, 1986). A detailed field and laboratory study of the effect of wind blown dust on rainwater chemistry in Northeast China concluded that dust particles containing $CaCO_3$ are carried by air masses from Siberia/Mongolia and Northwest China and dissolve in rainwater leading to acid neutralization over Northeast China (Zhang et al, 1993). A similar conclusion was reached in a modeling study by Mo et al (1988). They used a chemical equilibrium model that included concentrations of the major precursor gases. Including $CaCO_3$ in the aerosol resulted in significantly increased precipitation pH in their model output.

Wang and Wang (1996) compared precipitation chemistry from north and south areas of China for the years 1986 and 1993 and concluded that acid precipitation is observed in the south but not in the north. The surveying precipitation chemistry in the country scale is shown in Figure 2. The conclusion is that about 90% of sampling sites with an average pH < 5.6 are situated south of the Yangtze River (Zhao and Xiong, 1988; Zhao et al, 1988; Hao et al, 1998).

2.4. Vietnam

Vietnam has been experiencing rapid industrial development in recent years. In response to concerns over the potential environmental effects of this industrialization, the National Environment Agency (NEA) was established in 1993. One of the activities of NEA is the setting up and running of air quality and acid rain monitoring stations. The NEA operates several environmental monitoring stations in conjunction with other organizations, including a background air quality station located at the Cuc Phuong national reserve forest and a rainwater monitoring station located in Lao Cai province, near to the border with China. The rainwater monitoring station has been in operation since 1996 monitoring the following parameters: pH, electrical conductivity, SO_4^{2-}, NO_2^-, NO_3^-, Cl^-, NH_4^+, Na^+, K^+, Ca^{2+}, Mg^{2+}, and PO_4^{3-}. In addition, the Hydrometeorological Service operates 22 air monitoring stations as

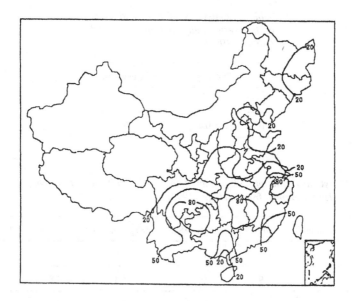

Figure 2. Contours of acid rain frequency corresponding to pH < 5.6 *(Hao et al, 1998).*

part of a larger network of environmental monitoring stations including river water quality monitoring stations (Anh, 1996).

The survey of rainfall acidity has been carried out at seven sites in the north of the country comprising three in-land sites, three coastal sites, and one background site (Khahn, 1993; Radojevic, 1998). This survey yielded the following results:

- the pH of rainwater was generally between 5 and 7;

- the pH values tend to be lower in winter, during the dry season with little rainfall, and higher in the summer, during the wet season when there is a large amount of rainfall;

- lowest pH values (< 4 to 5) were observed at a site located in an industrialized area, being from 6 to 15% from rain events.

Acid rains were also recently reported in Ho Chi Minh city (Thuc and Yen, 1997). At present 67% of rain events in Can Tho, the Mekong River Delta, has pH values less than 5.0, and 11%, pH 5.1–5.6 (Radojevic, 1998).

Air quality monitoring was initiated in 1995 at several sites throughout Vietnam. In Hanoi SO_2 concentrations varied between 0.001 and 0.04(mg/m^3) (Lan, 1996). Higher concentrations were observed close to industrial sources.

Table 2. Emissions of SO$_2$ and NO$_x$ in Indonesia (adapted from Kato and Akimoto, 1992).

Year	SO$_2$ (Tg/year)	NO$_x$ (Tg/year)
1975	0.20	0.33
1987	0.49	0.64

Table 3. Analytical results of bulk precipitation samples collected at Serpong, Tangerang, Indonesia in 1996 (Adapted from Saefudin, 1997).

Month	rainfall(mm)	pH	Conductivity (μS/cm)	SO$_4^{2-}$ (mg/L)	NO$_3^-$ (mg/L)	Cl$^-$ (mg/L)
June	29	6.4	43	2.5	2.2	1.9
July	12	5.9	16	5.3	2.8	1.7
August	41	6.2	20	3.3	3.1	1.3
September	81	5.3	23	3.3	2.9	3.5
October	98	5.0	26	3.6	3.4	1.8
November	4	6.1	32	1.9	2.6	1.9
December	146	5.8	13	0.4	0.2	1.4
Average	59	5.8	25	2.9	2.5	1.9

2.5. Indonesia

Indonesia, the most populous country in South East Asia, has experienced considerable industrial development over the last twenty years. Emissions of acidic precursors have doubled between 1975 and 1987, as shown in Table 2.

An automatic air quality monitoring station was set up at the Environmental Management Center (EMC), Serpong, Tangerang, about 40 km from Jakarta, in August 1993. This station monitors SO$_2$, NO$_x$, CO, non-methane HC, total-HC, O$_3$, PM$_{10}$ and meteorological data. Bulk precipitation sampling was started in June 1996, and an automatic acid rain analyzer was installed in November 1996. This measures: rainfall amount, temperature, pH, conductivity, SO$_4^{2-}$, NO$_3^-$ and Cl$^-$ automatically. Operational problems were encountered with the automatic analyzer and only data from the manual collector were available. The results obtained in 1996 are given in Table 3.

Table 4. Precipitation chemistry data from regional background air pollution monitoring station, Philippines (Radojevic, 1998 with additions).

Year	pH	Conductivity, μScm^{-1}	Chemical species, ppm				
			SO_4^{2-}	HCO_3^-	Ca^{2+}	Mg^{2+}	K^+
1994	6.9	11.4	6.80	48.5	0.63	0.30	0.30
1995	7.1	16.1	1.82	22.8	0.63	0.31	0.32
1996	6.6	11.4	0.48	16.7	0.75	0.04	0.12
1997	6.3	18.2	2.58	20.3	1.09	0.35	0.74
1998	6.2	13.7	3.20	10.1	2.75	0.28	.0.03

In a separate study, wet-only samples were collected at four sites in West Java (Jakarta, Serang, Cilegon, and Merak) over a 12 month period from June 1992 to June 1993, however, the samples were analyzed only for SO_4^{2-} and NO_3^- (Ayers et al, 1995). The authors conclude that these deposition rates of SO_4^{2-} and NO_3^- are significantly higher than natural fluxes, indicating significant anthropogenic acidification.

Some earlier data on precipitation pH determined in Jakarta has been reported. Samples were collected from January 1981 to July 1984 on a monthly basis and from December 1984 to October 1987 on a weekly basis. Volume weighted mean pH values were 4.79 for the monthly data and 5.33 for the weekly data (Ayers, 1991).

2.6. Philippines

A detailed source inventory of air pollutants, including SO_2 and NO_x, was prepared for Metro Manila in 1990 (Siador, Jr. and Calderon, 1997). A study of acid rain was conducted between April and November of 1986 and 1987 with 94 event samples collected at 15 sampling stations in Metro Manila and the provinces of Bulacan, Cavite, Laguna and Bataan. The pH of rain was found to vary between 3.7 and 7.7. The lowest value of 3.7 was reported in the central district of Metro Manila.

The dynamic background pattern of rainwater chemical composition is shown in Table 4.

Thus, we can conclude that acid deposition is not a regional problem in the Philippine Islands, however, the tendency to pH reducing is obvious.

The monitoring of acid rain precursor gases, SO_2 and NO_x, and other air quality parameters (O_3, CO, particulates) was initiated in Metro Manila in 1971 using manual methods to determine concentrations on a weekly basis at several stations (Siador, C. S., Jr., 1996). At present several automatic and mobile stations monitor pollutant levels in Metro Manila. The SO_2 concentration seems to have declined at most stations in Metro Manila between 1975 and 1993. Annual averages of 24 hour measurements of SO_2 ranged between 25 and 55 ppbv in 1975 at four sites. In 1993, annual averages of SO_2 ranged from 7 to 21 ppbv at four sites. NO_x concentrations tend to be below 25 ppbv.

2.7. Singapore

Before 1994 air quality monitoring was performed using semi-automatic methods. Since 1994, continuous monitoring of acid rain precursors, like SO_2, NO_x, O_3, CO and PM_{10} is carried out at 15 monitoring stations. Daily rainwater pH measurements have been carried out at some of the air quality monitoring stations since 1982. Between 1982 and 1994 rainwater was collected using bulk samplers. Since 1994 rainwater has been collected at four stations using an automatic precipitation collector. It is reported that the pH of rainwater is typically between 5.1 and 5.5 (Yong and Eng, 1997).

Precursor gases, SO_2 and NO_x, are routinely monitored at the air quality stations (Huan and Boo, 1996). Average annual SO_2 concentrations have been in the range of 10 and 35 ppbv since 1985. There appears to have been a slight upward trend in SO_2 concentrations at most sites since 1991. Annual average NO_2 concentrations have been in the range of 5.3 to 21 ppbv since 1985.

2.8. Malaysia

A detailed statistical analysis of the pH data obtained from the National Acid Rain Monitoring network prior to 1988 has been carried out (Leong et al, 1988). A seasonal variation in rainwater pH was observed. It was found that stations on the east coast of Peninsular Malaysia had significantly higher pH values during the North East monsoon (November–March) than during the South West monsoon (May–September) or the inter-monsoon periods (April and October). Some of the stations on the west coast had higher pH values during the South West monsoon. The explanation for these observations could be the direction of the wind. During the SW monsoon, the west coast experiences offshore winds which bring generally cleaner air, while during the NE monsoon the air mass traverses the Malayan peninsula bringing with it pollutants that lower the pH at these stations. The situation is reversed for the stations on the east coast; these experience offshore winds during the NE monsoon and hence the higher pH. These stations experience winds which traverse the peninsula during the SW monsoon bringing with them polluted air and lowering the pH.

The annual volume weighted rainwater pH values at various stations between 1985 and 1993 are given in Table 5.

It is apparent that Petaling Jaya, Perai and Senai consistently recorded the lowest pH values. The high acidity could be due to a high density of motor vehicles and industries that emit SO_2 and NO_x in these areas. Mersing in eastern Johore and Malacca also recorded relatively low pH values especially in later years, due to the expansion of industrial activity in these areas, and the transport of pollutants from other industrial areas. Central and eastern parts of Malaysia recorded relatively high pH values, due to the low level of industrial activity in these areas. At all stations, there was a decrease in rainwater pH between 1985 and 1988. The situation seems to

Table 5. Annual volume weighted mean pH of rainwater in Malaysia (Radojevic, 1998).

Site	1985	1986	1987	1988	1989	1990	1991	1992	1993
Petaling Jaya	5.05	4.54	4.46	4.44	4.51	4.50	4.63	4.41	4.47
Bayan Lepas	5.31	5.01	5.03	4.99	4.99	4.92	4.92	4.79	4.93
Senai	4.99	4.54	4.40	4.39	4.69	4.62	4.55	4.22	4.74
Alor Setar	5.58	5.49	5.53	5.38	5.42	5.67	5.74	5.08	5.19
Malacca	5.46	4.97	4.58	4.76	4.64	4.61	4.74	4.68	4.93
Mersing	5.35	4.89	4.84	4.65	4.78	4.90	5.05	4.70	4.93
Kota Bharu	5.51	5.47	5.47	5.18	5.37	5.37	5.28	5.14	5.31
Tanah Rata	5.64	5.23	5.21	5.08	5.22	5.30	5.11	4.95	5.33
Ipoh	5.43	5.23	5.21	5.10	5.16	4.97	5.13	4.97	5.09
Perai	4.71	4.51	4.61	4.52	4.66	4.53	4.63	4.60	4.69
Kuantan	5.35	5.22	5.13	5.10	5.09	5.09	5.23	4.85	5.10
Kuala Terengganu	5.44	5.33	5.27	5.16	5.19	5.17	5.26	5.12	5.24
Kuching	NR	NR	NR	5.25	5.34	4.84	5.50	5.35	5.56
Tawau	NR	NR	NR	5.54	5.64	5.75	5.95	5.92	5.94

Note: NR—no rain events.

have stabilized after 1988, but low pH values are still being recorded at many stations. The greatest increase in acidity was observed at Petaling Jaya and Senai. In terms of H^+ concentration, the acidity increased fourfold between 1985 and 1988. SO_4^{2-} and NO_3^- were also measured in rainwater at all sites. Mean annual SO_4^{2-} concentrations varied between 0.08 and 0.9 mg/L, while the mean annual NO_3^- concentrations varied between 0.01 and 0.25 mg/L. Petaling Jaya, Senai and Perai had the highest SO_4^{2-} concentrations. In general, these three sites also had the highest NO_3^- concentrations. A seasonal variation in the concentrations of SO_4^{2-} and NO_3^- was observed. Highest concentrations at all sites were generally observed during the months of May to August, which are the driest periods of the year. Also, the concentrations of SO_4^{2-} and NO_3^- increased significantly at Petaling Jaya and Senai between 1985 and 1987. The increasing acidity observed at the same sites during this period could be due to the increase in SO_4^{2-} and NO_3^- in rainwater.

It can be concluded that there is sufficient evidence to suggest increasing acidification of precipitation in industrial areas of Malaysia.

Data on precursor gas (SO_2, NO_x) concentrations are available from the air quality stations (Abdullah, 1996). In Petaling Jaya the average annual SO_2 concentration increased from 12 to 21 ppbv between 1992 and 1995. In nearby Kuala Lumpur the average annual concentration of SO_2 varied between 2 and 7 ppbv, with no consistent trend during this period. At both sites there appears to have been an upward trend in NO_2 concentrations during the same period. In Petaling Jaya the annual average concentration of NO_2 increased from 17 ppbv to 26 ppbv, while at Kuala Lumpur it increased from 16 to 24 ppbv in 1994, thereafter declining to 21 ppbv in 1995.

2.9. Brunei Darussalam

Brunei Darussalam, a small country with a land area of 5765 km^2 and a population of 261×10^3, is located on Borneo Island. The country has few anthropogenic sources of air pollution; circa 100,000 on-road motor vehicles, a small refinery, and one medical incinerator. Natural and man-made forest fires also contribute to local pollution, but most air pollution originates from transboundary sources. The first ever rainwater study was conducted in the capital city, Bandar Seri Begawan, in the period between 26 April and 3 November 1994 (Radojevic and Lim, 1995a). Bulk precipitation was analyzed for pH, conductivity, Cl^-, NO_3^-, SO_4^{2-}, Na^+, K^+, Ca^{2+} and Mg^{2+}. A statistical study of the interrelationship of various parameters was also carried out. The authors also conducted a study into the short-term variation of the concentration of selected ions within individual rainstorms (Radojevic and Lim, 1995b). This revealed that most of the ionic content of precipitation was deposited within the initial stages of the rainstorms. Although organic acids were not analyzed directly in this study, there was some evidence of the possible major contribution of organic acidity (formic and acetic) to the pH measured in these samples (Radojevic and Lim, 1995a).

An automatic, wet-only, collector was set up at the Brunei international Meteorological Station in Bandar Seri Begawan and this started sampling precipitation in July 1995 in conformity with the GAW protocol of the WMO (Radojevic et al, 1997). In the 185 samples analyzed between July 1995 and June 1996 it was observed that 91% of the samples had pH values less than 5.6 and could therefore be classified as "acid rain". Some 42% of the samples had pH values less than 5.

Prior to 1997 there was no monitoring of precursor gases in the country. A fully-equipped air quality monitoring station was set up in Bandar Seri Begawan in October 1997 to measure routinely SO_2, NO_2, O_3, CO and PM_{10} concentrations. A further 6 PM_{10} monitoring instruments were set up throughout the country. A survey of the distribution of average monthly SO_2, NO_2 and O_3 concentrations throughout the country using diffusion tubes was also carried out in October 1997 and February 1998 (Radojevic et al, 1998). Concentrations of SO_2 varied between 0.48 and 5.71 ppbv while concentrations of NO_2 varied between 0.3 and 19.68 ppbv. Concentrations of O_3 were between 5.2 and 57.2 ppbv. Legislation with regard to air quality in Brunei is still relatively rudimentary. Air quality standards similar to those adopted by Singapore have been suggested (Radojevic, 1998 with additions) but they have still not been enacted.

2.10. *Thailand*

At present, rainwater is routinely sampled at five sites in Thailand including urban, industrial and remote sites by the Pollution Control Department (Siriswasdi, 1997). Three of the sites measure pH, conductivity, SO_4^{2-} and NO_3^-, while the other sites measure only pH and conductivity. Samples are collected using wet-only automatic collectors.

In addition, precipitation has been collected on a daily basis at two rural sites in Thailand since mid-1991 in rural locations (Granat et al, 1996). Samples were analyzed for conductivity, pH, NH_4^+, Na^+, K^+, Mg^{2+}, Ca^{2+}, SO_4^{2-}, NO^{3-} and Cl^-. A high correlation was observed between H^+ and SO_4^{2-} suggesting that sulfur played an important role in acidification. Also, a strong interrelationship between Ca^{2+}, Mg^{2+}, Na^+ and Cl^- was found. The authors found a high correlation between NH_4^+ and NO_3^- but the reason for this relationship was not obvious. Bulk collectors gave results 10–30% higher than wet-only collectors.

Some earlier data on rainwater acidity in Thailand were summarized by Ayers (1991). These include 12 samples collected between November 1983 and July 1985 at a BAPMoN site on the island of Ko Sichang. The volume-weighted mean pH was 6.54, and NO_3^-, SO_4^{2-}, Ca^{2+} and K^+ were also measured. Some more data are available from four sites, Chiang Mai, Khon Kaen, Bangkok and Songkla, during 1987 (Ayers, 1991). The lowest pH measured was 4.5 and most samples had pH values > 5.6. Only < 10% of the samples had pH < 5.

Precursor gases are measured at 52 automatic air quality monitoring stations throughout Thailand, each of which measure CO, SO_2, NO_2, O_3, TSP, PM_{10}, non-methane hydrocarbons as well as relevant meteorological data as part of the Ambient Air Monitoring Network set up by the Pollution Control Department in 1992 (Charasaiya, 1996; Siriswasdi, 1997).

The sulfur deposition rates for years 1990 and 2020 based on RAIN/ASIA model are shown in Figures 3 and 4.

We can see that sulfur deposition rates and, accordingly, acid rain severity will be increased in the future, in spite of some reduction of coal-burning energy generation after the economic crisis of 1997–1999.

2.11. *Hong Kong*

Results for 1988 from the BAPMoN station of Yuen Ng Fan at Sia Kung are reported by Ayers (1991). Wet-only samples were collected on a weekly basis and the volume-weighted mean pH was 4.68. Data obtained between 1988 and 1993 are discussed by Ayers and Yeung (1996). The analysis suggested elevated levels of SO_4^{2-} and NO_3^-.

Results of daily precipitation samples collected from 1989 have been reported by Peart (1995). Results are summarized in Table 6. No apparent trend in pH was observed over the years. There was evidence that rainfall was less acidic during November, December and January than for the other months.

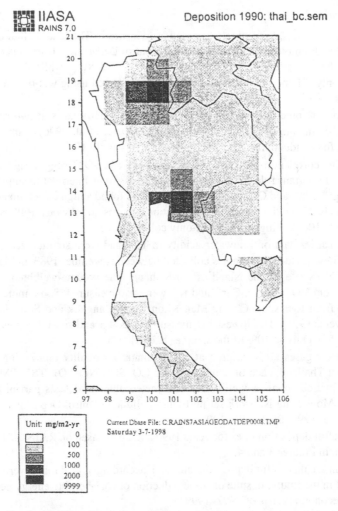

Figure 3. Predicted sulfur deposition rates for year 1990 using RAIN/ASIA model (Khum-mongkol, 1999).

No significant difference was observed between pH values measured in filtered and unfiltered samples in a rural site in Hong Kong (Sequeira et al, 1996). Hong Kong has a satisfactory air quality monitoring and management program in place.

2.12. South Korea

A number of separate groups monitor precipitation chemistry in Korea. Most of these measure the electrical conductivity, pH, SO_4^{2-}, NO_3^-, Cl^-, NH_4^+, Na^+, Ca^{2+}, Mg^{2+} and K^+. At most monitoring stations the volume weighted pH is generally around 5.0,

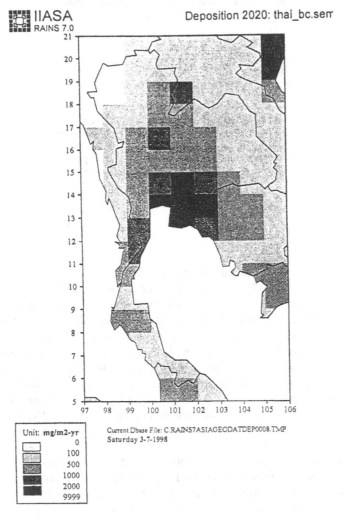

Figure 4. Predicted sulfur deposition rates for year 2020 using RAIN/ASIA model (Khummongkol, 1999).

although pH values as high as 6.8 have been reported during yellow sand periods in springtime (Shim et al, 1997).

Measurements of wet-only precipitation between August and November 1985 at 10 sites in the Seoul area were reported by Shin et al (1989). The volume weighted mean pH values ranged from 4.39 to 4.64. Concentrations of SO_4^{2-} varied between 1.4 and 3.3 mg/L, while NO_3^- was present at low levels < 0.37 mg/L. Measured levels of Ca^{2+} suggested that alkaline aerosol was important in the ion balance.

Table 6. pH of rainwater samples collected in Hong Kong (adapted from Peart, 1995).

Year	Minimum pH	Maximum pH	Median pH
1989	3.5	6.25	4.57
1990	3.45	6.78	4.69
1991	3.45	6.09	4.48
1992	3.81	6.56	4.54
1993	3.71	6.76	4.62
1994	3.56	6.92	4.59

A study of over four hundred rainwater samples collected during 1996 has been reported (Park et al, 1997). The most abundant anion was SO_4^{2-}, present at concentrations between 1.4 and 5.76 mg/L, and accounting for between 30 and 68% of total anions. Most of this was of non-sea salt origin. Sea salt contribution to SO_4^{2-} was <10%. The NO_3^- concentration ranged between 1.0 and 1.7 mg/L, while the Cl^- concentration varied from 0.28 to 8.86 mg/L. Comparison with results from previous years showed that the NO_3^- contribution rose between 1994 and 1996 due to increasing NO_2 emissions in the country. The volume weighted mean concentrations ranged from 4.5 to 5.2 with an annual mean of 4.8. Similar values were recorded in previous years. Some 2% of the individual precipitation events had pH values < 4. Deposition rates of SO_4^{2-} and NO_3^- were 2.0–6.9 g/m^2 and 1.0–2.0 g/m^2 respectively on the basis of studies over several years.

Chung et al (1996) report measurements at five stations over the period 1990–1993. The pH measurements are summarized in Table 7 indicating that acidity increased over the years at most sites. Long range transport of pollution from China was suggested as a possible factor in acidification of precipitation in Korea.

A recent review of rainwater chemistry has been conducted in 1996–1998 (Lee et al, 2000). The distribution of individual pH values over South Korea is shown in Figure 5.

Although the annual mean pHs do not differ sitewise, pH of individual precipitation showed a very large fluctuation, ranging from 3.4. to 8.0. Acidic precipitation appeared episodically; about 4% of the precipitation events showed pH of less than 4.0. About 70% of the precipitation had a pH below 5.6 and can be considered as acid rains. On the other hand, about 2% of the precipitation events had pH above 7.0, suggesting strong inputs of alkaline species to rainwater in this region. The input of Yellow sand is one of the important alkaline sources that effect the rainwater chemistry over South Korea.

Table 7. Annual volume-weighted mean pH values at five sites in South Korea (adapted from Chung et al, 1996).

Year	Koong-Hyun Ree	Chong-Woon	Tae-Ahn Peninsula	Seokeepo	Hahn-Ra Mountain
1990	5.22	4.73	NR	NR	NR
1991	4.82	4.67	4.80	5.04	4.61
1992	4.41	4.47	4.68	4.80	5.55
1993	4.39	3.99	4.42	4.51	6.08

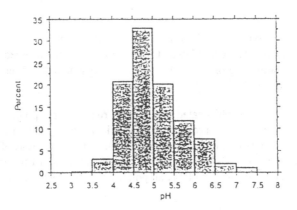

Figure 5. Distibution of individual pHs in South Korea (Lee et al, 2000).

2.13. Cambodia

Cambodia's population of 9 million is mainly (ca. 85%) engaged in agriculture. There are few industrial sources of air pollution in Cambodia. These include three power stations in Phnom Phen and approximately 70 factories (Sophy, 1997). In addition, there are some 47,000 motor vehicles and 170,000 motorcycles. The concentrations of pollutant gases are summarized in Table 8.

2.14. Taiwan

Although Taiwan is a highly industrialized country only a few studies of rainwater chemistry have been reported before the 1980's. Ayers (1991) reported the results from nine stations spread across Taiwan. Between 1985 and 1988, 300 rainwater samples were analyzed. Some 56% of the samples had pH values < 5.6. The latest

Table 8. The concentration of acid rain precursors in ambient air in Pnom Pehn City in 1999, ppbv (Sophal, 1999).

Pollutants	Dry season	Wet season
CO	4,666	1,958
SO_2	2.4	2.5
NO_2	28.7	33.1

monitoring showed that already > 70% of the precipitation in Taiwan exhibits a pH lower 5.6 and has to be considered as acid rains (Lin, 1998).

2.15. Acid in Asia: Conclusive Remarks

- Acid rain is observed regularly at many sites in East Asia, especially in China, South Korea, Hong Kong and Japan. Lower pHs tend to be observed in industrialized countries and regions while higher pH values tend to be observed in areas with little or no industries.

- Japan has the most sophisticated and longest running precipitation chemistry and air quality monitoring program in East Asia. Also, Japan is a world leader in air quality legislation, management and control. In many areas of air quality management it is on a par with the industrialized countries of Europe and North America, and in some areas, notably air pollution control technologies, it is more advanced.

- With the exception of Japan, other countries in East and South-East Asia have started monitoring acidic precipitation and air quality relatively recently. Monitoring programs vary from being relatively comprehensive (e.g., South Korea, Hong Kong, China) to rudimentary (e.g., Indonesia, Vietnam). Some countries have yet to set up air quality and acid rain monitoring programs of any kind (e.g., Myanmar, Laos, Cambodia). General air quality management, legislation, and control measures vary between countries of East Asia in the same proportion as air quality and acid precipitation monitoring.

- The quality of data produced by many of the monitoring networks in the region, with the possible exception of Japan, leaves much to be desired. There is an absence of quality control procedures in many of the networks. Much improvement could be achieved by the introduction of QA/QC schemes in acid rain and air quality monitoring networks. The use of certified reference materials (CRM) in rainwater analysis is to be encouraged and the setting up of international monitoring programs with a centralized QC/QA scheme would be desirable. This would greatly improve the quality of the data generated and allow for more meaningful comparisons to be made between different countries. Obviously, many of the earlier conclusions will have to be revised when more reliable data become available. The proposed international acid rain survey in East Asia would help by standardizing methods and procedures.

- Many of the poorer countries in East and Southeast Asia lack the technical expertise and instrumentation to set up adequate acid rain and air quality monitoring networks. Clearly, there is a need for international assistance in terms of providing training and financial resources to these countries.

- It is difficult to make reliable comparisons in pH and other measurements reported in the literature, in the absence of a uniform methodology strictly adhered to by all parties involved. Measurements of pH are especially sensitive to a variety of factors including: sampling bottle material, sampling period (event, daily, weekly, or monthly), type of sample (bulk or wet-only), and especially the time of pH analysis. The pH is not a static parameter, but is continually changing, during and after sampling, due to chemical reactions in the sample.

- From the measurements available so far it could be concluded that acid rain is still not a major problem in Southeast Asia. In Southeast Asia most rainwater pH measurements tend to be around 5.6, the pH of natural rainwater. Instances of pH < 5 are encountered at some sites, and these are mainly owing to localized industrial pollution. There is some evidence that pH values below 5 at unpolluted sites may be due to the contribution of weak organic acids, such as formic and acetic acids.

- It is difficult to make predictions about the future incidence of acid rain in East and Southeast Asia due to the unpredictable economic situation. Many countries in the region are experiencing a slowing down of their industrial development due to the recent currency crisis, which could in turn slow down the growth in pollutant emissions. On the other hand, the result could be fewer, but more polluting, industries, as there may be less financial resources available for introducing pollution control equipment. If the economic growth recovers to the rates experienced during the late 1980s and early 1990s, then we could expect to see a greater incidence of acid rain over a wider region.

- The haze phenomenon in Southeast Asia is a major air pollution problem and its impact on the chemistry of precipitation is still unknown. It has been speculated that forest fires could raise the acidity of precipitation. While it is known that forest fires produce acid precursor gases such as SO_2, NO and NO_2, the non-carbonaceous matter of particles produced by these fires contains Ca and K, which give rise to alkalinity in rainwater. There is a need for a research project to study both the emissions from the forest fires in Southeast Asia and their impact on precipitation chemistry and rainfall acidity.

- Most of the precipitation networks in East and Southeast Asia do not monitor weak organic acids, such as formic and acetic acids, in rainwater. At many tropical sites in Africa and South America it has been demonstrated that organic acids can contribute between 40 and 80% of the total acidity. In view of this it is important that precipitation surveys in East and Southeast Asia begin routine monitoring of organic acids at the earliest stage.

- It has been suggested some years ago that an acid deposition monitoring network should be set up in East Asia (Murano, 1997a, b). A uniform sampling and analysis protocol together with quality assurance/quality control (QA/QC) procedures has been proposed. This network is now in use and would produce more reliable data, which could be used to make more meaningful spatial and temporal comparisons.

3. CHEMISTRY OF ACID RAIN

We have seen that in the polluted regions the main causes of acid rain are sulfur dioxide and nitrogen oxides in the atmosphere. These gases are present in trace amounts (ppbv) even in natural, unpolluted air. However, they form acid rains only when they occur in higher than normal background concentrations owing to anthropogenic activities. Even so, the absolute amount of these gases in polluted atmosphere, as we have seen in Section 2 are very small, up to a few parts per million by volume, ppmv. Acid rain results when these gases are oxidized in the atmosphere and return to the ground dissolved in raindrops. SO_2 falls as H_2SO_3 and H_2SO_4 (reactions (2) and (3)), while NO_x falls as HNO_3 (see Chapter 3). Direct scavenging of NO_2 by atmospheric water (reaction (4)) is negligibly important, on account of the low solubility of NO_2 in water. A night time route to nitric acid is hydrogen abstraction from some suitable donor X-H by nitrate free radical (reaction (5))

$$SO_2 + H_2O \longrightarrow H_2SO_3, \tag{2}$$

$$SO_2 \xrightarrow{\text{oxidize}} SO_3 + H_2SO_4, \tag{3}$$

$$NO_2 + H_2O \longrightarrow 1/2\,HNO_2 + 1/2\,HNO_3, \tag{4}$$

$$NO_2 + O_3 \longrightarrow NO_3 \xrightarrow{XH} HNO_3. \tag{5}$$

In most acid rain areas of Asia, the sulfur oxides are the major contributor to the problem, but the nitrogen oxides predominate in Japan and Singapore. Sources of these gases are shown in Chapter 1.

Unpolluted rainwater, as we have already seen, has pH close to 5.6 as a result of equilibrium of raindrops with ca. 350–360 ppmv CO_2 in the troposphere. This yields the weak acid H_2CO_3 for which dissociation constant, $K_a = 4.2 \times 10^{-7}$ mol/L at 25 °C.

NO_2 and SO_2 are ultimately deposited in rains as nitric, sulfurous and sulfuric acids Nitric and sulfuric acids are strong acids, while sulfurous acid is rather weak,

$K_a = 1.7 \times 10^{-2}$ mol/L at 25 °C. However, we can consider SO_2(aq) and H_2SO_3 (aq) as interchangeable.

Since solubility of these three species, HNO_3, SO_2 and SO_3, is higher than CO_2, low concentrations of these acidic gaseous compounds have greater effects on pH of the rainwater than much higher concentrations of CO_2.

We can show this quantitatively.

For CO_2:

$$CO_2(g) + H_2O(l) \longleftrightarrow H_2CO_3(aq) \qquad K_H = 3.4 \times 10^{-2} \text{ mol L}^{-1} \text{ atm}^{-1}$$

$$H_2CO_3(aq) \longleftrightarrow H^+(aq) + HCO_3^-(aq) \qquad K_a = 4.2 \times 10^{-7} \text{ mol L}^{-1}$$

$$CO_2(g) + H_2O(l) \longleftrightarrow H^+(aq) + HCO_3^-(aq) \qquad K_c = 1.4 \times 10^{-8} \text{ mol}^2 \text{ L}^{-2} \text{ atm}^{-1}$$

For SO_2:

$$SO_2(g) + H_2O(l) \longleftrightarrow H_2SO_3(aq) \qquad K_H = 1.2 \text{ mol L}^{-1} \text{ atm}^{-1}$$

$$H_2CO_3(aq) \longleftrightarrow H^+(aq) + HCO_3^-(aq) \qquad K_a = 1.7 \times 10^{-2} \text{ mol L}^{-1}$$

$$SO_2(g) + H_2O(l) \longleftrightarrow H^+(aq) + HSO_3^-(aq) \qquad K_c = 2.1 \times 10^{-2} \text{ mol}^2 \text{ L}^{-2} \text{ atm}^{-1}$$

Summarizing, the equilibrium constant for the overall reaction is larger in the case of SO_2 than of CO_2 because SO_2 is more soluble in water than CO_2, and also because H_2SO_3 is a stronger acid than H_2CO_3. Consequently, a small concentration of SO_2(g) has a greater effect on the pH of rainwater than a much larger concentration of CO_2(g). For example, 0.12 ppmv of SO_2(g) in equilibrium with rainwater will produce a pH of 4.3, compared with pH 5.6 produced by 350 ppmv of CO_2(g).

3.1. Oxidation of Sulfur Dioxide

This process is very complex, because oxidation can occur in three quite different ways: heterogeneously on the surface of particles, homogeneously in the gas phase, and homogeneously in the aqueous phase of raindrops. The prevailing atmospheric conditions, especially the humidity and the concentration and composition of particulate matter, will determine the relative importance of these processes.

Heterogeneous oxidation on particulates

Accounting for the high SPM concentration in many Asian cities, that drives the SO_2 oxidation, heterogeneous oxidation of particles may predominant the process, especially under high urban pollution conditions.

The mechanism is probably similar to that involved in the industrial oxidation of SO_2 but still not known in detail

$$SO_2 + 1/2 O_2 \xrightarrow{\text{catalyst}} SO_3 \longrightarrow H_2SO_4 \text{ (or } SO_4^{2-}). \tag{6}$$

Salts of vanadium, manganese, and iron are all effective catalysts for this oxidation, since all can undergo redox reactions. These species are widely distributed in various environments; vanadium especially is found in particulate matter produced during coal burning (see more details for HM and SPM emissions from lignite-burning power plant in Chapter 12).

Heterogeneous gas phase oxidation

Hydroxyl radical is the important constituent in this SO_2 oxidation process, which may be written as follows (reaction (7))

$$SO_2 + OH \xrightarrow{\text{catalyst}} HSO_3. \tag{7}$$

The reaction constant, $k = 9 \times 10^{-13}$ cm^3 molec^{-1} s^{-1}. At the average global concentration of OH (about 8×10^5 radicals cm^{-3}), this reaction alone would be responsible for the generation of a half-life for SO_2 of about 10 days.

The HSO_3 formed in the following reaction (8) is a free radical species, not to be confused with the anion HSO_3^-, which is formed in the acid dissociation of H_2SO_3. HSO_3 is subsequently oxidized to SO_3 by reaction with molecular oxygen

$$HSO_3 + O_2 \longrightarrow HO_2 + SO_3. \tag{8}$$

Homogeneous aqueous phase oxidation

It has been shown recently that a hydroxyl radical may partition from the gaseous to the aqueous phase, or may be formed directly in the aqueous phase, and there effect the oxidation of sulfur dioxide. Sulfur dioxide can also be oxidized inside raindrops by hydrogen peroxide, H_2O_2, which is a minor atmospheric species formed principally by the disproportionation of HO_2 radicals

$$2\,HO_2 \longrightarrow H_2O_2 + O_2, \tag{9}$$

$$SO_2(aq) + H_2O_2(aq) \longrightarrow H_2SO_4(aq) \tag{10}$$

or

$$HSO_3^- + H_2O_2 \xrightarrow{H^+} HSO_4^-(aq). \tag{11}$$

Disproportionation of HO_2 radicals may occur in the gas phase, followed by dissolution of the H_2O_2 in the water droplets ($K_H = 10^5$ mol L^{-1} atm^{-1}); alternatively, the HO_2 radicals may first enter the aqueous phase ($K_H = 2 \times 10^3$ mol L^{-1} atm^{-1}), and than react together. This process is pH-dependent, since H_2SO_3 and its ionized form HSO_3^- react at different rates. Transition metals can catalyze this process as well. Other oxidants, like ozone, may also influence the process of sulfur dioxide oxidation. For example, ozone is more abundant than hydrogen peroxide in the troposphere, and although it is less soluble in water ($K_H = 1.3 \times 10^{-2}$ mol L^{-1} atm^{-1}) it reacts rapidly

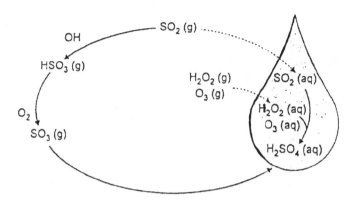

Figure 6. Summary of gas phase and aqueous-gas phase oxidation of sulfur dioxide in atmosphere (Bunce, 1994).

with SO^2 (reaction (12)). In addition, polluted air contains NO_2, which, in the aqueous phase, is an oxidant for SO_2, reaction (13)

$$HSO_3^- + O_3 \longrightarrow HSO_4^- + O_2, \tag{12}$$

$$SO_2 + NO_2 \longrightarrow SO_3 + NO. \tag{13}$$

The summary of gas and aqueous-gas phase oxidation is shown in Figure 6.

3.2. Oxidation and Deposition of Sulfur Dioxide

The chemistry of rain acidified by the sulfur oxides is complicated because the sulfur may be deposited in different forms. It may either precipitate as $H_2SO_3(aq)$ or it may be first oxidized to $SO_3(g)$ and precipitate as $H_2SO_4(aq)$. Deposition may occur either in the aqueous form (wet deposition) or in association with particulates (dry deposition), in which case much more of the sulfur will deposit in the form of sulfite or sulfate ions than the free acid (see Chapter 3).

Since the half-lives of SO_2 and SO_3 in the atmosphere are of the order of several days, acid precipitation can be expected over whatever distance is traveled by an air mass from the pollution source during this period. Assuming a wind speed of as little as $20 \, \text{km h}^{-1}$, such an air mass will travel nearly $3500 \, \text{km}$ over the course of a week. For example, sulfur acid-forming compounds can travel from Europe to Asia or from Asia to North America (see Chapter 1). Acid deposition is thus the transboundary pollution issue.

The rates of oxidation and deposition of SO_2 vary considerably with the conditions. Typical rates of oxidation are 1 to 10% per hour, i.e., pseudo first order constants 0.01 to $0.1 \, \text{h}^{-1}$. Rates down to 0.2% per hour have been recorded during dry seasons, when the homogeneous gas phase oxidation by OH radical is the predominant pathway

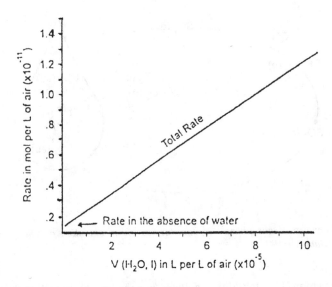

Figure 7. Rates of gas phase and aqueous phase oxidation of SO_2 *as a function of the liquid water content of the air (Bunce, 1994).*

in dry air. At the other extreme, rates up to 30% per hour have been reported for rainy seasons in the tropics, when the oxidation occurs mainly in the aqueous phase. The rates of both extreme mechanisms depend on solar intensity, because sunlight is needed for the formation of reactive oxidants like ozone or hydrogen peroxide. Figure 7 shows the proportion of SO_2 oxidation taking place in the aqueous phase as a function of the liquid water content of the atmosphere based on the assumption that the only prevalent mechanisms for oxidation are gas-phase attack by OH, and the aqueous phase oxidation by H_2O_2.

From the above discussion, we can say that there are three processes which take place following the release of SO_2 to the atmosphere. These are oxidation of SO_2 to SO_3, deposition of H_2SO_3/SO_3^{2-}, and deposition of H_2SO_4/SO_4^{2-}. We can draw the three reaction rate constants for these processes (reactions (14)–(16))

$$SO_2 \xrightarrow{k_1} SO_3, \tag{14}$$

$$SO_2 \xrightarrow{k_2} \text{deposition as } H_2SO_3 \text{ or } SO_3^{2-}, \tag{15}$$

$$SO_2 \xrightarrow{k_3} \text{deposition as } H_2SO_4 \text{ or } SO_4^{2-}. \tag{16}$$

In this simplified model, k_1, k_2, and k_3 are all pseudo first order rate constants, representing the sum of the several different mechanisms of oxidation or of deposition. Consequently, the values of the rate constants will vary according to the prevailing

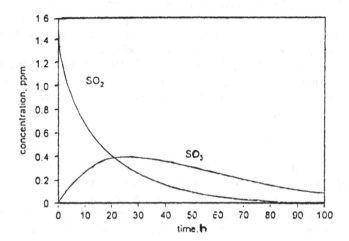

Figure 8. Concentration of $SO_2(g)$ *and* $SO_3(g)$ *as a function of the time elapsed since emission (Bunce, 1994).*

conditions. Such a model would apply to the emission from a single point source under atmospheric conditions that remains constant over the whole life-time of the emission plume. In real life, there might be multiple emission sources and changes in the weather conditions would undoubtedly cause these constants to change during the considered time period.

We can see the application of this model for the values of rate constants as the following: $k_1 = 0.08$, $k_2 = 0.025$, and $k_3 = 0.03$ h^{-1} (Figure 8). The concentration of SO_2 decays continuously with time, while that of SO_3 reaches a maximum and than falls again. At any time, the amounts of $SO_2(g)$ and $SO_3(g)$ will determined by the ratio of sulfite : sulfate deposited (Bunce, 1994).

If the wind speed were assumed to be constant., than the abscissa in Figure 8 could also depict the distance traveled by the plume from the emission source. Close to the source, most of the deposition should occur as sulfite, since little time for oxidation has elapsed. Farther from the source, the proportion of sulfur deposited as sulfate should increase and eventually predominate.

More complicated models of sulfur transport and deposition differ in details but typically use the same chemical reactions and mechanisms as shown above.

4. ECOLOGICAL CONSEQUENCES OF ACID RAIN IN ASIA

There are both local and regional impacts from acid rain in Asia. For example, damage of the historic Taj Mahal in Agra, India, is the result of the atmospheric pollution from local acid rain formed due to local industrial activity. Similarly, in China, acid rain has seriously corroded metal structures and concrete works in cities Chongging and Guiyan where SO_2 emissions are high. Conversely, depending on the height of smoke stacks and the prevailing wind conditions, emitted sulfur and nitrogen oxides can be carried hundreds and thousands of kilometers and kill trees, and acidify soil, lake and other water bodies, affecting biogeochemical cycles of important nutrients like nitrogen, phosphorus, sulfur and calcium. These consequences are different in various Asian countries and we consider them shortly. Other details can be found in Chapter 15 under discussion of critical loads.

4.1. Japan

During the last few decades the declines of several tree species and forests in different areas of Japan are noted; also European forests are suffering. However, they may be linked with different causes: forest succession dynamics, diseases and pests, meteorological extreme conditions, air pollutants, acidic deposition, and so on. Exposure experiments have been conducted to assess cause-effects relationships with acidic deposition and forest decline. Simulated acid rain or ozone in combination with sulfur dioxide was exposed to sixteen potted tree species for three growing seasons. Rain acidity below pH 4.0 could induce deleterious effects on some broad-leaved trees, however, coniferous trees did not show any significant growth reduction. Sulfur dioxide and/or ozone induced more complicated differential growth responses than the wet deposition did. Some indicated additive harmful effects of sulfur dioxide. Others showed synergistic adverse effects of ozone and sulfur dioxide. Experimental results suggest that an ambient level of ozone (or of the precursor of acid deposition connected with nitrogen species in the troposphere) may take an important role to reduce tree vitality, since ozone induced chronic changes in carbon allocation will be accelerated by increased nitrogen input.

Effects of acid rain on natural waters

A survey of the effects of acidic precipitation on inland waters and terrestrial ecosystems has also been initiated by JEA. Thirty-three lakes have been surveyed for pH, major ions, and other indicators of water quality. Soil analysis is carried out at 88 sites every three years. Visual investigation of vegetation in the surrounding area is carried out to look for any evidence of damage or decline. The pH of all lakes, with a few exceptions, was found to be in the range of 7 to 8. So far there is no clear evidence of adverse effects on aquatic or terrestrial ecosystems. There is some evidence of tree damage, but in most cases this was attributed to causes other than acid rain: weather, insect damage, magnesium deficiency, ozone, and water deficiency. In a separate study, a 17 year record of the pH of rivers and lakes in a mountainous region of central Japan was analyzed (Kurita et al, 1991). Decreases in pH of between 0.2 to 0.3 pH units were noted in some lakes and rivers with a high acid neutralizing capacity. In some lakes and rivers with a low acid neutralizing capacity the pH decreased by as much as 0.6 pH units over a 10 year period. However, it is suspected that factors other than acid precipitation are responsible for the observed acidification.

Acidification of natural waters is mainly a problem in areas where the underlying rocks provide poor buffering capacity. Rocks such as granites and gneisses offer little buffering protection. Chalk and limestone neutralize added acid, and so soils, lakes and streams in limestone areas are fairly insensitive to acidic precipitation

$$2\,H^+ + CaCO_3 \longrightarrow Ca^{2+} + CO_2 + H_2O.$$

The mechanisms underlying this reaction is that H^+ ions react with the HCO_3^- ions which are responsible for the alkalinity of the water, and solid $CaCO_3$ from bottom sediments dissolves to restore equilibrium. As a result, the pH in the aqueous phase is not changed significantly by the addition of the acidic rainwater.

Acidification of a body of water is generally accompanied by dissolution of metal ions from the underlying bedrock. These may include toxic metals ions such as Cd^{2+}, Pb^{2+}, and Hg^{2+}. In each case, the metals are solubilized because of the reaction of H^+ with the basic anion with which the metal is associated. Such a reaction is similar in every respect to the dissolution of $CaCO_3$ in lakes which overlie limestone. Consider, for example, the case of PS, for which $K_{sp} = 1 \times 10^{-28}$ mol^2 L^2. In pure water at equilibrium, [Pb^{2+}, aq] is calculated to be 1×10^{-14} mol L^{-1}, but at pH 4 the concentration of lead is over a billion times higher. Under some conditions, the presence of toxic metal ions in acidified water compromises its suitability for drinking (see Chapter 10).

Aluminum is a subject for attention because it is highly toxic to fish, and is major ion released upon neutralization of acid (Hedin et al, 1990). The aqueous chemistry of aluminum is quite complex, depending on the pH. Solutions of aluminum can contain Al^{3+}, $AlOH^{2+}$, $Al(OH)_2^+$ and numerous polynuclear aluminum-hydroxo species. At higher pH, the $Al(OH)_4^-$ anion becomes important, since Al_2O_3 and $Al(OH)_3$ are amphoteric, i.e., they dissolve in both acid and base. For our purposes, we will not consider the polynuclear complexes, as these form only at higher concentration of aluminum. We will also take the solid phases $Al(OH)_3$ and hydrated Al_2O_3 as interchangeable. Figure 9 shows the speciation of aluminum with pH accounting to this model (Bunce, 1994).

The hydrated Al^{3+} and $AlOH^{2+}$ ions are weak acids having pK_a about 5.5 and 5.6, respectively. Most natural waters have pH 5–8, so they will contain significant amounts of Al^{3+}, $AlOH^{2+}$, $Al(OH)_2^+$ and $Al(OH)_4^-$ when acidity mobilizes aluminum from bedrock

$$1/2\,Al_2O_3 \cdot nH_2O(s) + 1H^+ \longleftrightarrow Al(OH)_2^+(aq) + 1/2\,H_2O(l), \quad (17)$$

$$1/2\,Al_2O_3 \cdot nH_2O(s) + 2H^+ \longleftrightarrow AlOH^{2+}(aq) + H_2O(l), \quad (18)$$

$$1/2\,Al_2O_3 \cdot nH_2O(s) + 3H^+ \longleftrightarrow Al^{3+}(aq) + 1,5H_2O(l), \quad (19)$$

$$1/2\,Al_2O_3 \cdot nH_2O(s) + OH^- \longleftrightarrow Al(OH)_4^-(aq). \quad (20)$$

The explanation for the toxicity of aluminum towards fish is that in a lake of, say, pH 5 we can expect an aluminum concentration of roughly 10^{-5} mol L^{-1}. The fish blood is close to pH 7.4; its gill membrane is very thin in order to allow the efficient diffusion of oxygen from the water, and so a steep proton gradient is set up across the

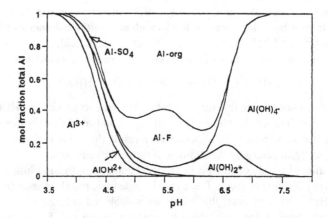

Figure 9. Speciation of aluminum over pH range from 3 to 9 (Bunce, 1994).

membrane. Figure 9 shows that the concentration of dissolved aluminum may change by 2–3 orders of magnitude for a pH change of two units. As a result, a gelatinous precipitate of $Al(OH)_3$ forms on a fish's gills, leading to death by suffocation.

The pH ranking in accordance with aquatic organisms depth is shown in Box 1.

Box 1. Loss of fish species and lake acidification (After Bunce, 1994)

The natural pH of surface waters is between 6.5 and 8.5 depending on type of water body, underlying geological rocks, water trophy levels, biogeochemical food web, etc. It has been recognized that a declining pH less than 6.0 is accompanied by undesired changes in biodiversity and even death of many aquatic species

Surface water pH	Aquatic organism lost
6.0	Death of snails and crustaceans
5.5	Death of salmon, rainbow trout and whitefish
5.0	Death of perch and pike
4.5	Death of eel and brook trout

Below pH ca. 4.0, the lake becomes a suitable habitat for white moss, which prefers an acidic environment. This plant forms a felt mat, which may grow to 0.5 or more thick, on the lake bottom. The mat prevents the exchange of nutrients between the water and the bottom sediments and also prevents the sediments from exerting any buffering action. The resultant lake waters are crystal clear, but this water supports very few forms of aquatic living organisms.

Table 9. Statistics of estimated parameters for acidity budget calculation in Japan, $mol_c\ ha^{-1}yr^{-1}$ (Shindo, 1998).

Parameters	Minimum	Maximum	Average
Base cation weathering $(BC_{we(E)})^*$	96	3564	1382
Base cation weathering $(BC_{we(R)})^*$	74	5198	1729
Base cation uptake $(BC_{gu})^{**}$	674	2839	1663
Nitrogen uptake $(N_{gu})^{**}$	337	1359	819
Acid production due to uptake $(BC_{gu}-N_{gu})^{**}$	337	1480	858
Deposition rates*** H^+	34	1343	335
NH_4^+	151	1690	407
$BC\ (Ca^{2+} + Mg^{2+} + K^+)$	226	2678	908

* Statistics for the whole of Japan (about 360,000 grid points).
** Statistics for the secondary or planted forests (about 180,000 grid points).
*** Statistics for the 29 monitoring stations.

Soil acidification in Japan

Accordingly, in Japan models were developed to evaluate soil acidification caused by acidic deposition and ecosystem sensitivity to it in different spatio-temporal scales. For the prediction of catchment scale acidification, a dynamic model was made. The model took rapid chemical reactions into consideration as the quasi-equilibrium processes and processes such as chemical weathering, nutrient uptake, nitrification etc. as element flux at a constant rate to the soil system. Application of this model to the soil acidification experiments using simulated acid rain showed that changes in soil chemistry could be well expressed by the model; the acid loaded into the soil was neutralized in the top 10 cm horizon and soil acidification occurred there. Against the urgent acidification, main buffering mechanisms were cation exchange and dissolution of Al hydroxides whose relative contributions differed according to soil characteristics and its acidification stage (Shindo, 1998).

For larger scale estimation, especially for the country scale, a steady state mass balance model was to be employed. Magnitudes of acid neutralizing capacity and acid production due to chemical weathering, base cation and nitrogen uptakes and base cation deposition were estimated for total Japanese Forest ecosystems based on existent data bases of geology, soil, vegetation etc., some measured data on soil properties and parameters derived from literature. It has been additionally stressed that the mineral weathering rate has the most significant effect on the neutralizing capacity of ecosystems. An indicator was proposed to evaluate the possible ecosystem impact by acid deposition based on the steady state model under the condition of current or predicted acid deposition rates (Table 9).

Relationship between proton input with acid rain and aluminum speciation in soils of Japanese Cedar ecosystems is considered in Box 2.

Box 2. Aluminum speciation in Japanese Cedar ecosystems (Sato et al, 1998)

The influence of acid rain on the environment is related to the various properties of different ecosystems. This influence varies depending upon the physico-chemical characteristics of soil, vegetation type, stemflow and throughfall interactions of rain waters with a canopy of different botanic species.

It is well known in Japan that soils close to the stems of Japanese cedar trees (*Cryptomeria japonica*) are strongly acidic. This is partly owing to the leaching of hydrogen ions from the stems. Since soil acidification owing to acidic deposition has not been observed in Japan, investigation of soils and soil solutions close to stems may be worthwhile to predict the situation that may occur after acidification due to acidic deposition. In a Japanese cedar forest in Gunma Prefecture, Japan, soil solutions have been collected in ceramic porous cups at a depth of 10 cm. Soil solutions far from stems (> 100 cm) are slightly acidic (pH ~ 5.8) and contain 1.0 μM of total Al in average. The speciation of Al, using cation exchange chromatography with fluorescence detection of the Al-lumogallion complex, shows that nearly 100% of the Al consist of species with a charge less than or equal to +1 (possibly, organically chelated Al). The molar ratios of BC (= $Ca^{2+} + Mg^{2+} + K^+$) to total Al are extremely high, ranging from 66 to 1050. In contrast, soil solutions close to a stem (10 cm) are markedly acidic (pH ~ 4.5) and contain 47 μM of total Al in average. Furthermore, more than 55% of the Al are in the form of Al^{3+}. The BC/T-Al ratios in winter decline to as low as 2, which is the closest to 1 that is known to be used as a critical value for starting soil acidification and appearance of harmful free aluminum ions.

4.2. China

The experimental results obtained by Chinese scientists have obviously shown that the areas suffering from acid rain in China have extended northwards from the south of the Yangtze River in 1986 to the whole of East China at present. The statistical results from the Acid Rain Survey in 82 cities from 1991 to 1995 indicate that the annual average pH value of the precipitation was lower than pH 5.6 in nearly half of these cities or in 87% of the southern cities which are located in the south to the Qingling Mountain and Huaihe River, and the lowest even reached pH 3.52 in Changsha, Hunan province. In addition, the frequency of acid rain was very high (higher than 60%) in one fourth of these cities. Up to now, acid deposition appears quite severe in wide areas south of the Changjiang (Yangtze) River and east of the Qingzang (Tibet) Plateau, covering 12 provinces/autonomous regions, with the formation of four core zones, namely Chongqing-Guiyang area in the southwest, Changsha-Nanchang area in the south, the southeast coastal area and the area near Qingdao in Shandong province.

The chemical composition of acid rain in China is generally different from that in Europe, with a lower pH value and a higher sulfate, calcium and ammonium concentrations. Another difference is that the concentration of calcium relative to sulfate is very high in China, while the nitrate concentration relative to other components is low. In some cases, the fluoride concentration in precipitation appears also high in China, owing possibly to the combustion of coal with high fluoride content. Besides, alkaline fly ash and soil dust build the capacity for acid neutralizing during washout.

Based on the mineralogy controlling weathering and soil development, sensitivity of the ecosystem to acid deposition is assessed with a comprehensive consideration of the effect of temperature, soil texture, land use and precipitation. The results show that the area most sensitive to acid deposition in China is a podzolic soil zone in the Northeast, followed by latosol, dark brown forest soil and black soil zones. The less sensitive area is ferralsol and yellow-brown earth zone in the Southeast, and the least sensitive areas are mainly referred to as the xerosol zone in the Northwest, the alpine soil zone in the Tibet Plateau, and the dark loessial soil and chernozem zone in central China. These regional different soil sensitivities to acid deposition can be attributed to differences in temperature, humidity and soil texture.

It has been shown that the assessment of ecosystem sensitivity to acidic loading depends strongly on the calculation of chemical weathering of soil base cations due to input of protons with depositions. There are two approaches for calculating this weathering. The use of a dynamic modeling approach has been identified as essential for predicting the time taken for the ecosystems to reach equilibrium between present and potential input of acidity. These models are much more complicated than so-called simplified steady-state mass balance (SSMB) equations. The first word indicates that the description of the biogeochemical processes involved is simplified, which is necessary when considering the large-scale application to East Asian ecosystems. The second word of the SSMB acronym indicates that only steady-state conditions are taken into account, and this leads to considerable simplification.

Damage to metal, concrete, trees and crops has been reported in areas seriously affected by acid rain (Zhao and Sun, 1986; Zhao and Xiong, 1988).

Let us consider the corrosion of metals due to acid rains. Iron and steel structures are highly susceptible to corrosion, and their protection with paint costs billions of US\$ annually. The chemistry of corrosion under atmospheric conditions is extremely complex and is catalyzed by hydrogen ions. This explains why acid deposition causes increased rates of corrosion. Unlike the tightly held oxide film that is formed on aluminum, protecting the metal underneath, iron oxide (rush) provides no such protection to iron and steel. Impurity sites in iron act as cathodes for the reduction of oxygen and the iron acts as the anode

$$2\,Fe(s) + O_2(g) + 4\,H^+(aq) \longrightarrow 2\,Fe^{2+}(aq) + 2\,H_2O(l). \tag{21}$$

The Fe^{2+} ions are oxidized to $Fe^{3+}(aq)$ and precipitate as rush (hydrated iron (III) oxide). Since the precipitate does not adhere tightly to the metal, it allows further corrosion to take place.

Steel protected with zinc (i.e., galvanized steel) is also subject to accelerated corrosion under acidic conditions, with dissolution of the protective layer of zinc.

Although zinc offers excellent protection to steel under dry conditions, 5-yr exposure tests indicate that in heavy industrialized areas, galvanized steel may last as little as 5–20 years.

4.3. India

As we have already mentioned, damage of the historic Taj Mahal in Agra, India, is the result of the atmospheric pollution from acid rains. This and similar deteriorating effects on constructions are connected with chemical reactions between limestone as a main ancient construction material, and acid rainwater.

Even under conditions of very clean air, limestone ($CaCO_3$) is subjected to slow attack, by the same chemical processes, which carve out caves and gorges in mountains (reaction (22))

$$CaCO_3(s) + H_2CO_3(aq) \longrightarrow Ca^{2+}(aq) + HCO_3^-(aq). \tag{22}$$

This process has a small equilibrium constant (5.3×10^{-5} mol^2 L^{-2} at 25 °C) and occurs exceedingly slowly (see Chapter 8 for more details). Acidic precipitation greatly increases both the equilibrium constant and the rate of dissolution

$$CaCO_3(s) + 4H^+(aq) \longrightarrow Ca^{2+}(aq) + HCO_3^-(aq). \tag{23}$$

The large value of K_c for reaction (23) results from the very strongly favored reaction of H+ with the basic anion CO_3^{2-}

$$CaCO_3(s) \longleftrightarrow Ca^{2+}(aq) + CO_3^{2-}(aq) \quad K = K_{sp} = 6.0 \times 10^9 \text{ mol}^2 \text{ L}^{-2}, \tag{24}$$

$$H^+(aq) + CO_3^{2-}(aq) \longleftrightarrow HCO_3^-(aq) \quad K = 1/K_a = 2.1 \times 10^{10} \text{ L mol}^{-1}, \tag{25}$$

$$CaCO_3(s) + 4H^+(aq) \longrightarrow Ca^{2+}(aq) + HCO_3^-(aq) \qquad K_c = 1.3 \times 10^2 \text{ mol L}^{-1}.$$

The damage done to historical monuments is mainly related to fine stone carving, as the outer layers of the stone flake off. This process, called "sulfation", involves the replacement of calcium carbonate by calcium sulfate, which is both more water soluble ($K_{sp} = 5.0 \times 10^{-4}$ mol^2 L^{-2}) v/v ($K_{sp} = 6.0 \times 10^9$ mol^2 L^{-2}) and has less structural strength

$$CaCO_3(s) + SO_2(g) \longrightarrow CaSO_4(s) + CO_2(g).$$

Thus the threat of acid deposition to the medieval construction in Europe that has been found in the 1980s is now of great concern for ancient Asian monuments like the Taj Mahal and the Great China Wall.

4.4. Taiwan

Assuming that fossil fuel energy consumption will continue at the current rate in China, sulfur emission will be doubled by the year 2010. Without any countermeasures for reduction of sulfur dioxide emission, it may induce possible adverse direct effects on the natural vegetation due to increasing concentration of sulfur dioxide rather than due to increasing wet deposition of sulfate. Exposure experiments suggest that differential sensitivity of plants to primary gaseous pollutants and its critical level will be a more important factor to explain forest decline than the soil acidification stress alone associated with wet acid deposition.

It is known that the geochemical and biogeochemical mobility and migration of the majority of heavy metals are increasing with the decreasing soil and water pH values that are occurring due to acid deposition. So it is of great scientific and public concern to study the influence of atmospheric acidity loading on the accumulation of hazardous metals in agricultural food production.

Acid deposition influence on the biogeochemical migration of heavy metals in the food web.

An interesting study of acid rain effects on the biogeochemical accumulation of heavy metals (Cd, Cu, Pb, and Zn) in crops has been presented by Chen et al (1998). The authors have compared the ratios of relative concentration of four heavy metals in brown rice and leaves of vegetables sampled from an acid rain affected area and a non-affected area. The data indicated that the ratios of relative concentration of Cd, Cu, Zn in brown rice and 19 vegetable species growing in an acid rain area (Lung-tang) and growing in an acid rain non-affected area (Lung-luan-tang) sampled from 1996 to 1997 are almost higher than 1, or higher than 3, except for Pb (Table 10). These results suggested that biogeochemical accumulation of heavy metals in brown rice seems not affected by long-term acid rains the contrary to vegetable species in northern Taiwan. Therefore, this accumulation is dangerous for humans eating vegetables produced in an acid rain affected area.

Table 10 also reveals that the mean concentration of Pb in brown rice and leaves of 19 vegetable species between acid rain affected area and non-affected areas are almost the same. In other words, the ratio is close to 1. This result indicated that acid rain does not influence the biological accumulation of Pb in brown rice and leaves of vegetables species sampled in Taiwan. Some studies have indicated that concentration of Pb in the crops was only affected when the concentration of Pb in the soils is higher than 500 mg/kg (Kabata-Pendias and Pendias, 1992). Sloan et al (1997) also indicated that the relative bioavailability of biosolids-applied heavy metals in agricultural soils was $Cd \gg Zn > Ni > Cu \gg Cr > Pb$, for the soils 15 years after biosolids application. It is quite consistent with the results from Chen et al's (1998) research. Thus, the phyto-availability of heavy metals caused by acid deposition followed the trend: $Cd > Zn > Cu \gg Pb$.

Forest ecosystems

As has been already mentioned, approximately 70% of the precipitation monitored in Taiwan is considered acid rain (exhibiting a pH lower than 5.6). Acid deposition shows large spatial variations with the highest acid deposition in northern Taiwan and the lowest deposition in southeastern Taiwan. The spatial pattern of acid deposition is

Table 10. The ratios of relative concentration of heavy metals in brown rice and the leaves of vegetable species growing in Lung-tang area (affected by acidic rains) and Lung-luan-tang area (non-affected by acidic rains) from 1996 to 1997 in Taiwan (Chen et al, 1998).

Rice and vegetable species	acid rain/non-acid rain affecting area (sampling number)	Ratio in acid rain/non-acid rain area			
		Cd	Cu	Pb	Zn
Rice					
Rice (*Oryza sativa Linn.*)	24/15	1.25	1.05	1.09	1.03
Vegetables					
Sweet potato (*Ipomoea bataus*)	14/9	1.00	1.45	1.07	1.11
Welsh onion (*Allium fistulosum*)	10/12	0.89	1.48	3.08	2.03
Pickled cabbage (*Brassica chineniss*)	3/10	5.03	1.23	—*	1.33
Chinese chives (*Allium tuberosum*)	7/5	4.97	0.70	0.08	1.56
Mustard (*Brassica juncea*)	2/4	—	1.59	—	2.19
Lettuce (*Lactuce sativa*)	6/8	3.73	1.87	1.00	1.97
Chickweed (*Alsine media*)	3/1	—	2.40	—	0.36
Garlic (*Allium sativum*)	6/7	0.85	2.44	—	4.64
Kohlrabi (*Brassica campestris*)	1/1	2.00	2.00	—	5.50
Cabbage (*Brassica oleracea*)	2/1	—	1.99	—	3.06
Tassel flower (*Amaranthus caudatus*)	6/2	0.97	2.23	—	1.47
Celery (*Apium graveolens*)	2/1	—	—	—	1.55
Spinach (*Spinacia oleracea*)	2/1	—	0.75	0.80	0.42
Coriander (*Coriandrum stivum*)	1/4	8.02	5.01	—	1.80
Basil (*Ocimum basilicum*)	1/3	—	8.05	—	0.36
Radish (*Raphanus sativus*)	4/2	—	2.76	—	1.08
Pepper (*Capsicum frutescens*)	3/4	1.97	2.04	3.92	0.88
Kidney bean (*Phaseolus vulgaris*)	3/10	2.07	1.78	1.09	1.44
Water convolvulus (*Ipomoea aquatica*)	6/3	0.28	1.97	3.50	0.66

* The ratios of relative concentration can not be calculated because the heavy metal contents of rice or vegetables growing in an acidic rain affected area or in a non-acidic rain affected area is lower than that of the method detection limit (MDL) of heavy metals.

correlated with the degree of urbanization and industrialization. The contribution of dry deposition to total acid deposition shows great spatial variations with precipitation patterns. Dry deposition is least and evenly distributed in northern and northeastern Taiwan where the annual precipitation amount is high while it becomes larger with high variability in southern Taiwan where the annual precipitation amount is low.

A RAINS-ASIA impact module is used to assess ecosystem sensitivity to acid deposition and to calculate the critical load of sulfur to six forest ecosystems in Taiwan. Results indicate that forest ecosystems in Taiwan are very sensitive to acid deposition owing to their low soil pH (< 5.5). Lowland subtropical forest ecosystems in Taiwan have low or moderatly low critical loads for S suggesting that they are vulnerable to acid deposition. Yet, many forest ecosystems are exposed to acid deposition far exceeding their critical loads. Although these forest ecosystems appear healthy, there may be a sudden detrimental change once the current buffering capacity is depleted.

4.5. Thailand

Cation leaching both from the forest canopy and forest soils is observed in some forest ecosystems. Continuous exposure to high levels of acid deposition can lead the forest to be in nutrient imbalance and thereby undermine forest health. The input of < 1 meq/100 g soil to the forest soil in the northern part of Thailand, with the organic content of 1.33 percent, has no changes in pH value owing to existing hydrogen buffering capacity. Simulated acid rain at pH 2 led to Al leaching from this soil naturally enriched by aluminum and iron, although there is no significant change of pH in the soil. The acid depositions cause also the reduction of bacteria, actinomyces and ammonifying microbes in the upper part of the soil. These could lead to a low nutrient cycling rate in the ecosystem as these micro-organisms play a significant role in organic matter decomposition in the soil. Owing to certain buffering ability of soils to acid depositions, the forest ecosystem can be sustainable in a short-time scale; however, some negative chemical and biological changes occur in soil. This will gradually decrease the ecosystem sustainability to acidic loading. More details on ecosystem sensitivity to acid deposition are considered in Chapter 15, Critical Loads.

5. MITIGATION OF ACID RAINS IN ASIAN COUNTRIES

5.1. Japan

Japan has an extensive air quality monitoring network. The concentration of acid rain precursor gases, SO_2 and NO_x, and other air quality parameters (CO, O_3, particulates) is monitored at more than 1,000 stations (Yanagisawa, 1989). In the area of air pollution control Japan has no rival; it has been the world leader in air pollution control technologies for quite some time, and it has the most stringent environmental legislation in the world. Japan started introducing SO_2 and NO_x control technologies at its power plants and other industries as far back as the early 1970s (Radojevic, 1996,1998). More than 1,000 power plants are equipped with flue gas desulfurization scrubbers, and all power plants are required to employ NO_x reduction technologies. The massive program of desulfurization and NO_x control at industrial plants initiated in the late 1960s and early 1970s, switches to low-sulfur fuels, and other air pollution control policies resulted in a significant decrease in mean annual SO_2 concentrations, from around 60 ppbv in 1967 to about 15 ppbv in 1981 (Hashimoto, 1989). Furthermore, Japan was the first in mass production of low emission motor vehicles, and it introduced strict vehicle emission standards well before the US. For example, the Japanese standard for NO_x was 0.4 g/mile in 1978, while the Californian and federal US standards in that year were 1.5 and 2.0 g/mile respectively (Nishimura and Sadakata, 1989). California only introduced the 0.4 g/mile standard in 1983, five years after Japan. Although the NO_x emissions per vehicle have been declining, the number of vehicles on Japanese roads grew steadily throughout the 1970s and 1980s. The average annual NO_2 concentration showed almost no change between 1973 and 1990; it has remained between 25 and 30 ppb throughout this period (Matsushita, 1996).

5.2. *China*

China is currently undertaking a massive pollution control program that will involve the fitting of desulfurization equipment to industrial plants. Also, the number of air quality monitoring stations in urban areas is growing.

The government of China attaches great importance to the pollution caused by acid rain and ambient sulfur dioxide. After the implementation of *The Suggestion on Controlling Acid Rain Development* in 1990, pollutant fees for SO_2 emission from industrial coal-combustion were charged in Guizhou and Guangxi province, and in nine cities: Liuzhou, Nanning, Guilin, Hangzhou, Qingdao, Chongqing, Changsha, Yichang, and Yibin since 1992, to set a standard for the comprehensive prevention and control of acid rain. These measures had a positive effect on the prevention and control of SO_2 pollution too. Thus, local sources of SO_2 should be firstly controlled in cities, and the SO_2 Pollution Control Areas should focus on cities. In brief, the SO_2 Pollution Control Areas should include the cities with an annual average concentration of ambient SO_2 higher than the second class of the national standard and the daily average concentration exceeding the third class, except the National Poverty Counties. The southern cities, where SO_2 pollution and acid rain are both serious, should be designated in the Acid Rain Control Areas.

The range of the Acid Rain Control Areas and the SO_2 Pollution Control Areas, as shown in Figure 10, is about 1.09 million km^2, and comprises 11.4% of the area of China.

The Acid Rain Control Area involves 14 provinces/autonomous regions/municipalities south to the Changjiang (Yangtze) River, with an area of 0.806 million km^2, while the SO_2 Pollution Control Area includes 63 cities north to the Changjiang (Yangtze) River, with a total area of 0.29 million km^2. The Acid Rain Control Area and the SO_2 Pollution Control Area comprise 8.4% and 3% of the Chinese land respectively.

In 1995, the SO_2 emission in all of China was about 23.7 million tons, 14 million tons (on about 6090) of which were in the Control Areas. Thus, the acid rain and SO_2 pollution control in China will not deteriorate if the SO_2 emission is well controlled in these areas.

The control measures are connected with implementations of different strategies of sulfur dioxide emission abatement.

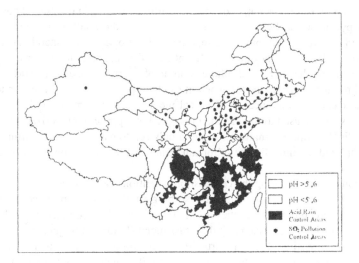

Figure 10. Sketch map of the acid control area and SO$_2$ pollution control areas in China (Hao et al, 1998).

Flue-gas desulfurization

In this scrubber process the SO$_2$ gas is removed by an acid-base reaction between it and calcium carbonate (limestone) or calcium oxide (lime) in the form of wet, crushed solid. The emitted gases are either passed through a slurry of the wet solid, or are bombarded by jets of the slurry. In some applications, fine grains of calcium oxide, rather than a slurry of calcium carbonate, are used to trap the sulfur dioxide from the emission gases. Up to 90% of the gas can be removed by this technique. The final product is a slurry of calcium sulfite and sulfate, or dry solid if granular calcium oxide is used, which then is usually buried in a landfill. In some modern technologies, notably imported from Japan, the product is fully oxidized by reaction with air, and the resulting calcium sulfate dewatered and sold as gypsum. The reactions with calcium carbonate are as follows

$$CaCO_3(s) + SO_2(g) \longrightarrow CaSO_4(s) + CO_2(g), \qquad (26)$$

$$CaCO_3(s) + O_2(g) \longrightarrow 2CaSO_4(s). \qquad (27)$$

In other operations, the sulfur dioxide is captured by use of slurries of sodium sulfite or magnesium oxide, and these original compounds that concentrated SO$_2$ gas are regenerated by thermally decomposing the product.

Clean coal technologies

Coal can be utilized in ways that are cleaner and often more energy-efficient than those used in the past. In the various technologies, the cleaning can occur precombustion, during combustion, postcombustion, or by conversion of the coal to another fuel.

In precombustion cleaning, coal has the sulfur contained in its mineral content— usually pyritic sulfur, FeS—removed so it cannot produce sulfur dioxide. The coal is ground to a very small particle size, effectively into separate mineral particle and carbon particles. Since they are of different densities, the particle types can be separated by mixing the pulverized solid in a liquid of intermediate density, and allowing the fuel portion to rise to the top where it can be separated. As an alternative to such physical cleaning, biological or chemical methods can be employed. Thus microorganisms can be used to oxidize the insoluble Fe^{2+} pyrite in pulverized coal to the soluble Fe^{3+} form. Alternatively, bacteria cultured to eat the organic sulfur in coal can be utilized. Chemically, the sulfur can be leached with a hot sodium or potassium solution.

In combustion cleaning the combustion conditions can be modified so as reduce formation of pollutants, or pollutant-absorbing substances can be injected into the fuel to capture pollutants as they form. In fluidized bed combustion, pulverized coal and limestone are mixed and than suspended (fluidized) on jets of compressed air in the combustion chamber. Virtually all the sulfur dioxide is thereby captured before it can escape. Since this procedure permits much-reduced combustion temperatures, it therefore greatly decreases the amount of nitrogen oxides formed and released.

Some of the advanced techniques used in postcombustion cleaning—such as the use of granular calcium oxide or sodium sulfite solutions—have already been described above. In the SNOX process, cooled flue gases are mixed with ammonia gas to remove the nitric oxide by catalytically reducing it to molecular nitrogen. The resulting gas is reheated and sulfur dioxide is oxidized catalytically to sulfur trioxide, which is subsequently hydrated by water to sulfuric acid, condensed, and removed.

In coal conversion the fuel is first gasified by reaction with steam, the gas mixture cleaned of pollutants, and the cleaned gas then burned in a gas turbine, which generates electricity; the waste heat of the combustion gases is used to produce steam for the conventional turbine and thus to generate more electricity. Alternatively, the gasified coal can be converted into liquid fuel suitable for vehicular use.

Fuel dioxide emissions from power plants can be also reduced by substituting oil, natural gas, or low-sulfur coal, but these fuels are usually more expensive than high-sulfur coal or lignite that are common in Asian deposits.

5.3. Singapore

Major sources of acidic precursors (SO_2 and NO_x) in Singapore include power stations, petroleum refineries, other industries and motor vehicles. Control of these pollutants is regulated by the Clean Air Act of December 1971. Emissions of SO_2 are controlled by restricting the sulfur content of automotive diesel and industrial fuel oil. The maximum permitted sulfur content of automotive diesel is 0.3%. As of July 1994 all new vehicles are required to be fitted with catalytic converters in order to comply with vehicular emission standards.

FURTHER READING

1. Bunce N., 1994. *Environmental Chemistry.* 2nd edition. Wuerz Publishing Ltd, Winnipeg, Canada, Chapter 6.

2. Radojevic M., 1998. Acid rain monitoring in East And South-East Asia. In: Bashkin V. N. and Park S.-U. (Eds.), *Acid Deposition and Ecosystem Sensitivity in East Asia*, Nova Science Publishers, New York, 95–124.

3. Bashkin V. N. and Park S.-U. (Eds.), 1998. *Acid Deposition and Ecosystem Sensitivity in East Asia*, Nova Science Publishers, Ltd, 427 pp.

4. Radojevic M. and Bashkin V. N., 1999. *Practical Environmental Analysis.* RSC, UK, 44–73.

5. Bashkin V. and Radojevic M., 2001. Acid check in Asia. *Chemistry in Britain*, No 6, 38–42.

WEBSITE OF INTEREST

1. Air pollution information network,
http://www.york.ac.uk/inst/sei/APIN/welcome.html
http://www.york.ac.uk/inst/sei/SEI/welcome.html

QUESTIONS AND PROBLEMS

1. Give the definition of acid rain. Present the relative ranking of acid rain depending on pH values.

2. Present a review of the acid deposition problem in various countries of Asia. Why is this region of the most environmental concern at present?

3. Describe the methods of acid rain monitoring that are employed in various Asian countries. Compare these methods with those used in your country.

4. Characterize the chemistry of acid rain formation accounting for both sulfur and nitrogen compounds.

5. Present a quantitative explanation of why nitric and sulfuric acids have more effects on acid rain formation than carbonic acid.

6. Present different mechanisms of sulfur dioxide oxidation and explain what mechanism will be dominant in your region.

7. Discuss oxidation and deposition of sulfur dioxide. Describe the processes which happen in sulfur dioxide plume depending on time.

8. What sulfuric species will be deposited under different weather conditions? Present examples.

9. Characterize the general ecological effects of acid rain in the Asian countries.

10. Discuss the chemistry of natural water acidification and the role of aluminum in these processes.

11. Discuss the acid rain problems in South East Asia. Focus your attention on acid deposition effects on biogeochemical cycling in Tropical Rain Forest ecosystems.

12. Discuss the interactions between acid deposition and accelerating migration of heavy metals in biogeochemical food webs. Present examples.

13. Present a discussion of chemical processes that lead to the damage of construction materials in Asian cities.

14. Discuss the mitigation strategy for sulfur dioxide emissions in various countries. Pay attention to advanced technologies.

15. Characterize the methods for reduction of sulfur dioxide formation during coal combustion.

CHAPTER 5

HAZE POLLUTION

1. INTRODUCTION

The most devastating environmental issue events in Southeast Asia are the forest fires. The recent experience and insight on the causes of catastrophic forest fires provides a strong push to review land use politics throughout this sub-region. The indiscriminate clearing of land for pulpwood and oil palm plantations is fueled by the high demand for paper and palm oil products in the World. Moreover, the traditional way of clearing land in most of the sub-region is by fire. Here we should mention that forest fire is not a new event. In 1887, Lady Annie Brassey noted that "the bay itself is surrounded by vast forests, and not long ago a steamer was prevented from entering the river for three days, in consequence of a fierce forest fire, the dense volumes of smoke from which completely obscured the entrance".

The above quotation from the 19th century travel book (Brassey, 1889) is a direct reference to Brunei Bay and Brunei River, and it demonstrates that forest fires and the associated smoke pollution have been taking place during the whole period of forest clearing in this area. This natural event is enhanced by anthropogenic activities and many effects on human health and ecosystem behavior have been recorded.

In this chapter we will consider the geographical distribution of recent forest fires in Southeast Asia, chemical composition of haze and synergetic effects of haze pollutants and anthropogenic air contaminants on human health.

2. SOURCES AND TRANSPORT OF HAZE POLLUTANTS

2.1. Geographical Distribution of Forest Fires in Southeast Asia

Incidents of haze (severe smoke pollution) have occurred from time to time in many parts of the Asian region. A particularly lengthy and severe episode, resulting from uncontrolled forest fires mainly in the Indonesian states of Kalimantan and Sumatra, affected several countries of Southeast Asia from July to October 1997. The long duration of the episode was associated with the fact that the arrival of the rain-bringing monsoons was considerably delayed, believed to be linked to the El Nino phenomenon. In early January 1998, fires once again ravaged these parts, which lasted until April 1998. Figure 1 shows the extent of haze in forest fires of 1994, 1997 and 1998.

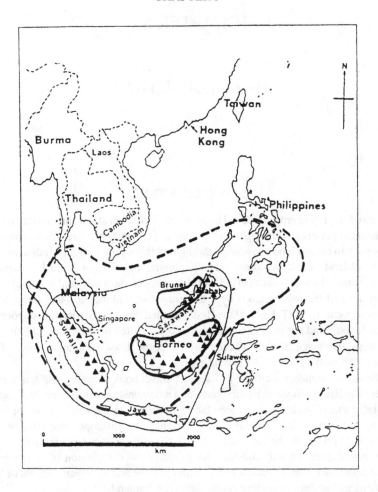

Figure 1. Approximate location of forest fire hot-spots and area affected by regional haze in Southeast Asia (Radojevic and Tan, 2000).

In Indonesia, the 1994 fires consumed an area greater than 50,000 km². The 1994 regional haze episode affected Brunei between the middle of August and the beginning of October due to the long-range transportation of pollution from the Kalimantan region of Borneo and from Sumatra. The prevailing wind direction was generally southerly and southwesterly, characteristic of that time of year. It should be noted that the 1994 episode was considerably less severe in Brunei than the 1997 and 1998 episodes.

In Brunei Darussalam, measurements taken during the dry weather period from February to April 1998 showed that the Pollution Standard Index (PSI) readings exceeded 100, which is already unhealthy, and sometimes went as high as 250.

This caused the disruption of daily activities, closure of schools, and changes in governmental working hours.

Twelve provinces of Indonesia had forest fires during the 1997 haze episode and an estimated 165,000 hectares were burnt. During the peak period of September–October 1997, significant increases in asthma, bronchitis, and ARI were observed in eight Indonesian provinces. As estimated 12.4 million persons were affected by the haze, and about 1.8 million cases of the diseases mentioned above were reported. Levels of total suspended particulates (TSP) exceeded the national standard by 3–15 times during the peak air pollution period of October 1997.

In the Philippines, the impact of the haze associated with the forest fires was noted mainly on the southern islands. The visibility in these areas was reduced to 4–5 kilometers.

The PSI in Singapore exceeded 100 for 12 days during the 1997 haze episode, reaching a maximum of 138. About 94% of haze particles were found to have a diameter less than 2.5 microns, and thus easily breathed in. Hospital visits for all haze-related illness increased by about 30%.

The Southern provinces of Thailand were affected by the haze from Indonesian forest fires. The PM_{10} concentrations in the city of Hatyai increased by about $20 \, \mu g \, m^{-3}$. The first haze peak episode lasted from 22 to 29 September, with a maximum concentration of $211 \, \mu g \, m^{-3}$. The second haze peak episode lasted from 6 to 8 October 1997 (ESCAP, 2000).

2.2. Economical and Ecological Consequences

A joint study by WWF Indonesia and the Economy and Environmental Program for Southeast Asia concluded that the 1997/1998 fire and haze cost a total US$4.4 billion (Table 1).

Of the total, US$1.4 billion were health costs (e.g., sore eyes, skin rashes), affecting around 70 million people in three countries (Schweithelm, 1999). Timber losses account for US$493 million. In agriculture (plantation and smallholdings), the cost is US$470.4 millions. Other costs are related to direct forest benefits, i.e., raw materials and food benefits, indirect forest benefits related to hydrology and soil conservation, and losses of international funding of biodiversity conservation.

The tragedy is that even at the conservative estimate of US$4.4 billion, the cost of the fire and haze would have provided basic sanitation and water sewage services to all of Indonesia's poor. The loss is huge in terms of lost opportunities for sustainable development.

3. CHEMICAL COMPOSITION OF HAZE

3.1. Smoldering Combustion

A fire that is 100% efficient in the combustion of organic material emits all carbon as carbon dioxide. Most forest fires in the Southeast Asian sub-region are not this efficient and emit carbon in other forms, such as carbon monoxide, as a result of

Table 1. Direct and indirect costs of forest fire episodes of 1997/1998 in Southeast Asia (ESCAP, 2000).

Parameters	Cost, million US$
Health cost	1,400
Timber losses	493
Agriculture losses (plantations)	470.4
Direct forest benefit losses	705
Indirect forest benefit losses	1,077
Indirect losses for biodiversity studies	30
Indirect losses due to carbon release	272.1
Cost of international fire fighting help	13.4
Total	4,460.9

Table 2. Expressions for describing combustion efficiency (McKenzie, 1999).

Expression	Smoldering combustion, %
$CO_2/1834^a$	< 90
$CO_2/(CO + CO_2)^b$	< 92
$CO/CO_2{}^b$	> 8

Notes: [a] Maximum possible CO_2 emission $(g\ kg^{-1})$, assuming a vegetation carbon content of 50%;
[b] Molar ratios.

incomplete combustion. Therefore the completeness of forest biomass combustion or combustion efficiency can be expressed by comparing carbon dioxide emissions to emissions of incomplete combustion products. The dominant presence of visible radiation from flames and black smoke characterizes flaming fires, whereas sparse or no flames and the presence of dense white smoke characterize smoldering fires or the smoldering phase of fire. Thus we can say that forest haze is mainly the product of smoldering combustion of forest biomass.

There are several expressions that have been used to describe combustion efficiency (Table 2).

Physical and chemical processes that produce emissions during smoldering combustion are complex.

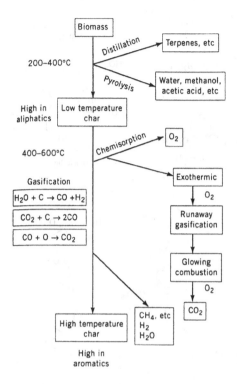

Figure 2. Physical and chemical processes in smoldering combustion (McKenzie, 1999).

Smoldering combustion is not a process in and of itself, but rather a composite of several physical and chemical processes (McKenzie, 1999). These processes, summarized in Figure 2, include volatilization, pyrolysis, oxidation, and reduction.

During a biomass fire, all smoldering composition processes may be occurring simultaneously, along with flaming combustion processes. Several emissions result from more than one process (e.g., methane, methyl halides, carbon dioxide). Periods of oxygen depletion or thermal quenching of oxidation reactions, characteristic of large forest fires, may result in an increase in smoldering combustion emission.

The smoldering of a fuel such as duff (the forest floor subsurface layer consisting of partially decayed organic matter) occurs by way of relatively slow advance of a glowing combustion front. As this front advances, biomass adjacent to the front gains heat, plant products such as terpenes are distilled off, and the biomass is converted to char. This process produces mostly water and occurs in a relatively oxygen-deficient environment because of the oxidative processes in the adjacent glowing combustion front. Thus, the low temperature (200–400 °C) char formation phase produces volatile pyrolysis products. These pyrolysis products are known to arise from the pyrolysis of polysaccharides and lignin products, which compose biomass (Figure 3).

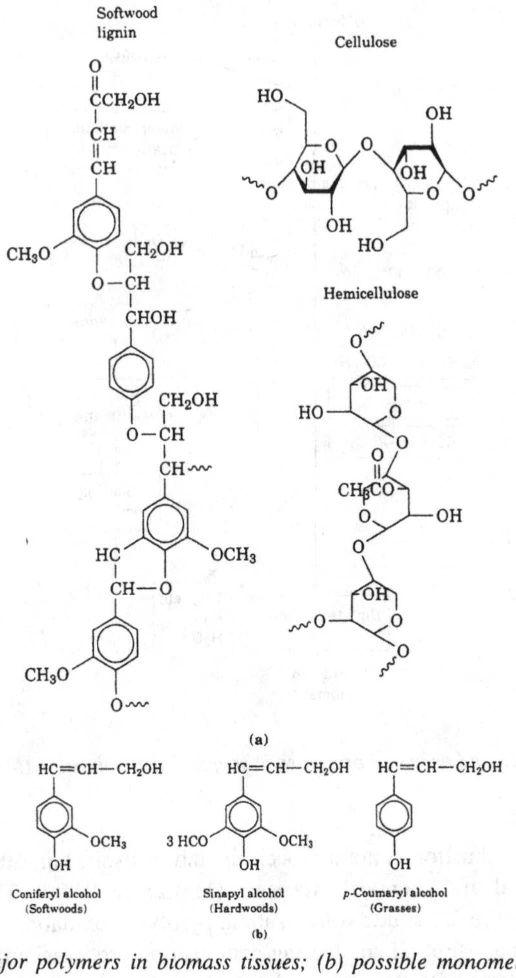

Figure 3. (a) Major polymers in biomass tissues; (b) possible monomeric units in lignin (McKenzie, 1999).

For example, acetol is a major pyrolysis product from a range of polysaccharides, especially in the presence of inorganic components, and guaicol is a major product of the pyrolysis of lignin. Both acetol and guaicol have been observed in the emissions of smoldering combustion. Secondary reactions of pyrolysis products catalyzed by the low temperature char may also occur, such as methyl halide formation. Subsequently, the char layer is heated further as the front advances, reaching 450–500 °C, where oxygen chemosorption on the char is maximized, especially in the presence of inorganic compounds. This chemosorption is the major exothermic process in smoldering combustion of the char and results in the generation of carbon monoxide, carbon dioxide, hydrocarbons and PAHs.

3.2. *Monitoring of Haze Emission Products*

Emissions from forest haze have been measured in several different laboratory experiments and field monitoring during haze episodes in the Southeast Asian sub-region. These labor-consuming studies were carried out by an international team of scientists headed by Dr. Miroslav Radojevic and the results are published (see further reading list).

Experimental studies

An experimental study has been carried out with peat samples from the forest area of Brunei Darussalam. We should note here that the measurement of emission products requires comprehensive analytical equipment. Hydrocarbons (C_1–C_4) are determined by gas chromatography with flame ionization detection (GC/FID), CO_2 and O_2 are analyzed by gas chromatography with thermal conductivity detection (GC/TCD), and CO, by gas chromatography with electron capture detection (GC/ECD). Aldehydes and polynuclear aromatic hydrocarbons (PAHs) are determined by gas chromatography with mass spectrometry (GC/MS).

Results of the experimental study of peat smoldering combustion at 480 and 600 °C are shown in Table 3.

As we can see from Figure 2, a smoldering stage is expected to release large amounts of incompletely oxidized compounds like CO, VOCs, PAH, which are potentially more harmful than substances emitted during the light temperature flaming stage of combustion. The most abundant C-containing combustion product is CO_2, followed by CO in the second place and CH_4 in the third place. The concentration of alkanes generally decreases with increasing molecular mass. The proportion of the alkanes of lower molecular mass (methane, ethane) increases as the temperature is increased and the concentration of long chain alkanes decreases. The concentrations of CO and C_2H_4 decrease significantly at higher temperature. The emission factors of CH_4 agree well with those of between 5.7 and 19.4 earlier reported for smoldering combustion in moderate climate. Emission factors between 1 and $4.2 \, g \, kg^{-1}$ were reported for flaming combustion.

Aldehydes and PAHs were not detectable in the emission products. This fact does not imply that they were absent; they may have been present at levels below the detection limits. The presence of benz(a)pyrene was reported in particulate matter emitted from the combustion of pine needles in one experimental study in a moderate climate.

Field monitoring

Smoke from forest fires may be collected by various means for subsequent laboratory analysis. The most applicable approach is to trap emissions from biomass burning on absorbent material. The absorbent material is then either solvent-extracted for GC analysis or directly desorbed onto a GC. For example, smoke is pulled with an air pump at a rate of 10–$50 \, mL \, min^{-1}$ through triple-layer glass cartridges with separate

Table 3. Emission of chemical species from smoldering combustion of peat (Muraleedharan et al, 2000).

Component	Temperature, °C			
	450		600	
	%, volume/volume	mg kg^{-1} peat	%, volume/volume	mg kg^{-1} peat
Moisture	51.4	—	51.4	—
Loss-on-ignition	75	—	77	—
Ash	25	—	23	—
CH_4	1320	5785	2.761	11338
C_2H_6	0.148	1320	0.434	3122
C_3H_8	0.146	1960	0.001	11
C_4H_{10}	0.049	875	0.014	203
C_2H_4	0.087	721	0.017	113
CO	4.159	37134	2.122	15279
CO_2	16.36	185,000	13.216	149,591
O_2	6.52	—	5.81	—
HCHO	< dl	—	< dl	—
CH_2CHCHO	< dl	—	< dl	—
Other aldehydes	< dl	—	< dl	—
PAHs	< dl	—	< dl	—

layers of Carbosieve S-III, Catbotrap, and Tenax TA. The cartridges are then placed in a desorption oven in line with a liquid nitrogen cold trap and a gas chromatographer (GC) in series.

Various types of glass fiber filters can be also used to collect haze from smoldering combustion. Glass fiber filters can either be heated (400–750 °C) or solvent-rinsed to remove any possible organic contaminants. After sample collection, the fibers are either Soxhlet or ultrasonically extracted with various combinations of organic solvents. The fine aerosols in haze can be collected by first passing the haze through several parallel cyclone separators, and then through several parallel quartz fiber filters. The filters are then combined and ultrasonically extracted in hexane and benzene/2-propanol mixture. The extracts are concentrated, filtered and analyzed by GC/MS for oxygenated organic compounds.

Table 4. Concentration of organic micropollutants in Brunei during haze episode of 1998, μgm^{-3} (Muraleedharan et al, 2000).

Species	Date of sampling				
	15/4/98	16/4/98	17/4/98	18/4/98	19/4/98
Acetic acid	< dl	7.5	3.9	4.2	12.1
Formaldehyde	6.2	6.3	6.3	5.0	16.8
Acetaldehyde	2.5	2.3	2.2	3.6	14.6
Propionaldehyde	2.0	2.3	5.5	3.0	17.0
Butylaldehyde	51.2	35.0	71.6	35.6	37.8
Benzene	< dl	2.4	5.0	4.4	< dl
Toluene	1.8	8.2	9.0	6.8	0.9
Ethyl benzene	0.14	0.57	0.43	0.43	1.62
Xylenes	7.5	11.0	14.3	14.0	7.9
Phenol	< dl	< dl	< dl	0.07	0.12
PAHs	33.8	20.8	8.5	4.2	5.6
Naphthalene	29.8	20.6	7.4	5.1	5.6
TPH	15.1	21.6	89.1	100	307

Notes: dl—detection limits.

You can see further analytical details in McKenzie (1999), Radojevic and Bashkin (1999) and Muraleedharan et al (2000).

We will discuss the chemical composition of forest haze using the examples from field monitoring during the haze episode of 1998 in Brunei Darussalam.

Organic micro-pollutants. Concentrations of various organic micro-pollutants are shown in Table 4.

As can be seen, haze contains a range of substances, which are also present in wood smoke, and it may be reasonable to state that those compounds which were not determined were probably present as well. Samples were also analyzed for cresol and cresol was not detected in any of the samples; if it was present its concentration would have been below the detection limit ($< 0.01 \mu gm^{-3}$). Concentrations of TPHs were rather influenced by industrial emissions from local pollution sources (the oil terminal and refinery) than originated from biomass smoldering. For the other organic micropollutants, like BTEX, and some aldehydes, the concentration pattern was different. During haze development, the concentrations of most aldehydes were increased but none of them exceeded both existing standards and recommendations from WHO

Table 5. Suspended particle (PM_{10}) concentrations as 24 hour average at three sampling stations in Brunei during the haze episode of April 1998 (Muraleedharan et al, 2000).

Date	PM_{10} concentrations, μgm^{-3}		
	1st station	2d station	3d station
15/4/98	66	96	126
16/4/98	47	45	101
17/4/98	41	38	49
18/4/98	77	130	228
19/4/98	149	176	372
US EPA standard	50		
WHO guideline	70		

and EC (Radojevic and Bashkin, 1999). For other compounds, only benzene and PAHs exhibited levels exceeding the available recommendations. In the case of PAHs, however, the major component was naphthalene, which is not one of the more harmful PAHs, as indicated by the high value of recommended guidelines ($1000 \mu gm^{-3}$).

Nitrogen dioxide (NO_2) and ozone (O_3). Concentrations of NO_2 and O_3 were in the range of 1.5–27.0 and 0.1–11.0 μgm^{-3}, respectively. These levels are well below WHO guidelines and accepted air quality standards even though forest fires were actively burning in the vicinity (2–5 km) of the sampling sites.

Particulate matter (PM_{10}). The observed concentrations of inhalable particles, PM_{10}, at the three monitoring stations in Brunei are presented in Table 5.

Some of the observed concentrations exceed accepted air quality guidelines and standards. Owing to the enhanced effect of small particles on the lungs, more stringent standards are being set for smaller particles ($< 2.5 \mu m$), which can penetrate deeper into lung system. For example, the US EPA has set a primary National Ambient Air Quality Standard (NAAQS) of $150 \mu gm^{-3}$ (24 h average) for particles of size $10 \mu m$ or less (PM_{10}) but it is proposing a further stringent standard of $65 \mu g\,m^{-3}$ (24-h average) for particles $< 2.5 \mu m$ ($PM_{2.5}$). Particle size distribution for this haze episode is shown in Figure 4.

There is a pronounced peak in the particle size distribution at around $0.8 \mu m$ diameter. It is obvious that more than 99% of the particles of haze were $< 2.5 \mu m$ and these could have major impacts on the respiratory system.

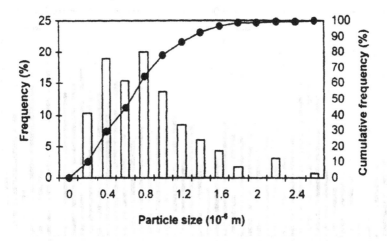

Figure 4. Size distribution of haze sampled on 11 April 1998 (adopted from de Jong and Hollander, 1998).

We can see also the substantial variations in the particulate concentration over the 24 h period. These variations can be explained by fire hotspots and prevailing wind direction during the monitoring period.

The chemical composition of aerosol particles was also determined. Analysis of one typical haze sample gave the following results in % of aerosol mass: C = 70.0, H = 4.0, N = 2.8, S = 4.7, Cl = 1.0, K = 19.6, Na = 3.3, Ca = 5.4, Mg = 0.05, Fe = 0.24, and Zn = 19.4. Some heavy metals like As, Cd, Ni, V, and Hg were also present but their mass percentages were generally < 0.01 of the total aerosol mass. This chemical composition of haze reflects the relevant chemical composition of burning biomass in the tropical forests.

4. SYNERGETIC EFFECTS WITH OTHER ANTHROPOGENIC POLLUTANTS

4.1. Acid Deposition

As we have discussed in Chapter 4, acid rains are of great environmental concern in the Southeast Asian sub-region and the pH of rainwater drops often below 4.5 in many countries of question. Over the last 10 years, a number of authors have suggested that forest fires in the tropics may have an acidifying effect on rainwater with potential consequences to tropical ecosystems (see, for instance, Levine J. S., Global Biomass Burning, 1991, in the list of further reading).

Biomass fires are inversely related to precipitation; they generally occur during dry seasons when the precipitation amounts are very low or non-existent since heavy rains tend to put the fires out. Therefore, one would anticipate that even if the acidity

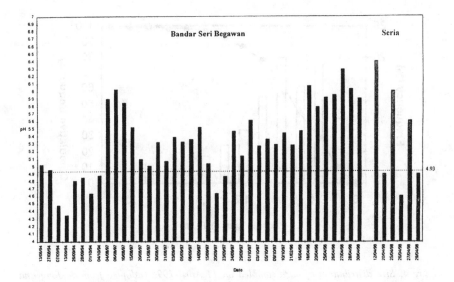

Figure 5. pH *of rainwater samples collected in Brunei Darussalam during episodes of biomass fires and haze in 1994, 1997, and 1998. Line is VWM* pH *of 228 rainwater samples collected in the absence of biomass fires and haze (Radojevic and Tan, 2000).*

and the ionic content of individual rainwater samples, that is when there is any rain, may be high, the wet deposition rates tend to be low simply because rainfall amounts are so low during the burning seasons. Most wet deposition takes place during the wet season when precipitation amounts are exceedingly large. It is the deposition rates that should be related to ecological impacts, rather than the composition of individual rainwater samples.

Here we will consider the appearance of synergetic effects of acid rain and haze in some places of Southeast Asia. The example will be shown from a case study in Brunei Darussalam (Radojevic and Tan, 2000).

Measurements of pH *values of rainwater during haze episode*

During both non-haze periods and haze episodes of 1994, 1997, and 1998 (see map on Figure 1), samples were analyzed for pH immediately after collection. Bulk samples were used to collect participation during the 1994 haze episode. Wet-only samples were collected during 1997 and 1998 haze episodes using an automatic wet-only collector. The details of analytical procedure are very important in this measurement and described (Radojevic and Bashkin, 1999).

The pH values of all the rainwater sampled in Brunei during the haze episodes of 1994, 1997, and 1998 are given in Figure 5.

The Volume Weighted Mean (VWM) pH of rainfall determined during periods unaffected by biomass fires and haze is shown as a line for comparison. The data does

Table 6. Wet deposition of H^+ *in Brunei Darussalam during episodes of forest fires and haze and percentage of contribution to total annual wet deposition hydrogen ion (Radojevic and Tan, 2000).*

Haze period	Rainfall amount, mm	VWM pH	Deposition of H^+, μeq m^{-2}	Percentage of total annual deposition of H^+
August 1994	35.3	5.02	334	0.73
September 1994	104.3	4.63	2460	5.35
October 1994	4.4	4.88	58	0.13
Total 1994 episode	144.0		2852	6.21
August 1997	64.0	5.14	459	1.00
September 1997	273.3	5.17	1830	3.98
October 1997	189.9	5.35	853	1.85
Total 1997 episode	527.2		3142	6.83
February 1998	18.2	5.28	95	0.21
March 1998	0.0	—	0	0.00
April 1998	115.2	5.91	141	0.31
Total 1998 episode	133.4		236	0.52

not demonstrate any significant acidifying effect of biomass burning on rainwater, contrary to what has been postulated in the literature from other regions of the World, mainly North and South America, and Africa (Crutzen and Andrea, 1990; Lacaux et al, 1993; Cachier and Ducret, 1991). Although biomass burning produces many acidifying compounds, for example SO_2, NO_2, and organic acids such as formic and acetic acid, it also produces ions such as Ca^{2+}, Mg^{2+} and K^+, which may act to neutralize the acidity.

Deposition rates of acidity during haze episodes

The wet deposition of H^+ can be considered as a measure of acidity deposition rates. This is a value of the most important ecological concern regarding the acidification loading and effects on the ecosystems (see Chapter 4).

During the period without any significant biomass burning (1995–1996) the total annual wet deposition of H^+ was 0.046 eq m^{-2}. Taking this value as a reference, the contribution of H^+ deposition during periods of biomass burning to the annual wet deposition of H^+ was calculated (Table 6).

The monthly contributions during episodes of biomass fires and haze to the total annual wet deposition varied between 0.0 and 5.25%. Even assuming that all the H^+ during haze episodes was due to biomass fires (which it obviously is not) the results suggest that forest fires are a minor source of acidity in wet deposition.

The results from Southeast Asia agree with studies of rainwater composition in Venezuela during periods of biomass fires. In these studies higher concentrations of all ions were observed than during periods when there were no biomass burnings. Higher pH values were observed during periods of vegetation burning in one study, while in the other it was concluded that biomass burning had no significant effect on rainwater acidity (Sanhueza et al, 1992). In other investigations carried out in Africa, South America (Brazil) and Australia, the acidity of rainwater was increased during biomass burning periods. Here we should point out that the chemical composition of haze is closely related to the chemical composition of burning vegetation that in turn, is connected with chemical composition of soils. In other words, the ratio between base cations (Ca^{2+}, Mg^{2+} and K^+) and acid forming compounds such as SO_2, NO_2, and organic acids (formic and acetic acid) in biogeochemical cycling is the most important factor influencing the chemical composition of haze and its influence on the acid deposition.

4.2. Photochemical Smog

Globally, both chemical and meteorological evidence suggests biomass fires as the source of photochemical smog layers over the Atlantic Ocean and African tropical forests. The global ozone production from biomass smoldering is estimated to be compared in strength to the stratospheric input (McKenzie, 1999). Locally elevated ozone is significant owing to its toxicity to humans and plants (see Chapter 11). Ozone can cause inflammation of the eyes and upper respiratory tract and tissue damage in plants. Sharp increases in ozone with altitude have been observed in the tropical regions. Therefore the likelihood of vegetation damage by ozone from biomass burning would be especially possible in the mountainous areas of the tropical belt.

The emissions from smoldering combustion contain all the necessary compounds needed for the formation of photochemical smog. We have mentioned in Chapter 3 that four main components are necessary for the initiation of this tropospheric phenomenon: sunlight, temperature above 18 °C, hydrocarbons and nitrogen oxides, and all of them are present during forest fires in Southeast Asia.

Diurnal variation of air pollutants during haze episode

We have seen that formation of photochemical smog depends on the diurnal variation of important chemical species. Sunlight is also among the most important parameters since all chemical reactions are initiated by photons.

Accordingly, here we will consider the diurnal variations of gaseous air pollutants such as SO_2, CO, NO, NO_2 and O_3 in Brunei's capital, Bandar Seri Begawan, during the 1998 haze episode, for one haze day (Figure 6).

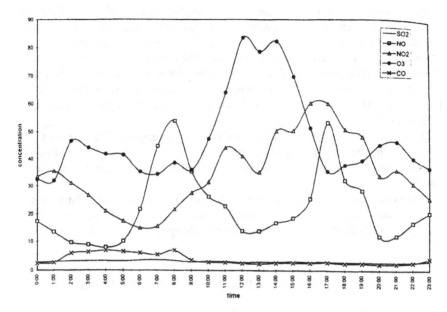

Figure 6. Diurnal variation in gaseous air pollutants at Bandar Seri Begawan on 20 February 1998. All concentrations are in μgm^{-3} except CO in $mg\ m^{-3}$ (Radojevic and Hassah, 2000).

Although there was considerable day to day variation in the precise concentration patterns, the general trends were quite similar on many occasions. Concentrations of NO and NO_2 generally exhibit peaks during the morning and afternoon rush hours and O_3 concentrations usually peaked at around noon. Concentrations of CO also exhibited diurnal variation, generally rising to a maximum during the early morning rush hour. This picture looks like that depicted in Chapter 3 for the diurnal variation of gaseous air contaminant during photochemical smog formation (see Figure 5, Chapter 3). This is very characteristic for oxidation of NO to NO_2, followed by the photochemical formation of O_3 with increasing sunlight, and a decline to lower levels at night.

Such pattern suggest that in urban conditions even during haze episodes gaseous criteria pollutants originate mainly from motor vehicles and that biomass burning is not a major contributor to their atmospheric concentrations. However, emissions of unsaturated hydrocarbons and nitrogen oxides from smoldering biomass burning play the most important role in photochemical smog formation in remote areas around the Southeast Asian sub-region.

Regional and global consequences

Emissions from biomass burning could disrupt the oxidizing potential of the atmosphere. The atmosphere has a self-cleaning mechanism, which photochemically oxidized pollutants, such as hydrocarbons, to carbon dioxide, preventing their buildup in the atmosphere. The key oxidizer in the troposphere is the hydroxyl radical. We can remember the characteristic reactions of OH radical (see Chapter 3) and conclude that the hydroxyl radical provides a major sink for methane and carbon monoxide. When nitrogen oxide concentrations are low, OH radicals are consumed as they oxidize these pollutants through a series of chain reactions. An increase in global hydrocarbon, methane, and carbon monoxide concentrations due to regional forest fires in Asia, Africa, North and South America, could result in a greater decrease of hydroxyl radicals.

However, this effect reverses at localized biomass burning sites. We have discussed previously that biomass smoldering emits nitrogen oxides. Hydrocarbon oxidation in the presence of large amounts of nitrogen oxides creates additional hydroxyl radicals and ozone.

The consequences of local peculiarities for global atmospheric chemistry are still uncertain but if the levels of hydroxyl radical would be decreased, many pollutants that are normally oxidized by OH radical could build up, leading to significant relevant changes.

5. HUMAN AND ENVIRONMENTAL IMPLICATIONS

5.1. Human Health

In the Southeast Asian sub-region, people come into contact with smoke from smoldering combustion in several ways. They may be exposed to these emissions while cooking or heating with wood stoves, clearing agricultural residues or land, or fighting fires. During haze episodes residents in the vicinities of large biomass fires may be also exposed to the emissions of smoldering combustion. Moreover, as we have discussed in Section 2, transboundary air pollution during haze episodes has been monitored in the vast Southeast Asian area, often far away from forest fire locations. The exposed people may experience significant eye and respiratory tract irritation, which may cause a slight decrement in lung function and could lead to increased incidents of respiratory illness.

Table 7 lists exposure levels to some of the compounds produced during smoldering combustion of biomass.

Most of the air contaminants from smoldering biomass emissions have irritant effects but some of them are also carcinogens and immunodepressants. Several PAHs identified in smoke from smoldering combustion are also known or suspected human carcinogens, such as benz(a)pyrene. Ames Salmonella tests have been performed with PAHs extracted from wood smoke. The tests were positive for mutagenicity, which increased in tests with extracts from oxygen-starved fires since oxygen-starved fires produced higher concentrations of PAHs.

Case study in Brunei

The harmful effects of regional haze episodes have caused considerable concern to the government authorities and the public in Brunei. Effects of the haze include a drastic

Table 7. Exposure levels for compounds found in smoldering biomass emissions (after McKenzie, 1999).

Compound	Exposure level, ppb	Effect
2-Furaldehyde	1.0–4.2	Suspected carcinogen, irritant
5-Methylfuraldehyde	5.4–14.6	Irritant
2-Acetylfuran	1.13–1.87	Irritant
Phenol	6–28	Irritant
o-Cresol	4.25–5.55	Irritant
m/p-Cresol	7.5–10.9	Irritant
Guaiacol	13.5–18.5	Irritant
4-Methylguaiacol	6.8–10.6	Irritant
Vanillin	1.0–1.2	Irritant
Acetol	55–75	No toxicological data
Vinyl acetate	100–200	Suspected carcinogen, irritant
2-Cyclopenten-1-one	2.6–5.2	Irritant
Acetic acid	530–870	Irritant
Formic acid	64–156	Irritant
Propionic acid	74–98	Irritant
Methyl-3-oxobutanoate	11–23	Irritant
Formaldehyde	500	Carcinogen, irritant
2-Propenal	70	Irritant
Benzene	60	Carcinogen, anemia, immune system depressant

reduction in visibility, closures of airports and schools, increase in transportation accidents, physical destruction of biota by fires, increased incidents of illness, lower productivity, restriction of normal recreational and work-related everyday activities, psychological effects, and economic cost. The health effects of the haze are a major concern in Brunei, as well as elsewhere in the region. Observed symptoms and illness include: upper respiratory tract infections, asthma, conjunctivitis, bronchitis, eye and throat irritation, coughing, breathlessness, blocked and runny noses, skin rashes,

Table 8. Changes in the respiratory morbidity in the south (haze) and upper northern Thailand (control) and the net health impacts from 1997 haze episode, September–October 1997 (Phonboon, 1998).

Illness	South	North	% of net haze impacts	P-value
ODP visits				
All respiratory	26	18	8	< 0.01*
IPD admissions				
All respiratory	33	26	7	< 0.01*
Pneumonia	36	18	18	< 0.01*
Bronchitis	40	28	12	0.01*
Asthma	12	9	3	NS

* significant.

and cardiovascular disorders. During the September 1997 haze episode there was a high association between the air pollution index (API) and daily attendance at clinics. However, no deaths have yet been attributed to the haze in Brunei, Singapore, or Malaysia.

Case study in South Thailand

During a two month period covering the haze episode from September–October 1997, a substantial increase in respiratory morbidity of both outpatients (ODP) visits and inpatient (IPD) admissions was observed in the study areas of southern Thailand in comparison with the control area of northern Thailand where haze episodes have not been registered. The increases were: 26% vs. 18% (south vs. control area) for all respiratory visits, 33% vs. 26% for all respiratory admissions, 36% vs. 18% for pneumonia admissions, 40% vs. 28% for bronchitis/COPD admissions, and 12% vs. 9% for hospitalized asthma although this category is not significant (Table 8).

Hence, the net health impacts from the 1997 haze are 8% and 7% increases in respiratory visits and admissions, respectively. It is interesting to observe that the percentage of net haze impacts is higher in two specific respiratory diseases, pneumonia and bronchitis/COPD. From this finding and the monthly report of respiratory morbidity, the increase during the 1997 haze would be approximately 45,000 visits and 1,500 admissions in southern Thailand.

Regression analysis demonstrates the significant associations between all categories of monthly respiratory admissions and monthly PM_{10} levels. The effects of monthly PM_{10} (for each $1 \mu gm^{-3}$) were about 85, 28, 13, and 13 monthly admissions for all respiratory, pneumonia, bronchitis/COPD, and asthma, respectively. The proportion of variance of illness that is accounted for by the predictor variables of the models varied from 45% in bronchitis/COPD to 80% for pneumonia cases.

5.2. Ecological Impacts

Disturbance of biogeochemical cycles

Biomass burning volatilizes nitrogen from ecosystems. Over 50% of the volatilized nitrogen are in the form of molecular nitrogen and is not redeposited into ecosystems, resulting in a significant loss of nutrient nitrogen through pyrodenitrification. Long-range transfer of nitrogen emitted in other forms, such as ammonia and nitrogen oxides, also depletes this nutrient from ecosystems that are repeatedly burned. Long-term perturbations of nitrogen biogeochemical cycling also occur as a result of biomass burning, with elevated soil nitrogen oxide and nitrous oxide emissions continuing for up to six months after burn. Other nutrients such as potassium, phosphorus, magnesium, and sulfur may also be depleted from ecosystems through repeated forest fires.

Biomass burning results in the sequestration of carbon as charcoal, constituting a significant sink for atmospheric carbon dioxide and therefore a source of atmospheric oxygen. The charcoal is not subject to microbial oxidation. On the geological time scale atmospheric oxygen levels may increase, which increases the risk of fires and thus establishes a positive feedback loop for the buildup of atmospheric oxygen.

Climate change

Methane and carbon dioxide, as the products of biomass burning, are known greenhouse gases. In addition to direct greenhouse effects of carbon dioxide, indirect effect of forest fires is connected with decreasing sequestration of CO_2 into growing biomass. Biomass burning competes with industry as a source of carbon dioxide, contributing from 20 to 40% to global greenhouse effect (Andrea, 1993). Biomass burning and smoldering combustion in particular are significant sources of methane, contributing about 6% of the global flux.

The reverse effects are related to the formation of aerosols during forest fires in tropical areas. Aerosol lifetimes are on the order of days and their climate effects are initially on the regional scale. They form nuclei for the formation of cloud droplets, which then form clouds. The large number of aerosols in biomass-burning plumes results in the formation of many small cloud droplets, which are effective in reflecting solar radiation. Significant daytime temperature drops have been observed during haze episodes in Southeast Asia.

Stratospheric effects

Biomass smoldering emits methyl chloride and methyl bromide and may contribute up to 50% and 30% of their total atmospheric budget. These compounds have significant ozone-depleting potential and may contribute to the catalytic reactions of ozone destruction (see next Chapter 6).

FURTHER READING

1. Levine J. S. (Ed.), 1991. *Global Biomass Burning*. Cambridge, MA, MIT Press.

2. McKenzie L., 1999. Smoldering biomass fuels, measuring and modeling. In: Meyers R. A. (Ed.), *Encyclopedia of Environmental Pollution and Cleanup*, John Wiley, NY, 1536–1548.

3. Muraleedharan T. R., Radoejeic M., Waugh A. and Caruana A., 2000a. Chemical characterization of the haze in Brunei Darussalam during the 1998 episode. *Atmospheric Environment*, 34, 2725–2731.

4. Muraleedharan T. R., Radoejeic M., Waugh A. and Caruana A., 2000b. Emissions from the combustion of peat: an experimental study. *Atmospheric Environment*, 34, 3033–3035.

5. Muraleedharan T. R. and Radoejvic M., 2000. Personal particle exposure monitoring using nephelometry during haze in Brunei. *Atmospheric Environment*, 34, 2733–2738.

6. Radoejvic M. and Tan K. S., 2000. Impacts of biomass burning and regional haze on the pH of rainwater in Brunei Darussalam. *Atmospheric Environment*, 34, 2739–2744.

7. Radoejvic M. and Hassah H., 2000. Air quality in Brunei Darussalam during the 1998 haze episode. *Atmospheric Environment*, 33, 3651–3658.

WEBSITE OF INTEREST

1. Air pollution information network,
   http://www.york.ac.uk/inst/sei/APIN/welcome.html
   http://www.york.ac.uk/inst/sei/SEI/welcome.html

QUESTIONS AND PROBLEMS

1. Characterize the main reasons for forest fires in the Southeast Asian sub-region and their geographical distribution.

2. Describe the ecological and economic consequences of forest fires in the whole Southeast Asian sub-region and in your country, if any.

3. Discuss the chemical processes which occur during smoldering combustion and present the corresponding example of resulting chemical species.

4. Discuss the experimental approaches to estimating the chemical composition of flumes from smoldering combustion.

5. Characterize the analytical approaches to determination of various chemical species in emissions from smoldering combustion.

6. Present examples of field monitoring of haze chemical composition in Southeast Asia.

7. Describe the possible effects of tropical forest fires to the acidity of rainwater and acid deposition fluxes.

8. Characterize inputs of haze compounds into photochemical processes in the whole Southeast Asian sub-region and in your particular area.

9. Discuss the impacts of haze from forest fires on human health in various countries of the region.

10. Describe the ecological consequences of forest fires on local and global scales and give relevant examples.

CHAPTER 6

STRATOSPHERIC OZONE DEPLETION

1. INTRODUCTION

Ozone in the stratosphere lies between 12 and 50 km above the surface of the earth, forming a protective layer, which shields the Earth from the harmful effects of the Sun's highly energetic ultraviolet radiation. The chemistry of ozone formation is tightly related to the dynamic equilibrium between oxygen and ozone and this equilibrium is affected by man-made chemicals known as chlorofluorocarbons, or CFCs. Some other halogen-containing compounds such as the bromine species or halons used in fire extinguishing equipment also play a large role in the depletion of ozone concentration in the stratosphere. Ozone depletion is the real global problem.

The term "ozone layer" being very popular in the mass media is misleading, since it implies a distinct region of atmosphere in which ozone, O_3, is a major atmospheric constituent. In reality, ozone is found in both the troposphere (see Chapter 3) and the stratosphere and its concentration depends on latitude and season. The graphic way of picturing the ozone content in the atmosphere is in the "ozone shield" image. If we could imagine compressing all the atmospheric ozone into a single layer, the thickness of this layer would be just 3 mm.

Therefore even ozone is a very important player in stratospheric chemistry, it is a mistake to consider this species as a major component of this part of the atmosphere. In reality, the major species in terms of numbers of molecules are still nitrogen (78%), oxygen (21%), and argon (1%).

2. STRATOSPHERIC CHEMISTRY OF OZONE

Ozone is distributed through the atmosphere from the ground up to about 100 km. The maximum content is only 10 ppmv (4.7×10^{12} molecules cm^{-3}) at the height of about 22 km, so this is a minor constituent albeit a very important one (Figure 1). The most chemical processes involving ozone are carried out in the stratosphere, the atmospheric layer from 15 to 50 km.

2.1. Light Absorption by Molecules

The chemistry of ozone depletion is driven by energy associated with sunlight. Accordingly, we will start the consideration of stratospheric chemistry of ozone by

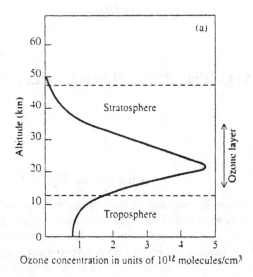

Figure 1. Variation with altitude of ozone concentration for various regions of the lower atmosphere (Baird, 1999).

investigating the relationship between light absorption by molecules and the resulting activation, or energizing, of the molecule that enables them to react chemically.

The spectrum of solar light that is considered from a biological point of view can be divided in ultraviolet range (wavelength, $\lambda < 280$ nm is known as UV-C, 280–320, UV-B, and > 320, UV-C), visible range, 400–750 nm (from violet to red light) and infrared (> 750 nm).

The absorption maximum of oxygen molecules lies above the stratosphere and O_2 gas filters from sunlight most of the UV light from 120 to 220 nm; the rest of the light in this region is filtered in the stratosphere. Ultraviolet light, having wavelengths shorter than 120 nm, is filtered in and above the stratosphere by O_2 and other constituents of air such as N_2. Thus no UV light having wavelengths shorter than 220 nm reaches the Earth's surface, thereby protecting our skin and eyes, and in fact protecting all life, from extensive damage by this part of the Sun's radiation. Molecular oxygen also filters some but not all of the UV light in the 220–280 nm range.

Ultraviolet light in the 220–320 nm range is filtered from sunlight mainly by ozone molecules with maximum of 250 nm. Ozone, aided to some extent by O_2 at the shorter wavelengths, filters out the Sun's ultraviolet light in the 220–290 nm range, which overlaps the 200–280 nm region known as UV-C. However, ozone can absorb only a fraction of the Sun's UV light in the 290–320 nm range. The remainder of the sunlight of such wavelengths, 10 to 30% depending upon the latitude, penetrates to the Earth's surface. Thus ozone is not completely effective in shielding us from light

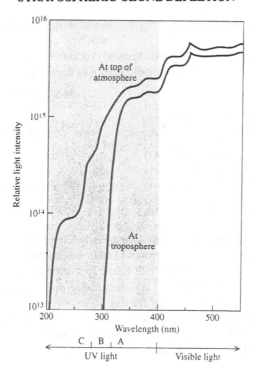

Figure 2. The intensity of sunlight in the UV and in part of the visible region measured outside of the atmosphere and at the Earth's surface (Chameides and Davis, 1982).

in the UV-B region. Because neither ozone nor any other constituent of the clean atmosphere absorbs significantly in the UV-A range, i.e., 320–400 nm, most of this, the least biologically harmful type of ultraviolet light, does penetrate to the Earth's surface (Figure 2). Thus, under natural conditions, combining screening of O_2 and O_3 has protected the biosphere from the Sun's damaging radiation.

2.2. Formation and Destruction of Ozone

Under the influence of sunlight, oxygen (O_2) is continually being changed into ozone (O_3) and ozone is likewise converted back to ordinary oxygen. Each day 350,000 tons of ozone are made and destroyed in the atmosphere. It would be a mistake to think the natural process is only conversion of molecular oxygen to ozone and that the reverse process, ozone to oxygen, is only because of air pollution. If that were so, all the O_2 in the atmosphere would long since have been changed to O_3. What certain air pollutants can do is to speed up the rate of ozone loss so that its steady state concentration (equilibrium between formation and destruction of ozone) declines (see Section 4).

The principal reactions in the photochemical process of formation and destruction of ozone are reactions (1)–(4), collectively known as the Chapman mechanisms

$$H\Delta°, \text{ kJ mol}^{-1}$$

$$O_2 \xrightarrow{h\nu,\ \lambda < 240\,\text{nm}} 2O \qquad\qquad 495 - E\,(\text{photon}), \qquad (1)$$

$$O + O_2 \xrightarrow{\ M\ } O_3 \qquad\qquad -105, \qquad (2)$$

$$O_3 \xrightarrow{h\nu,\ \lambda < 325\,\text{nm}} O_2 + O \qquad\qquad 105 - E'\,(\text{photon}), \qquad (3)$$

$$O + O_3 \longrightarrow 2O_2 \qquad\qquad -389. \qquad (4)$$

Reactions (1) and (2) represent ozone formation, while (3) and (4) are the reverse process. Oxygen atom (O) is very reactive, and consequently has a short lifetime in the stratosphere. Sunlight drives both ozone formation and destruction and this means that all four reactions will halt at sunset, and so during the nighttime the (O_3) is essentially the same as at the end of the day.

We see that different wavelength ranges are involved in formation and destruction of ozone. This corresponds to the different absorption spectra for molecular oxygen and ozone as it is shown in Figure 2.

2.3. Catalytic Reactions in Oxygen—Ozone Cycling

Until the mid-1960s the Chapman's mechanism was thought to be a complete description of the chemistry of the O_2/O_3 system in the stratosphere. Research since that time has greatly improved our understanding of this process. Using the improved estimates of reaction rate constants (reactions (1)–(4)), the steady state concentration of ozone was calculated to be twice to three times larger than what was observed experimentally by direct measurements in the stratosphere. This means that there must be other natural sinks, which had not been discovered, by which stratospheric ozone is destroyed. These additional mechanisms are catalytic processes, each of which is the propagation cycle of a free radical chain reaction. They are shown in generalized format as reaction (5) and reaction (6), with X as a catalyst

$$X + O_3 \longrightarrow XO + O_2, \qquad (5)$$
$$XO + O \longrightarrow X + O_2. \qquad (6)$$

Adding (5) and (6) together gives equation (4)

$$O + O_3 \longrightarrow 2O_2.$$

Therefore the sequence (5) and (6) is another way of carrying out reaction (4), and increases the rate of ozone destruction.

Table 1. Activation energies for individual steps of catalytic reactions for ozone destruction, kJ mol^{-1} (Bunce, 1994).

X	$X + O_3 \rightarrow XO + O_2$	$XO + O \rightarrow X + O_2$
Cl	2.1	1.1
NO	13.1	≈ 0
H	3.9	≈ 0
OH	7.8	≈ 0

Note: Uncatalyzed reaction, equation (4) has activation energy equal 18.4 kJ mol^{-1}.

Four separate catalytic cycles have been discovered, all with catalyst X an add-electron species. X can be a chlorine atom (Cl), NO, OH, or H. The chain sequence with $X = Cl$ is shown below

$$Cl + O_3 \longrightarrow ClO + O_2, \tag{7}$$

$$ClO + O \longrightarrow Cl + O_2. \tag{8}$$

We can see that the chlorine atom consumed in reaction (7) is regenerated in reaction (8), allowing the numerous recycling. Eventually termination reactions (e.g., equation (9)) remove the odd electron chlorine species

$$Cl + HO_2 \longrightarrow HCl + O_2. \tag{9}$$

Following the discovery of these catalytic cycling reactions, the concentrations of the reactive intermediates such as Cl and ClO have been measured experimentally at several different altitudes. The rates of the various steps can be computed using these concentrations along with the rate constants for the propagation steps (5) and (6), which have been determined in laboratory experiments at different temperatures and whose activation parameters are presented in Table 1.

At ≈ 30 km altitude the relative rates of the possible reactions for ozone destruction are as follows:

$$NO/NO_2 \text{ cycle} > \text{uncatalyzed reaction} \approx$$
$$\approx Cl/ClO \text{ cycle} > OH/HO_2 \text{ cycle} \gg H/OH \text{ cycle}.$$

These data show that we cannot conclude that two-step cycles will always be faster just because they have smaller activation energies: the rates of chemical reactions depend on concentrations of reactants as well as magnitudes of rate constants. For the same reason, all four catalytic cycles have reaction (6) as the rate limiting reaction, even though it has the smaller activation energy.

The H/OH cycle is not important at 30 km because reaction (10) is so fast, but becomes more significant at higher altitude, where the concentrations of O_2 and M are smaller (p(total) is less), and the reaction with O_3 can compete more effectively

$$H + O_2 \xrightarrow{M} HO_2. \tag{10}$$

More details on ozone destruction over 30 km altitude can be found in Box 1.

Box 1. The nature of noctilucent clouds and the Earth's ozone layer (Nikolaev and Fomin, 1997)

It has been revealed theoretically that hydrogen self-ignites at a height of 120 km and burns up nearly completely at heights of 165–200 km. Still higher, its concentration falls off in proportion to pressure because of burning and, therefore, hydrogen does not leave the terrestrial atmosphere. The water vapor which forms during hydrogen combustion descends and changes, under definite temperature conditions, to very small pieces of ice at a height of about 85 km, the number of which is enough to form nocticulent clouds. These clouds are easily observable in twilight. Atomic oxygen forms during the same process of hydrogen combustion and combines with molecular oxygen to form ozone. The authors note that the power of this source of ozone is proportional to the hydrogen concentration near the Earth's surface and can be higher than that of all other sources. Apparently, this circumstance will make it necessary to develop a new approach to clarify the sources of appearance of the ozone layer and ozone "holes". The energy released during hydrogen combustion is sufficient to account for the existence of the termosphere.

The catalytic destruction of ozone by NO is the most important process that occurs in the middle and upper stratosphere. We should stress here that this process is possible even in unpolluted atmosphere since small amounts of nitrous oxide, N_2O, from biological denitrification have always been present in the stratosphere, which is the precursor of nitric oxide, NO, in reaction with atomic oxygen (reaction (11)). Most collisions with atomic oxygen form N_2 and O_2, but a few form NO (see also Section 4)

$$O + N_2O \longrightarrow 2NO. \tag{11}$$

Another source of NO in the upper troposphere and low atmosphere is flights of planes. The air heating due to fuel combustion produces nitrogen oxide (reaction (12))

$$O_2 + N_2 \longrightarrow 2NO. \tag{12}$$

The NO molecules, which are the products of the above reaction catalytically destroy ozone by extracting oxygen atoms in ozone and forming nitrogen oxide, NO_2, i.e., they act as X in the mechanisms (5) and (6)

$$NO + O_3 \longrightarrow NO_2 + O_2, \tag{13}$$

$$NO_2 + O \longrightarrow NO + O_2. \tag{14}$$

Adding (13) and (14) together gives equation (4)

$$O + O_3 \longrightarrow 2O_2.$$

In natural conditions, the main source of nitrous oxide is biological denitrification (see more details in Chapter 13). This natural process is significantly increased in the whole World and in Asia as well by massive application of synthetic nitrogen fertilizers. Denitrification of fertilizer nitrogen leads to an increase of N_2O concentration in the stratosphere (see Chapter 13 for details).

As we have mentioned already, the Cl/ClO cycle is also important in stratosphere. The natural source of atomic chlorine, Cl, is methyl chloride gas, CH_3Cl, produced at the Earth's surface, mainly in the oceans as a result of the interaction of chloride ions with decaying vegetation. Only a portion of methyl chloride gas is destroyed in the troposphere. When intact molecules of it reach the stratosphere, they react photochemically decomposed by UV-C or attacked by OH radicals (reactions (15) and (16))

$$CH_3Cl \overset{UV\text{-}C}{\longrightarrow} Cl + CH_3, \tag{15}$$

$$CH_3Cl + OH \longrightarrow Cl + \text{ other products.} \tag{16}$$

At present, the concentration of chlorine atoms is mainly determined by destruction of synthetic chlorofluorocarbons, CFCs (see Section 3).

As with methyl chloride, large quantities of methyl bromide, CH_3Br, are also produced naturally and some of it eventually reaches the stratosphere, where it decomposes photochemically to yield atomic bromine. Like chlorine, bromine atoms can destroy ozone by the mechanism shown in reactions (5) and (6). These reactions are as follows:

$$Br + O_3 \longrightarrow BrO + O_2, \tag{17}$$

$$BrO + O \longrightarrow Br + O_2. \tag{18}$$

Almost all the bromine in the stratosphere remains as the radicals Br and BrO, since the inactive non-radical forms, hydrogen bromide, HBr, and bromine nitrate, $BrONO_2$, are efficiently decomposed photochemically.

The percentage of stratospheric non-radical bromine species is lower than that of chlorine because the slower speed of this reaction and because of the efficiency of the photochemical decomposition. For this reason, stratospheric bromine is more efficient in destroying ozone than is chlorine. The estimated factor is from 40 to 50, but the concentration of active bromine species is much less than chlorine species.

We should remember here that each average chlorine atom can catalytically destroy 10,000 ozone molecules and bromine atoms up to 40,000–50,000. However, there are additional reactions in the stratosphere that significantly decrease the rate of these and similar processes both in natural and polluted stratospheres.

2.4. Additional Reactions in Stratospheric Chemistry

The reactions shown in Table 1 do not account completely for the chemistry of the stratosphere. Complicating reactions are connected with:

- Temporary reservoirs of active species;

- Interactions between catalytic cycles;

- Null cycles;

- Initiation and termination reactions.

We will consider these reactions for the most important cycles in the low and middle stratosphere, NO/NO_2 and Cl/ClO.

Temporary reservoirs
Catalytically, active species NO_x and ClO_x can be converted into substances which reduce their instantaneous concentrations and from which they can be regenerated, as follows:

For NO_2:

$$OH + NO_2 \xrightarrow{M} HNO_3. \tag{19}$$

$HNO_3(g)$ acts as a temporary reservoir from which NO_2 can be regenerated by photolysis. We can see that at any time, as much as half is temporarily inactivated in an $HNO_3(g)$ reservoir and excluded from ozone destruction reactions.

For Cl:

$$Cl + CH_4 \longrightarrow HCl + CH_3. \tag{20}$$

Here HCl is the temporary reservoir; Cl can be regenerated by the reaction of HCl with OH

$$HCl + OH \longrightarrow Cl + H_2O. \tag{21}$$

Interaction between cycles
We can expect that various catalysts will inter-react. The example of this reaction in NO/NO_2 and Cl/ClO cycles is shown below

$$ClO + NO_2 \longleftrightarrow ClONO_2. \tag{22}$$

Chlorine nitrate is formed by the combination of chlorine monoxide and nitrogen dioxide; after a few days or hours, a given $ClONO_2$ molecule is photochemically decomposed back to its components, and thus catalytically active ClO is re-formed.

2.4.1. Null (do-nothing) cycles

Such a cycle results in no net chemical change; generally, sunlight is converted to kinetic energy. We can see examples in the following reactions:

$$NO_2 \xrightarrow{h\nu,\ \lambda < 325\,nm} NO + O, \tag{23}$$

$$O + O_2 \xrightarrow{M} O_3,$$

$$O_3 + NO \longrightarrow NO_2 + O_2. \tag{24}$$

When these reactions are added together, all the chemical species drop out on both sides. The previous three reactions may alternatively be considered as the photochemical pseudo-equilibrium, reaction (25)

$$NO_2 + O_2 \underset{thermal}{\overset{h\nu}{\longleftrightarrow}} NO + O_3. \tag{25}$$

Initiation and termination reactions

The overall rates of the chain reactions referred to in Table 1 depend on the rates of initiation and termination reactions. This is because the average number of catalytic cycles propagated depends upon the balance between the rates at which chains are initiated (generation of free radicals from non-radical precursors) and terminated (conversion of free radicals to non-radical products). The initiation of photochemical reactions and their rates depends on the intensity of sunlight and falls to zero at night. We can consider characteristic examples as follows:

$$HNO_3 \xrightarrow{h\nu} NO_2 + OH, \tag{26}$$

$$NO_2 \xrightarrow{h\nu} NO + O, \tag{27}$$

$$O_3 \xrightarrow{h\nu} O_2 + O, \tag{28}$$

$$CH_3Cl \xrightarrow{h\nu} Cl + CH_3,$$

$$H_2O + O \longrightarrow 2\,OH, \tag{29}$$

$$N_2O + O \longrightarrow 2\,NO. \tag{30}$$

We can see that reactions (29) and (30) are not photochemical, but one reactant, atomic oxygen is formed in reaction (3), which drives the reaction photochemically.

Termination reactions remove radicals from the system. Generally, such reactions involve the combination of two radical (i.e., odd electron) species. Some termination reactions are shown below

$$2\,ClO \longrightarrow ClOOCl, \tag{31}$$

$$H_2O_2 \longrightarrow H_2O_2 + O_2, \tag{32}$$

$$NO_2 + Cl \longrightarrow NO_2Cl. \tag{33}$$

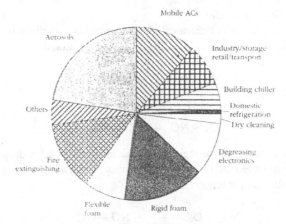

Figure 3. World consumption of CFCs and Halons by major application in 1986 as weighted value of ozone depletion potential (ESCAP, 1995).

Thus these additional reactions in the stratosphere decrease significantly the rate of ozone destruction processes, especially under contamination of the atmosphere by chlorofluorocarbons and nitrous oxides as the catalysts of ozone destruction reactions.

3. ROLE OF CHLOROFLUOROCARBONS IN STRATOSPHERE OZONE DEPLETION

Chlorofluorocarbons, or CFCs, are substances used in refrigeration, plastic foam production, the electronic industry and many other applications. CFC and related chemicals have been used from 1929 and their World consumption in 1986 is shown in Figure 3.

In the 1980s about 1 million tons of CFCs were released annually into the atmosphere. These compounds are nontoxic, nonflammable, nonreactive, and have useful condensation properties (suiting them for use as coolants, for example); on account of these favorable characteristics they found a multitude of applications. There were several CFCs of commercial importance, like CFC-11, CFC-12. The code numbers for CFC compounds are shown in Box 2.

Box 2. CFC nomenclature

The chemical formulas for individual CFCs, such as CFC-11, can be deduced from their code numbers by adding 90. This is so-called "Rule-90". The resulting digits correspond, respectively, to the number of carbon, hydrogen and fluorine atoms present in one molecule. The digit "0" means the lack of this atom in the chemical formula. For example, adding 90 to 11 gives 101, so it follows that CFC-11 contains one carbon, zero hydrogen, and one fluorine atom. Since the total number of noncarbon atoms in substituted alkane hydrocarbons adds up to $2n + 2$, where n is the number of carbons, the number of chlorines can be deduced by difference, that is, by subtracting from $2n + 2$ the number of hydrogen plus fluorine atoms. Thus the rest is represented by chlorine atoms. For example, for CFC-11, $2n + 2 = 2(1) + 2 = 4$, so the number of chlorine atoms is $4 - (0 + 1) = 3$. The formula CFC-11 contains one atom of carbon, zero of hydrogen, one of fluorine and three of chlorine, i.e., $CFCl_3$. Another example: CFC-141: $141 + 90 = 231$, so this molecule contains 2 carbons, 3 hydrogens, 1 fluorine, and hence 2 chlorines. However, you cannot tell from the formula $C_2H_3FCl_2$, which isomer is involved, and the letters a, b, etc. are used to differentiate them.

3.1. Characteristic Examples of CFC Applications

The compound CFC-11 with chemical formula $CFCl_3$ is a liquid, which boils near room temperature. CFC-11 was used to blow the holes in soft foam products such as pillows, carpet padding, cushions, and both seats and padding in cars. It is also employed to make the rigid urethane foam products used to insulate refrigerators, freezers, and some buildings. The utilization of insulating foam products rose in the last quarter of the 20th century due to emphasis on energy conservation.

CFC-12, which is pure CF_2Cl_2, is a gas at room temperature but is readily liquefied under pressure. From first formulation at the beginning of the 1930s, it was used as the circulating fluid in refrigerators, replacing the toxic gases like ammonia and sulfur dioxide. It was produced by reacting carbon tetrachloride, CCl_4, with gaseous hydrogen fluoride, HF (reaction (34))

$$CCl_4 + HF \longrightarrow HCl + CF_2Cl_2. \tag{34}$$

Until recently CFC-12 was used extensively in automobile air conditioners, from which much was released (about 0.5 kg per vehicle annually) to the atmosphere during use and servicing. Today special equipment is used to capture the CFCs (and their modern replacements) when air conditioners in cars are serviced.

CFC-12 has been widely used also for producing rigid plastic foams. The characteristic example is the formation of foam sheets such as those used in white trays of the sort to package fresh meat products and in hamburger "clamshell" cartons used by fast-food restaurants.

Both CFC-11 and CFC-12 were employed extensively as propellants in aerosol spray cans. This application has greatly decreased in some developed countries (USA, Canada, Norway, and Sweden) but still utilization of CFCs in spray cans continues elsewhere, especially in Asia.

The other CFC of major environmental concern was CFC-113, $C_2F_3Cl_3$. It was used extensively for cleaning the glue, grease and solder residues from electronic circuit boards after their fabrication, up to $2 \, kg \, m^{-2}$.

Brominated analogs of CFCs are in commercial use as fire extinguishers, under the name of Halons. These compounds are very valuable in fighting electrical fire, for example at computer installations. Their mode of action, besides smothering the fire with a heavy vapor, involves cleavage by heat of the weak C − Br bond. The bromine atoms thus formed act as terminators for the radical chain reactions which take place in flame. Some bromine compounds, like H1211 and H1301, are effective in stratospheric ozone depletion (see Section 2.2).

3.2. CFC Degradation in the Stratosphere

CFCs have no tropospheric sinks, since they are very resistant to any type of chemical reactions in lower parts of the atmosphere. So all of their molecules eventually rise to the stratosphere. In the middle and upper layers of the stratosphere they decompose under the influence of high energetic UV-C from sunlight, releasing chlorine atoms. For example, reaction (35) shows this decomposition for CFC-12

$$CF_2Cl_2 \xrightarrow{UV\text{-}C} CF_2Cl + Cl. \tag{35}$$

CFCs do not absorb components of sunlight with wavelengths greater than 290 nm, and generally require wavelengths of 220 nm or less for photolysis. Because vertical migration of CFCs in the stratosphere is slow, their atmospheric lifetimes are long: 60 years on average for CFC-11 molecules and 105 years for CFC-12. Furthermore, CFC-11 is photochemically decomposed easier at lower altitudes than in case of CFC-12, and is more able to destroy ozone at low stratospheric altitudes, where the concentration of ozone is greatest.

3.3. Stratospheric Ozone Depletion and Antarctic Polar Hole

The first warning that CFCs might be responsible for depleting stratospheric ozone was made in 1974. At that time, it was not known that reactions (7) and (8) occurred naturally in the stratosphere; chlorine atom assisted destruction of ozone was suggested to occur only as a consequence of the release of CFCs to the atmosphere. It was not known that Cl/CLO chain proceeds even in the unpolluted atmosphere. At present we can state that the release of CFCs does not introduce a brand-new sink for ozone. This is only an enhancement of the natural process of ozone destruction (Figure 4).

There are three reasons for environmental concern about CFCs

1. CFCs increase the sink strength for stratospheric ozone with no possibility of compensation by increasing the source strength, which, in turn, is fixed by the intensity of sunlight. This lowers the steady state concentration of ozone. We should remember that steady state concentration is the balance between sources and sinks.

2. CFCs are extremely long-lived pollutants and their destruction is very slow, even in the stratosphere, since molecular oxygen and ozone having wavelength less than 250 nm, absorb radiation much more efficiently than CFSs.

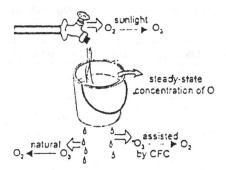

Figure 4. Release of CFCs boots the rate of ozone removal—i.e., it increases the strength of existing sink (Bunce, 1994).

Table 2. Concentrations of organochlorine species in the atmosphere (Bunce, 1994).

Species	Concentration, ppbv	Estimated increase from 1950, times
CH_3Cl	0.6	Natural
CCl_4	0.6	4
CH_3CCl_3	0.5	3
CFC-11	0.8	3
CFC-12	1.0	2
CFC-113	0.2	3

3. Owing to the long-life period of existence in the stratosphere, environmental damage caused by CFCs is a problem that could persist for many generations (see below).

The atmospheric concentration of all organochlorine species now stands at $\sim$ 4 ppbv, having risen from an estimated < 1 ppbv in 1950 (Table 2).

3.3.1. Ozone depleting potential

The ozone depleting potential (ODP) is the propensity of the substance to destroy stratospheric ozone, integrated over the life of the compound. By definition, ODP for CFC-11 = 1. For other CFC compounds, both the total ODP and time scale of the effect can vary (Table 3).

We should point out here that definite evidence for loss of stratospheric ozone is hard to obtain. The concentration of ozone varies greatly depending on daytime and season and the pattern is not exactly reproducible from year to year. However, since

Table 3. Properties of some chlorinated species and their Ozone Depleting Potentials (Bunce, 1994).

Compound	CFC code	Lifetime, yr	% increase per year in late 1970's	ODP	ΔO_3 (%)
CH_3Cl	11	70	6	1	2.0
CF_2Cl_2	12	1100	6	0.86	2.1
CHF_2Cl	22	25	> 10	0.05	0.03
$CF_2ClCFCl_2$	113	90	—	0.80	0.5
CH_3CCl_3	—	< 10	9	1.15	0.5
CCl_4	—	≈ 10	2	1.10	0.6

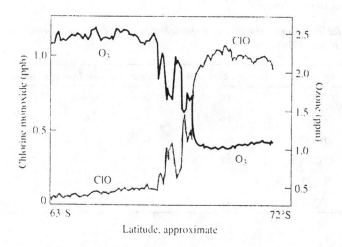

Figure 5. ClO and O_3 concentrations over Antarctica at 18 km altitude as a function of latitude (Baird, 1999).

the mid-1980s clear evidence of stratospheric ozone loss has been obtained for the Antarctic Polar region, and then, from about 1990, for mid-latitudes and near 40 km in altitude (McElroy and Salawich, 1989). See also the websites at the end of this chapter for updated monitoring results.

3.3.2. Polar ozone hole

In 1985 the correlation between concentration of chlorine atoms and ozone in the stratosphere of high latitudes (> 70 ° S) was recorded (Figure 5).

Local depletions up to 50% in stratospheric ozone were developed in the late winter and this phenomenon was called an Antarctic ozone hole. During the winter, the Antarctic is a unique location in that the air circulated is entirely circumpolar, with almost no admixture of air from lower latitudes. Since ozone formation is a sunlight-dependent reaction, no ozone is generated during the long dark winter. In the late winter/early spring, when the temperature of the polar stratosphere is lowest, crystals of $HNO_3 \cdot 3H_2O$ (at T $< -70\,°C$) and water ice (T $< -85\,°C$) are formed. Temporary reservoirs of chlorine such as $ClONO_2$ and HCl are broken down heterogeneously on the surface of these crystals, releasing active chlorine radicals, such as Cl_2 and $HOCl$. Immediately after sunrise, these species are cleaved by visible or near-UV light, releasing chlorine atoms with the promotion of catalytic ozone destruction (see reactions (7) and (8)). This process is as follows:

$$HCl + ClONO_2 \xrightarrow{h\nu} Cl_2 + HNO_3, \tag{36}$$

$$H_2O + ClONO_2 \xrightarrow{h\nu} HOCl + HNO_3, \tag{37}$$

$$HCl + HOCl \xrightarrow{h\nu} Cl_2 + H_2O. \tag{38}$$

Two factors lead to the restoration of ozone in the stratosphere over the Antarctic in spring.

- The intensity of sunlight increases and more ozone is formed;

- Increasing stratospheric temperatures sublimate the polar stratospheric clouds, releasing $HNO_3(g)$, which soon cleaves to NO_2, trapping some of the ClO in the temporary reservoirs of chlorine nitrate.

This phenomenon, however, has not been recorded in the Arctic Polar region. Although elevated concentrations of active chlorine species were detected and significant amounts of ozone were destroyed, the infusion of air from lower altitudes prevented large scale loss of ozone in the high Arctic.

4. ROLE OF NITROUS OXIDE IN STRATOSPHERE OZONE DEPLETION

The modern global pattern of nitrous oxide emission is shown in Table 4. If current trends in N_2O emission continue, anthropogenic emissions could more than double during the 21st century.

Based on various estimates, the Asian contribution to global N_2O emission might be from 45 to 70%, i.e., from 8 to 12 million tons per year (ESCAP, 2000).

N_2O and stratospheric ozone depletion
Most N_2O is destroyed in the higher atmosphere (stratosphere), although some uptake at the Earth's surface may take place. Of the N_2O that reaches the stratosphere, about 90% are photolyzed

$$N_2O \xrightarrow{h\nu} O + N_2. \tag{39}$$

Table 4. Global sources of nitrous oxide in 1990 and expected increase (Oonk and Kroeze, 1999).

Source	1990 $N-N_2O$ emission, 10^6 tons per year	Expected increase in 21st century, %
Natural	11 (8–14)	No change
Anthropogenic		
Agriculture	3 (1–9)	+80
Stationary combustion	0.5 (0.3–1.1)	+100
Traffic	0.4 (0.2–0.6)	+1275
Industry	0.5 (0.3–0.7)	No change
Wastes, biomass burning	1.8 (0.4–2.9)	Uncertain
Total anthropogenic	6 (2–14)	+65–233
Total	17 (10–28)	+24–77

The remaining 10% are oxidized

$$N_2O + O \longrightarrow O_2 + N_2 \quad (42\%), \tag{40}$$

$$N_2O + O \longrightarrow 2\,NO \quad (58\%). \tag{41}$$

The NO produced through reaction (41) allows the chain of reactions leading to catalytic destruction of stratospheric ozone to be initiated. N_2O is the main source of NO_x in the stratosphere and is, therefore, the important natural regulator of stratospheric ozone. An increase in N_2O most probably affects stratospheric ozone concentrations. At present, there is too much uncertainty to predict the extent of this destruction but in any case, the role of the Asian countries in magnification of this process is significant.

5. CONSEQUENCES OF OZONE DEPLETION

Ozone depletion has a number of consequences for human health and agriculture. These include increased rates of skin cancer and eye cataracts, weakening of immune systems, damage to crops, reductions in primary producers (plankton) in the ocean and increasing air pollution.

The relationship between UV-B radiation and skin cancer is by now well documented and accepted. The US EPA has estimated that each 1% decrease in stratospheric ozone concentration will result world wide in a 2% rise in cutaneous malignant melanoma. Among the Asian and Pacific region, people of many countries, such as Australia, New Zealand, South Pacific Islands, Indonesia, Malaysia, Philippines, Thailand, etc, are the most vulnerable to the exposure of UV-B radiation. For example, between 1980 and 1991, when the growth of CFCs production was the maximum, melanoma cancer registration rates in New Zealand increased by 22% (ESCAP, 1995).

Eyesight is also affected by UV light and the incidence of cataracts, already the leading cause of blindness around the World, is expected to grow with ozone depletion. The US EPA estimates that 1% in stratospheric ozone content will result in an extra 100,000 to 150,000 cases of cataract induced blindness each year.

As UV-B radiation weakens the immune system, the decreasing ozone concentration in the stratosphere is expected to cause increased outbreaks of measles, chicken pox, malaria, leishmaniasis, tuberculosis and leprosy. The immune depressing effects of UV-B radiation will be felt most severely in the Himalayan regions such as Nepal, Bhutan, India and the Tibetan plateau in China. Due to the high altitudes, the population of these regions is already exposed to greater amounts of ultraviolet light.

In general, many plants exposed to increased amounts of UV-B radiation display stunted growth and reduced population and germination. The drop of crop yields is changing from 5% for wheat, 20% for potato and soyabean and even 90% for squash (UNEP/GEMS, 1992). Other crops which may be affected negatively include cotton, peas, melons and cabbage.

The single most important agricultural product in the Asian region, rice, may not be directly affected by increased UV-B radiation, but it has been found that ultraviolet light can inhibit the activity of nitrogen-fixing micro-organisms, which play an important role in nitrogen nutrition of rice. A reduction of this natural process will require the expanded use of mineral fertilizers to maintain current yields. The production of nitrogen fertilizers in China only is already about 27 million tons annually (see Chapter 13). This, in turn, will enhance the biological denitrification of fertilizer nitrogen and increase the input of nitrous oxide into the stratosphere with subsequent destruction of ozone molecules.

The decline in phytoplankton productivity in the South Pacific sub-region has been connected to increasing UV radiation, which would also adversely affect fisheries. It has been estimated that a 16 per cent ozone depletion would cause a 5% decrease in amount of plankton leading to a 6 to 9% drop in fishing yields. Coral reefs and mangroves, which are important habitats for fish and other marine organisms, might be stressed or even killed as a result of high UV radiation (ESCAP, 1995).

The increased amount of UV-B radiation reaching the troposphere will exacerbate the air pollution problems, which are already severe in many cities of the Asian region (see Chapter 3). UV-B reacts with pollutants emitted from various sources to produce photochemical smog and increased ground level ozone. Ironically, decreased levels of stratospheric ozone can result in increased levels of tropospheric ozone, known urban as air pollutant. Each per cent reduction in stratospheric ozone is expected to increase tropospheric ozone by 0.3%.

5.1. Sunscreens

Human skin contains various amounts of a natural pigment, melanin, which absorbs radiation, especially in the UV-B range. Melanin is found in specialized cells called melanocytes. The production of melanin is stimulated by exposure to UV-B, although a delay occurs between exposure and melanin formation. The pigmentation is found mostly in the surface layers of the skin, which consist entirely of dead, keratinized cells. Melanin filters out UV-B radiation in the keratinized cells, inhibiting it from reaching the living cells underneath. Fair-skinned and dark-skinned people differ not in the total number of melanocytes, which they produce, but rather in the amount of melanin produced by each melanocyte (Bunce, 1994). Thus black-skinned people are less sensitive to UV-B radiation.

A sunscreen is a chemical substance, which is applied to the skin to carry out the same function as melanin, namely to prevent high energy radiation from penetrating to the living cells in the lower layers of the skin. Sunscreens have to absorb in the proper region of the spectrum, and to be cosmetically acceptable they should be colorless, i.e., their absorption should not extend into the visible. Substances possessing the benzene or other aromatic ring as part of their structure are used most commonly as sunscreens.

6. CFCs EMISSION ABATEMENT STRATEGY

6.1. Political Implementations

Political acceptance of scientific predictions of stratospheric ozone depletion occurred in North America before any adverse consequences had actually been observed. This is contrary to most other environmental problems which global society had faced during the last part of the 20th century.

Thus the prediction of a threat to stratospheric ozone by CFCs was made in 1974, and as early as 1978 the use of CFCs as aerosol propellants was banned in North America. International consensus on the need to restrict CFC emissions emerged with the discovery of an Antarctic ozone hole (see above) that correlated with an increase in stratospheric chlorine.

Awareness of the ozone depletion problem was reflected by signing in 1987 of the "Montreal Protocol on substances that deplete the ozone layer". This international treaty initially set targets for CFC production to be cut back to 1986 baseline levels by mid-1989, cut to 80% of baseline by 1993, and to 50% of baseline by 1998. Subsequent information suggested that these cuts in CFC production would be insufficient to prevent substantial loss of stratospheric ozone over the first half of the 21st century. Consequently, amendments have been made to strengthen the original terms of the Montreal Protocol, so that a complete phase-out of the "hard" CFCs such as CFC-11 and CFC-12 (Figure 6) will be effected.

Without the Protocol, ozone depletion would have risen by 50 % in 2050 in the Northern Hemisphere's mid-latitudes, and 70% in the Southern Hemisphere's mid-latitudes. The implications would have been dramatic as there would have been 19 million more cases of non-melanoma cancer, 1.5 million more cases of melanoma cancer, and 130 million more cases of eye cataracts.

In fact, the production of CFCs and Halons has declined by 86% in the last ten years. However, it has been mainly on the account of developed countries whereas the developing countries and newly developed countries of Asia have still increased

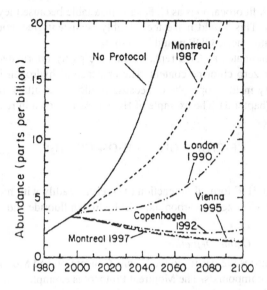

Figure 6. Impact of Montreal Protocol on ozone depletion (ESCAP, 2000).

the CFCs production (ESCAP, 2000). It is estimated that in the Asian region approximately 100,000 tons of CFCs were used annually in the beginning of the 1990s. Japan was the second World largest producer, following the United States. At present, these countries have cut their CFCs production sufficiently, however, China, India, Indonesia, Iran, South Korea, Malaysia, Pakistan, the Philippines and Thailand still produce and consume significant amounts of CFCs, especially with high ODP.

6.2. CFC Replacement Compounds

Because CFCs have important uses in modern society as the working fluids in refrigerators and air-conditioners, and as the blowing agents for plastic foams, the response to phasing them out has been an intensive search for alternatives which retain their desirable properties in terms of volatility and low toxicity, but which are environmentally friendly. The environmental problems to be overcome in finding a CFC replacement are the following:

- CFCs are so stable chemically that they do not break down in the troposphere and hence over time can migrate to the stratosphere;

- They contain chlorine, which catalyzes the ozone destruction in the stratosphere.

The ideal CFC replacement molecule should therefore be somewhat reactive in the lower stratosphere so that it will be oxidized before it has time to reach the stratosphere, and should also contain few, or ideally no, chlorine atoms. Totally fluorinated compounds, such fluorocarbons as C_xF_y, are unsuitable because they are even more inert than CFCs. This would make them highly persistent greenhouse gases, even though their stratospheric ODP would be zero.

The chief candidates as CFC replacements are partly fluorinated hydrocarbons with minimal or zero chlorine content. The presence of hydrogen in the molecule confers reactivity in the troposphere, because it allows the attack of OH upon the molecule (see Chapter 3). The example of HCF-134a is shown in reaction (42)

$$CF_3CH_2F + OH \longrightarrow H_2O + CF_3CH_2F. \tag{42}$$

The radical CF_3CH_2F formed in reaction (42) reacts readily with molecular oxygen and is ultimately oxidized to carbon dioxide, hydrogen fluoride and water.

6.3. Responses to Ozone Depletion

Most Asian countries are parties to the Vienna Convention and Montreal Protocol on ozone depleting compounds. The Montreal Protocol is exemplary in the sense that it truly embodies the principles of common but differentiated responsibility between the industrial and developing countries. It recognizes the fact that industrialized countries are responsible for the bulk of emissions of CFC. Moreover, they have the financial and technological resources to find proper replacements for these substances. Meanwhile, the developing countries are given a period of grace before they must start their phase-out schedules.

The other strong feature of the Montreal Protocol and its various amendments is the financial mechanism for implementation. The Protocol established the Multinational Fund (MFMP) in 1990. The Fund is used to pay the incremental cost incurred by developing countries in phasing out their consumption and production of CFCs.

The easy availability of substitutes is a factor that permitted compliance to the Protocol. Science and industry have been able to develop and commercialize alternatives to the CFCs and other ozone-depleting compounds. In fact, industrialized countries like Japan and Singapore ended their dependence on CFCs at less cost than was anticipated.

Examples of CFC replacements are shown in Table 5.

Despite the financial and technical feasibility of CFC alternatives, the task of phasing out ozone-depleting substances in developing countries of the Asian region is still not simple. Weak economic and technological bases do not allow these countries a smooth transition to alternatives. However, the establishment of MFMP is a milestone towards promoting alternatives to CFCs.

Effective sanctions are provided in the Protocol to promote participation and honest compliance. The Protocol provides for trade sanctions. The objective is to encourage countries to participate in the Protocol by preventing non-participating countries from gaining competitive advantages. Each Party is prohibited from importing controlled substances from states not party to the Protocol.

Table 5. Abatement percentage and cost data of some substitutes for ozone-depleting substances (ESCAP, 1995).

Use	Ozone-depleting substance replaced	Substitute	Abatement ozone depletion, %	Cost	
				US$/kg of ODP saved	Increment cost of alternative as per cent of cost of CFC replaced, %
Aerosols	CFC-11 and/or CFC-12	Hydro-carbons	100	From −0.76 to 1.96	−11 to +45
Solvents	CFC-113	Aqueous process	100	16.84	241
	MCF	Aqueous process	100	From −6.6 to 8.9	−54 to +73
Foam-blowing: rigid poly-urethane	CFC-11	Reduced CFC option	30–50	2.24	97
Foam-blowing: polystyrene/ polyethylene	CFC-11	HCFC-22	95	0.92	21
Foam-blowing: flexible slabstock polyurethane	CFC-11	Methylene chloride	100	2.2	50
Foam-blowing: polystyrene	CFC-12	Hydro-carbons	100	—	—
Refrige-rators	CFC-12	HFC-134a	100	—	—
Mobile air conditioners	CFC-12	HFC-134a	100	—	—
Fire extingui-shers	Halons	CO_2, foam, dry powders	—	—	—

FURTHER READING

1. Bunce N., 1994. *Environmental Chemistry*. second edition, Wuerz Publishing Ltd, Winnipeg, Chapter 2.

2. ESCAP, 1995. *State of the Environment in Asia and Pacific*. United Nations, NY, Chapter 6.

3. Baird C., 1999. *Environmental Chemistry*. second edition, W. H. Freeman and Company, NY, 17–84.

4. Oonk H., 1999. Nitrous oxide emissions and control. In: Meyers R. A. (Ed.), *Encyclopedia of Environmental Pollution and Cleanup*, John Wiley, NY, 1054–1069.

5. ESCAP, 2000. *State of the Environment in Asia and Pacific*. United Nations, NY, Chapter 21.

WEBSITES OF INTEREST

1. NASA: Ozone contour images (both poles), plus links to useful UK and European sites,
   http://jwocky.gsfc.nasa.gov/

2. USEPA: Information on CFC substitutes, the methyl bromide, the UV index etc.
   http://www.epa.gov/docs/zozne/index.html

3. World Meteorological Organization (UN): Regular bulletins on the current Antarctic ozone hole, links to data for Northern Hemisphere depletions,
   http://www.wmo.ch/web/arep/ozobull.html

4. NOAA: Contour maps of total column ozone for both hemispheres, daily,
   http://nic.fb4.noaa.gov/80/products/stratosphere/tovsto/

QUESTIONS AND PROBLEMS

1. Explain why "ozone layer" is a misleading definition of ozone stratospheric distribution.

2. Show the profile of ozone concentrations in the stratosphere and present the explanation of such a distribution.

3. Characterize light absorption by molecules and its role in stratospheric chemistry of ozone and other reactants.

4. Describe the Chapman mechanism for ozone formation-destruction processes in the stratosphere.

5. Does the Chapman mechanism present the correct description of ozone chemistry in the stratosphere and what are the limitations of this mechanism?

6. Describe the role of catalytic reactions in ozone destruction. Do these reactions describe natural processes or are these brand new reactions related to anthropogenic pollution?

7. Characterize the known catalytic cycles and their relative significance in different layers of the stratosphere.

8. Discuss the sources of catalytic species that serve as ozone-depleting substances in the stratosphere.

9. What are the chemical and physical properties of chlorofluorocarbons? Describe the CFC nomenclature.

10. Characterize hydrocarbons containing bromide and their chemical reactions with ozone molecules.

11. Describe the chemical reactions that drive the destruction of various chlorofluorocarbons in the stratosphere.

12. Define the ozone depletion potential and present the examples of ODP for different species.

13. Discuss the role of nitrous oxides in stratospheric chemistry of ozone. Discuss the relative importance of this species in ozone destruction.

14. Discuss the contributions of the Asian countries to production and consumption of substances with ozone-depleting potentials.

15. Present the mechanism of ozone hole formation at the Antarctic pole. Is a similar phenomenon recorded at the North Pole?

16. Characterize the ecological consequences of excessive UV-B radiation in the Earth's biosphere in general and in the Asian region in particular.

17. What were the reasons for phasing out the application of CFCs and what international treaties do you know that manage these processes?

18. Discuss the CFCs emission abatement strategy. Why is this strategy significantly different for developed and developing countries?

19. Characterize the properties of ideal replacement for CFC compounds and explain the chemical processes behind these replacements.

20. Discuss the list of possible replacements for different "hard" CFCs and explain the difficulties of their implementation in developing countries of the Asian region.

PART II

SOIL AND WATER POLLUTION

CHAPTER 7

SOIL POLLUTION

1. INTRODUCTION

We can recognize two categories of human-induced soil degradation and pollution: 1) resulting from erosion processes with displacement of soil material, mainly through water erosion and wind erosion; and 2) resulting from internal physical and chemical-biological deterioration. The manifestation of these processes depends on many natural and human-induced factors and varies significantly in the vast Asian region. Water erosion will be considered in Chapter 8 and the main attention in this chapter will be given to land and soil degradation due to various sources of pollution. Many Asian soils have specific natural biogeochemical features and most are very sensitive to pollutant inputs.

2. CHARACTERISTIC BIOGEOCHEMICAL FEATURES OF SOILS AND NATURAL ECOSYSTEMS IN ASIA

2.1. General Principles

The knowledge of biogeochemical cycling is a key to understanding many problems of environmental pollution in general and soil pollution in particular. The combination of soil and ecosystem parameters with the quantitative assessments of biological, geochemical and hydrochemical turnover gives an opportunity to calculate the rates of biogeochemical cycling and coefficients of biogeochemical uptake, C_b, for different ecosystems and varying chemical elements (Bashkin and Howarth, 2002). In addition to the coefficient of biogeochemical uptake, we will also apply the active temperature, C_t, coefficient. We know that in northern Asian areas the real duration of any process (biochemical, microbiological, geochemical, biogeochemical) depends strongly on the temperature regime. These processes are depressed annually for 6–10 months and the influence of acid forming compounds, as well as any other pollutant, occurs during summer. A process duration term has been derived as the active temperature coefficient, C_t, which is the duration of active temperatures $> 5\,°C$ relative to the total sum.

Soil-biogeochemical mapping can serve as a measure of soil sensitivity to pollution loading (Bashkin and Park, 1998). Let us consider biogeochemical properties and

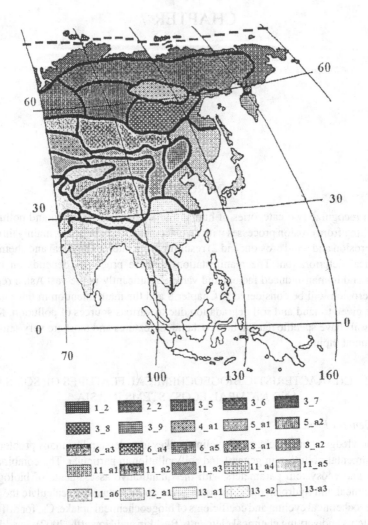

Figure 1. Soil-biogeochemical mapping of the Asian region (Bashkin and Park, 1998).

cycling of various elements in predominant ecosystems of the Asian region as shown in Figure 1 and Tables 1–3.

The values of C_b for each geographical region were ranked in order to determine the type of biogeochemical cycling and these ranks are shown in Table 2. Five types of biogeochemical cycling were divided: very intensive, intensive, moderate, depressive and very depressive.

The corresponding values of active temperature coefficients ranked in accordance with the main climatic belts are shown in Table 3.

Table 1. The values of biogeochemical cycling (C_b) and active temperature (C_t) coefficients in various soil-ecosystem geographical regions of East Asia.

Ecosystem	Main FAO soil type	Geographic region	Index, Figure 1	C_b	C_t
Arctic Deserts	Lithosols, Regosols	Eurasian	1_2	10.0	0.06
Tundra	Cryic Gleysols, Histosols	Eurasian	2_2	18.0	0.15
Boreal Taiga Forest	Podzols, Podsoluvisols, Spodi-Distric Cambisols, Albi-Gleyic Luvisols, Gelic and Distric Histosols, Rendzinas and Gelic Rendzinas, Andosols	North Siberian	3_6	9.5	0.25
		Central Siberian	3_7	9.3	0.30
		East Siberian	3_8	7.5	0.20
		Kamchatkian-Aleutian	3_9	5.0	0.25
Taiga Meadow-Steppe	Gleysols, Planosols	Central Yakutian	4_a1	10.0	0.35
Subboreal Forest	Podzols, Dystric and Eutric Cambisols, Umbric Leptosols, Podsoluvisols	East Asian	5_a1	2.6	0.67
		East Chinese	5_a2	1.5	0.81
Forest Meadow-Steppe	Luvic Fhaeozems, Chernozems	South Siberian	6_a4	2.0	0.42
		Amur-Manchurian	6_a5	1.5	0.65
Steppe	Chernozems, Kashtanozems, Solonetzes	Mongolian-Chinese	8_a2	0.8	0.61
Desert-Steppe and Desert	Xerosols, Regosols, Arenosols, Yermosols, Solonetzes, Solonchaks	Pamir–Tibetan	11_a2	0.6	0.62
		Hindukush-Alayean	11-a5	0.4	0.86
		Tan-Shanean	11-a6	0.6	0.60
Savanna, Tropical Forest	Luvi-Plinthic Ferrasols, Luvisols, Vertisols, Subtropical Rendzinas, Ferralitic Cambisols, Nitosols, Arenosols, Ferralitic Arenosols	South Asian	12_a1	0.3	1.00
Subtropical and Tropical Wet Forest	Subtropical Solonchaks, Ferrasols, Eutric Subtropical Histosols, Gleyic Subtropical Podsols, Plintic Gleysols, Nitosols	Southeast Asian	13_a1	0.2	1.00
		Himalayan	13_a2	0.4	0.80
		Malaysian	13_a3	0.1	1.00

Table 2. *The ranks attached to biogeochemical cycling data to assess the migration capacity of soil-ecosystem types.*

Ranks	Biogeochemical cycling	Biogeochemical cycling coefficient, C_b
1	Very intensive	< 0.5
2	Intensive	0.5–1.4
3.	Moderate	1.5–3.0
4	Depressive	3.1–10.0
5	Very depressive	> 10.0

Table 3. *The ranks attached to temperature regime data to assess the duration of active biogeochemical reactions.*

Ranks	Temperature regime	Active temperature coefficient, C_t
1	Arctic	< 0.25
2	Boreal	0.26–0.50
3	Sub-boreal	0.51–0.80
4	"Mediterranean"	0.81–0.99
5	Subtropical and tropical	1.00

2.2. Tundra Ecosystems

The tundra ecosystems are represented in Asia by the Eurasian geographical region with primitive humid podzols, cryic gleysols, histosols, lithosols and regosols. The biogeochemical processes in these ecosystems are characterized by a small heat quantity, a short but very intensive period of active temperatures (the mean value of C_t is equal to 0.15), wide distribution of permafrost, a low precipitation, low biological and microbiological activity and a low rate of chemical weathering. The mean C_b values are equal to 18 (15–50), which correspond to a very depressed type of biogeochemical cycling. However, the long winter period enhances the accumulation of various pollutants in snow cover, with a sharp increase of their rapid influence on different components of ecosystems during the short summer period.

2.3. Boreal Taiga Forest Ecosystems

These are represented in East Asia by North Siberian, Central Siberian, East Siberian and Kamchatkian-Aleutian soil-biogeochemical geographical regions. The predominant land use/ecosystem categories are Dense Needle Evergreen Forest, Open Needle Evergreen Forest, Dense Mixed Evergreen Forest and Open Mixed Evergreen Forest. In spite of the differences in species composition, these coniferous forests are characterized by a depressed type of biogeochemical cycling; the C_b values vary from 5.0 (Kamchatkian-Aleutian geographical region) to 9.3–9.5 in Central and North Siberian regions. The predominant soils are podzols, podzoluvisols, and histosols, which have low pH, low base saturation and low cation exchange capacity. The mean values of C_t range between 0.25–0.35. Under these cold climate conditions the additional stress of acidic deposition on exposed plants may tend to make the vegetation in the taiga forest ecosystems more sensitive to changes caused by acidification.

2.4. Taiga Meadow Steppe Ecosystems

In East Asia, these ecosystems are represented in the Central Yakutian geographical region with planosols. The biogeochemical features of these soils are connected with their localization in the inner part of Northern Asia where the climate is the most severe and driest and soils develop under insufficient atmospheric deposition. The ratio of precipitation to potential evapotranspiration (P : PE) is equal to 0.45–1.00 and drops during the summer period down to 0.20–0.45. The permanent permafrost and an abundance of carbonate salinization of parent material lead to the formation of planosol-solonchak-solod-solonetz complexes. Under the long severe winter period and hot, dry short summer, the biogeochemical cycling is depressed; mean C_b is equal to 10.

2.5. Sub-Boreal Forest Ecosystems

These ecosystems develop in a monsoon climate with predominant distribution of such land-use categories as Dense Deciduous Forest, Dense Deciduous Broad Leaf Forest, and Open Deciduous Broad Leaf Woodland. Two geographic regions are presented: East Asian and East Chinese.

The East Asian geographical region is situated in the continental part of the Far East (Russia, China, Korea) and island parts (Russia, Japan). These are characterized by different subtypes of cambisols (spodi-distric, spodi-distric cryic, humid, orti-distric, distric) and podzols, especially on Hokkaido and in the Manchurian and Sikhote-Alin mountains under Dark Needle Forest ecosystems. In the plains, cambisols are located in the most drained areas. The type of biogeochemical processes in these ecosystems is moderate with mean values of C_b equal to 2.5 and C_t, 0.67, which is favorable to soil acidification under the input of sulfur and nitrogen acid forming compounds. This process can be especially enhanced in ecosystems with predominant vitric andosols where porosity favors fast chemical and biogeochemical weathering with allophane-kaolinite formation processes. The abundance of free iron and aluminum oxides under acid soil reactions, reinforced by acidic deposition, leads to a release of Al^{3+} ions and toxic influence on the fine roots of trees.

The East Chinese geographical region has almost the same spoil-biogeochemical features as above-mentioned but due to some climate characteristics the mean C_b value is 1.5 and the mean C_t-0.81

2.6. Forest Meadow Steppe Ecosystems

The East Asian part of these ecosystems is represented by South Siberian and Amur-Manchurian geographical regions having predominant luvic phaeozems and luvic chernozems with Open Deciduous Forest natural land use categories. In the wide space of the low plains of the Amur-Sunguri drainage basin of limno-alluvial origin, one can observe paleohydromorphic soil features as well as modern parameters of the given pedogenic process. This determines the biogeochemical cycling of elements in the above mentioned regions as moderate, characterized by mean values of C_b, 1.5 and C_t, 0.65. The more continental climate of the South Siberian geographical region favors an accumulation of various organomineral compounds in a humus biogeochemical barrier; the biogeochemical coefficient C_b is 2. In local depressions, salinization processes develop with formation of meadow-steppe solod and solonets complex and even peaty-swampy solonchaks. The acidification processes are more pronounced in those parts of the South Siberian geographical region, which occupy a very complex orographic area of large mountain ranges and vast intermountain depressions. This region occupies the central part of the Eurasian continent on the border between the Boreal Taiga Forest ecosystems of Siberia and the Dry Steppe and Desert ecosystems of Central Asia. It presents a contrast in climatic conditions: the northern and western slopes of mountain ranges have significantly more precipitation than the southern and eastern slopes, inter-mountain depressions and hilly areas. In accordance with climatic conditions, the different soil types from podzols to kashtanozems develop but the predominant soils in the slopes of hills and low mountains are luvic phaeozems and calcic chernozems, whereas in the highest mountain sites podzols and podzoluvisols are widespread. Thus the biogeochemical processes can be characterized as moderate in depressions and as semi-intensive in high mountain forest ecosystems with cambisols; the average C_b is equal to 2 and C_t to 0.42. However, the mountain forest ecosystems are very sensitive to acidic deposition.

2.7. Steppe Ecosystems

The main characteristic features of these ecosystems are connected with continental climate and insufficient precipitation, P : PE ranges between 0.6–0.3. The annual maximum precipitation is during summer, but due to high temperature and evapotranspiration the values of moisture coefficients are in the minimum range during this time. In accordance with the given climatic conditions, the soils of steppe ecosystems (chernozem, kashtanozem, solonets) are characterized by the presence of a few biogeochemical barriers such as humus, carbonate, and gypsum that makes them insensitive to actual and potential acidic precipitation. In the East Asian part of the world, these steppe ecosystems are represented in the Mongol-Chinese geographical region. The biogeochemical cycling is moderate: mean C_b values are between 0.7–0.8 and mean C_t values are between 0.57–0.61.

2.8. Desert-Steppe and Desert ecosystems

These ecosystems are widespread in the Asian continent and they coincide with sub-boreal and subtropical climatic belts of very strong aridity. There are three geographical regions in East Asia: Pamir-Tibetan, Hindukush-Alai and Tien-Shan (Table 1, Figure 1) The main soil types are xerosol, arenosol, yermosol, solonets, regosol etc. and all of them are characterized by high buffering ability, high pH values, and a low ratio of P : PE. Thus, in spite of an intensive, and even very intensive, type of biogeochemical cycling, C_b values are in limits of 0.3–0.6; these soils and corresponding ecosystems are insensitive to actual and potential acidification.

2.9. Xerofitic Savanna and Tropical Monsoon Forest Ecosystems

These ecosystems are represented in Asia by the South Asian geographical region with predominant ferrosols, vertisols, subtropical rendzinas, ferralitic arenosols, etc. (Table 1, Figure 1). In spite of very intensive biogeochemical cycling (C_b-0.3; C_t-1.00), most soil/ecosystem combinations in the Indostan peninsula are insensitive to acidic precipitation. These biogeochemical features are complicated in the Sri Lanka Islands and in plain and low plain areas of the Mekong and Menam river basins. These sub-regions are characterized by a monsoon climate with wet summer (1200–1300 mm) and dry winter periods. The natural land use categories are represented by Dense Drought Deciduous Forest, Open Drought Deciduous Woodland, Tropical Broad Leaf Forest, Tropical Savanna and other subtropical and tropical vegetation types. In the eastern part of the given Mekong-Manam geographical sub-region, the nitosols and rhodic ferrasols characterized by very intensive biogeochemical cycling and very high buffering capacity are predominant. These soils have low sensitivity to acidic deposition. On the other hand, luvi-plinthic and xantic ferrasols, subtropical albi-gleyic luvisols with plinthite are widespread in accumulative low plains of river deltas. The combination of these moisture conditions with very intensive biogeo-chemical cycling (C_b is 0.3) leads to the formation of ecosystems which are very sensitive to actual and potential acidic depositions reinforcing the release of free Al^{3+} in soil-water system.

2.10. Subtropical and Tropical Wet Forest ecosystems

These ecosystems are represented by monsoon subtropical and tropical forests (South-east Asian and Himalayan geographical regions) and equatorial wet forests (Malaysian geographical region). The main characteristic features of the given regions are a very old type of soil parent materials, which are transformed by very intensive geochemical and biogeochemical weathering leading to the destruction of all primary minerals except quartz and the accumulation of new formed minerals such as kaolinite, hematite, gibbsite, hydrogillite, etc. The predominant soils are ferrasols characterized by very low buffering capacity, abundance of free Al^{3+} and Fe^{3+}, acid reaction of soil depth, lack of accumulative biogeochemical barrier, very intensive biogeochemical cycling of all elements and especially such nutrients as N, P, K, S, Ca, Mg, etc. The combination of these features with a monsoon and equatorial climate leads undoubtedly to a shift in the original equilibrium towards acidification under the increasing input of acid forming sulfur and nitrogen compounds.

The *Southeast Asian Geographical Region* occurs in the northern part of this zone and is characterized by predominant distribution of acric ferrasols. The biogeochem-ical cycling is very intensive (mean C_b is 0.2) but definite differences exist in this cycling between hilly plains and low mountains up to 400–500 m a.s.l. and middle elevation mountains (up to 1000 m a.s.l.) where the humus biogeochemical barrier emerges in the profiles of podzolized ferrasols.

The *Himalayan Geographical Region* is situated in the eastern part of Tibet and the Chino-Tibetan mountains that determine very complex biogeochemical cycling in vertical biogeochemical catenas of different soils, such as evergreen broad leaf forests on mountainous acric ferrasols (1400–2000 m a.s.l.), evergreen and deciduous forests on transitive ferrasol-cambisol soils (2000–2700 m a.s.l.), mixed coniferous/deciduous forest on cambisols (2500–2800 m a.s.l.), dark coniferous forest on histosols, gleysols and cambisols (2700–3000 m a.s.l.) and mountainous meadows on mountainous phaeozems (3000–3200 m a.s.l.). The given soil–ecosystem consequences are connected with hydrothermal differences and accompanied by biogeochemical cycling forwarding from very intensive in the lowest parts ($C_b < 0.2$) to moderate in the highest ecosystems ($C_b > 0.5$). The mean values of the biogeochemical cycling coefficient are equal to 0.4 and active temperature coefficient, 0.8. The sensitivity of these ecosystems to acidic deposition varies but the majority of them, especially in the highest elevations, are very sensitive and should be protected to avoid the irreversible destruction of their biogeochemical cycling.

The *Malaysian Geographical Region* is situated in the Malaysian peninsula and the Indonesian and New Guinea islands. The predominant ecosystems are Wet Equatorial Tropical Forest with ferrasols. In accordance with very intensive biogeochemical cycling ($C_b < 0.1$) and natural acid features of soils, all components of these ecosystems are very sensitive to actual and potential acidic deposition.

Resuming the analysis of biogeochemical turnover in East Asian ecosystems, one should take into account that these general assessments are complicated by inadequacy and uncertainty of initial information that is required to carry out the correct quantitative parameterization and characterization of biogeochemical cycling even for such elements as nitrogen and sulfur.

3. MODERN LAND USE IN ASIA

Humankind throughout history has derived its sustenance from soil and corresponding land. Despite great technological advancement, modern man has not been able to develop synthetic substitutes for food and will have to depend heavily on soil and land use for food, fiber, fuel and shelter. In Asia, demand for the necessities of life increases with tremendous population growth, resulting in pressure on soil resources and relevant land use, which under improper perennial management has resulted in soil pollution, land degradation and lost productivity of both natural and agricultural ecosystems.

In Asia, there are three important uses of land: arable or croplands, permanent pastures and grazing land, and forest and woodlands. The land use pattern in the region has undergone a major change over the years with a sharp increase in cropland but a marked decline in forest. The growth in cropland is characterized by three phases (Figure 2).

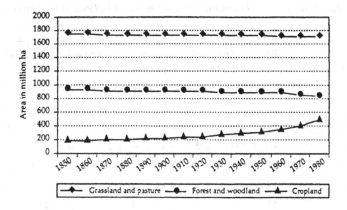

Figure 2. Land use trends in Asia and the Pacific during the last 150-year period (ESCAP, 2000).

During the first phase, rate of expansion of cropland was moderate followed by a sharp increase in the second phase (1950–76). In the post-1976 phase, the expansion of cropland leveled off. We should note here that the proportion of the Asian land area free from soil related constraints on agricultural production is only 14 per cent.

In 1993 cropped land accounted for 19 per cent of the total land area of developing countries of the region. This indicates that significant inroads have already been made into land with low production potential, which include unirrigated areas with arid and semi-arid climates, or unreliable rainfall, areas with steep slopes or poor soil drainage or a combination of these features. Such areas tend to be ones where environmental degradation is most severe.

The land use in selected countries and sub-regions is given in Table 4. Amongst the sub-regions, Southeast Asia has substantial land (49%) under forest most of which are natural forests. All other sub-regions have depleted natural forest cover due to excessive cutting. Northeast Asia also has substantial land under forest and woodlands but much of these have been contributed through plantations. Deforestation and over exploitation of natural vegetative cover are the major causative factors of human-induced soil degradation in forest and woodland of the Asian region.

Amongst various sub-regions, the biggest area under crops is in South Asia followed by Northeast Asia. Land use details for Central Asia are not available but the region has about 24,200 ha of croplands and 37,100 ha under rangelands (Glazovsky, 1997).

3.1. Land and Soil Degradation in Asia

Land use changes in Asia, particularly those related to agricultural expansion and yield growth on a massive scale were associated with considerable degradation of soil resources in terms of erosion, nutrient depletion and pollution (Table 5).

Table 4. Land use in selected countries of the Asian region in 1996, million ha (FAO, 1996).

Country/region	Total land area	Arable and permanent crops land		Permanent pasture		Forest and woodland		Other lands	
		area	%	area	%	area	%	Area	%
Bangladesh	13.0	9.7	74	0.6	5	1.9	15	0.8	6
Bhutan	4.7	0.1	3	0.3	6	3.1	66	0.1	25
India	297.3	169.7	57	11.4	4	68.5	23	47.8	16
Iran	163.6	18.1	11	4.4	27	11.4	7	90.1	55
Maldives	30	3	10	1	3	1	3	25	83
Nepal	13.7	2.4	17	2.0	15	5.8	42	3.6	26
Pakistan	77.1	2.1	28	5.0	6	3.5	5	4.7	61
Sri Lanka	6.5	1.9	29	0.4	7	2.1	32	2.0	32
South Asia total	575.9	223.2		63.7		96.2		192.7	
Cambodia	17.7	3.8	22	1.5	8	12.2	69	0.1	1
Indonesia	181.2	30.2	17	11.8	7	111.8	62	27.4	15
Lao	23.1	0.9	4	0.8	3	12.6	54	8.8	38
Malaysia	32.9	7.6	23	0.3	1	22.3	68	2.7	8
Myanmar	65.8	10.1	15	0.3	1	32.4	49	22.9	35
Philippines	29.8	9.2	31	1.3	4	13.6	46	5.7	19
Taiwan									
Thailand	51.1	20.8	41	0.8	2	13.5	26	16.0	31
Vietnam	32.5	7.0	21	0.3	1	9.6	30	15.6	48
China	932.6	95.8	10	400.0	43	130.5	14	306.4	33
Japan	37.7	4.4	12	0.7	2	25.0	66	7.6	20
North Korea	12.0	2.0	17	0.05	0	7.4	61	2.6	22
South Korea	9.9	2.1	21	0.09	1	6.5	65	1.3	13
Mongolia	156.7	1.3	1	117.2	75	13.8	9	24.4	16
Northeast Asia total	309.5	108.6		518.0		183.1		342.2	
Asia total	2159	418.4	16	598.8	34	507.4	24	634.3	26
Rest of the world	10044	981	10	2369	24	3432	34	3264	32
World	13045	1450	11	3395	26	4138	32	4061	31

Table 5. Changes in agricultural land use and soil degradation in Asia (Scherr, 1999).

Land type	Main changes	On-site soil degra-dation and pollution	Other kinds of degradation
Irrigated lands	Increase in irrigated area increased multi-ple cropping	• Salinization and water-logging • Nutrient constraints under multiple cropping • Biological degradation due to agrochemical ap-plication	• Nitrate and phos-phate pollution of sur-face and ground waters • Pesticide pollution • Waterborne disease • Water conflicts
High-quality rainfed lands	Transition from short fallow to continuous cropping, high yield crops, mechanization	• Nutrient depletion • Soil compaction and physical de-gradation from machinery cultivation • Acidification • Removal of natural vegetation, perennials from landscape • Soil erosion • Biological degradation due to agrochemical application	• Pesticide pollution • Deforestation of commons
Densely populated marginal lands	Transition from long to short fallows or continuous cropping; cropping in new landscape niches	• Soil erosion • Soil fertility depletion • Removal of natural vegetation, perennials from landscape • Soil compaction, physical degradation from overcultivation • Acidification	• Loss of biodiversity • Watershed degradation
Extensively managed marginal lands	Immigration and land-clearing from low input agriculture	• Soil erosion from land-clearing • Soil erosion from crop & livestock production • Soil nutrient depletion • Watershed degradation • Weed infestation • Biological degradation from top soil removal	• Deforestation • Loss of bio-diversity
Urban and peri-urban agricultural lands	Rapid urbanization; diversification of urban food market; rise in urban poverty	• Soil erosion from poor agricultural practices • Soil contamination from urban pollution • Overgrazing and compaction	• Water pollution • Air pollution • Human disease vec-tors

(million ha)

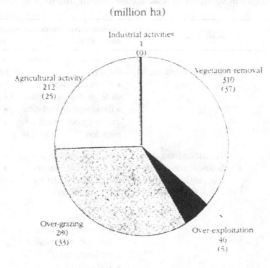

Source: Oldeman (1990).
Note: Figures in parenthesis indicate percentage share.

Figure 3. Area affected by various causes of soil degradation in Asia (ESCAP, 1995).

In Asia, the degraded soil area constitutes just about a quarter of the total land area. Amongst various categories of land uses, degradation of cropland appears to be most extensive over a third, pasture about a fifth and that of forest about a fourth. Thirteen per cent of the degraded area has suffered severe soil degradation and 62% light degradation. In the whole Asian region, 747 million ha of soil cover are degraded (Figure 3).

4. CHEMICAL DETERIORATION OF SOILS

It has been estimated that about 75 million ha of soil deteriorated chemically in the region during the last 45 years (ESCAP, 1995). Among the most dominating processes is salinization, which has spread over an area of 54 million ha. About 15 million ha suffers from nutrient loss, while acidified and polluted soils account for 4 and 2 million ha respectively.

4.1. Soil Salinization and Alkalization

The effect of salt on soils is measured in terms of salinity (soluble salts in soil solution) and sodicity or alkalization (sodium adsorbed or bound to soil particles). More details are shown in U.S. Soil Salinity Lab Stuff (1954).

4.1.1. Salinization

Salinity defines the concentration of ions dissolved in soil water, and is measured directly by electrical conductivity (EC). Soil water is held under tension or suction and as such is not available for analysis. To compensate for this, the standard procedure is to measure soil EC on a water extract. Soil scientists use what is called a saturated paste soil water extract (for more details see Radojevic and Bashkin, 1999).

The saturated paste method provides an estimate of salt in soil at normal field moisture equivalents. Saturation percentage is about four times the 15-atmosphere moisture equivalent or permanent wilting point (irreversible plant desiccation), and approximately two times the 0.33 atmosphere moisture equivalent or field capacity (moisture held in soil against drainage by gravity). Soil with a saturated paste EC of 4-desiSeimens per meter (dS/m) is considered saline by agricultural standards and can inhibit growth of sensitive plants (Deuel, 1999).

4.1.2. Alkalization

Two parameters are used to measure and evaluate the effects of excess sodium in soils. These are the sodium adsorption ratio (SAR) and the exchangeable sodium percentage (ESP). SAR is a parameter used to describe the composition of irrigation water or existing soil solution and is used to predict the ESP. SAR is defined by the equation

$$SAR = Na^+/[(Ca^{2+} + Mg^{2+})/2]^{1/2},$$

where cation concentrations are expressed in milliequivalents/liter (meq/L).

ESP is either measured directly as the exchangeable sodium divided by the cation-exchange capacity ((CEC) $\times$ 100), or estimated from SAR by the following equation:

$$ESP/(100 - ESP) = 0.015 \times SAR.$$

This equation assumes an equilibrium status and may not reflect the true conditions, if soil is not in equilibrium with the irrigated water or soil water extract. A soil is considered sodic if the ESP exceeds 15% of the CEC. CEC refers to the negative charge of soil colloids (clay size particles) and the attraction of this negative charged surface for the positively charged cations. Capacity refers to the total number of cations that must be adsorbed to achieve electrical neutrality. This is a surface phenomenon and the cations adsorbed are continually being desorbed and replaced by other cations in soil solution, thus the term cation-exchange capacity. The dominance of a given cation in soil solution favors its adsorption in an exchange reaction. Too much sodium in soil solution results in a sodic soil.

Excess sodium is toxic to plants and causes deterioration of soil structure. Deterioration of soil structure is closely associated with the process of swelling and dispersion. Calcium-saturated clay particles are comprised of four or more parallel clay particles called tactoids. Tactoids clump together to form particles and aggregates. As Na replaces Ca on the external exchange sites, particles within aggregates separate from each other and become dispersed. Dispersed particles can move from one layer to the next, becoming lodged in finer pores and eventually clog up the soil. As sodium

Table 6. Soil salinization in South and Central Asia, 000 ha (ESCAP, 1995; 2000).

Country	Degree of salinization			Total	Total as per cent of agr. land
	Light	Moderate	Severe		
Afghanistan	1,271	0	0	1,271	3
India	0	2,111	2,033	4,144	2
Iran	10,099	14,272	8,301	32,672	55
Pakistan	3,457	377	0	3,834	15
Sri Lanka	47	0	0	47	2
Kazakhstan	3	2	0.5	5,500	10.2

increases and replaces calcium in the interlayer, the tendency to swell increases. Multivalent cations provide electrostatic bridges between platelets. Sodium is not only monovalent but also has a large hydrated radius that further increases the distance between platelets. The deterioration of tactoid structure is generally complete with an ESP of about 50%. This condition provides the greatest challenge in reclamation of salt-affected soils (Deuel, 1999).

4.1.3. Modern soil salinization areas in Asia

Dryland salinity is a serious problem in Afghanistan, India, Iran, Kazakhstan, Thailand, Turkmenistan, and Uzbekistan. Replacement of forest vegetation by agricultural crops, especially in Thailand, has brought about hydrological changes through reduced evapotranspiration, allowing salts to concentrate in the low lying parts of the landscape. Excessive application of irrigated waters with high content of dissolved mineral salts is the main reason of soil salinization in Central Asia (see Chapter 8).

The figures of salt-affected soils reported by different researchers in many countries of South Asia show wide variation. Among eight countries in South Asia (Table 6), Iran has the greatest area of salt affected soils and strong salinization occurs only in this country and India (Yadav, 1986).

In Southeast Asia, the worst affected country is Indonesia, with 2.2 million ha of salt affected soils (Dent et al, 1992). In Malaysia, about 0.23 million ha of saline marine soils are found in the Peninsular region and Sarawak (Aminuddin et al, 1994). Some of these soils have been reclaimed by construction of coastal bunds, check gates, and drains for production of crops like coconut, oil palm, cocoa, coffee and paddy.

In Myanmar, approximately 1.4 per cent of the land is affected by saline and sodic soils, occurring mainly in the coastal belt, deltaic and arid areas (Tha Tun and Swe, 1994). In the Philippines, an estimated area of 0.4 million ha is affected by salinity along the 18,000 km coastline.

In Thailand, saline soils occupy 0.58 million ha in the coastal area and parts of the central plain, while potentially saline soils cover over 3.04 million ha. Moreover, the severely affected coastal saline soils are used for salt making, which involves heavy fuelwood consumption, thereby leading to extensive deforestation and further environmental deterioration (Potisuwan, 1994). In Vietnam, saline soils along 2,500 km coastline occupy 2,0 million ha (Phien and Siem, 1994).

Saline soils in China are mainly distributed in arid and semi-arid and coastal zones, occupying an area of 81.8 million ha. Of these, recently formed saline soils account for 36.93 million ha, and relic saline soils 44.87 million ha (Zitong, 1994). The latter soils occur mostly in the desert areas. About 6.24 million ha of saline soils occur in cultivated fields. In addition, there are 17,33 million ha of potentially saline soils with salic horizon in the deeper layers, which are prone to secondary salinization after irrigation (ESCAP, 1995).

In Northeast Asia, the worst sufferers from land degradation are the Kalmykia and Astrakhan regions (6 million ha) in Russia as well as south Siberian steppe soils. In the Siberian steppe at least 25% of arable land is subject to erosion. About 12 million ha have been salinized and waterlogged with development of soil alkalization. The latter process also proceeds in the northeast Kazakhstan steppes.

4.1.4. Remediation of saline soils

As stated previously, a saline soil is one with saturated paste EC in excess of 4 dS/m. Reclamation is indicated when the salt level limits or alters the intended land use. The only way to reclaim a saline soil is to leach it with fresh water. Soluble salts are carried with the water to a maximum depth of penetration that will prevent the return of salt by capillary action.

Soil structure is generally not a limiting factor for saline soils. However, a textural or structural impediment to leaching requires some form of drainage if the restriction is above the required depth of penetration. The quantity of water needed for reclamation is dependent on the required depth of penetration and texture, which control the effective pore volume. The requirement may be as great as 1.5 to 2 pore volume. For a heavy clay soil, the depth of water required for leaching is roughly equivalent to the required depth of penetration. In arid Asian regions a heavy clay soil could require up to 4 m of fresh water per hectare of affected land.

Sometimes a moderately saline soil (EC < 25 dS/m), comprised of coarse or medium structure, can be reclaimed to support salt-tolerant crops, by adding up to 10% low salt manure. The addition of organic matter increases the cation-exchange capacity of the affected soil. A 10% treatment can absorb up to 8000 kg of salt per hectare (Deuel, 1999).

4.1.5. Remediation of sodic soils

Remediation of sodic soil is far more difficult than reclaiming a saline soil. The condition of sodicity (ESP > 15%) requires the addition of calcium amendments, time for them to react with the soil, and percolation of water through the soil to remove excess sodium. Whether or not this can even be done depends on the degree of dispersion and swelling.

Table 7. Area affected by soil fertility decline in South Asia (000 ha).

Country	Degree of soil fertility decline			Total	Total as per cent of agr. land
	Light	Moderate	Severe		
Bangladesh	6,367	0	0	6,367	65
India	26,200	0	3,183	29,383	16
Pakistan	5,200	0	0	5,200	20
Sri Lanka	693	731	0	1,424	61
Total	38,460	731	3,183	42,374	11

The usual practice is to apply sufficient calcium to replace sodium from the top 30 cm of the impacted soil. Calcium is normally obtained from either high ionic strength, soluble calcium sources (e.g., $CaCl_2$ or $Ca(NO_3)_2$), or gypsum ($CaSO_4 \cdot 2H_2O$). Acids are also effective in combination with dolomitic lime ($CaMgCO_3$). Lime alone is of no benefit unless the soil is acidic. This would be an anomaly since most sodic soils are strongly alkaline with a pH in excess of 8.5. Dry powder or granular amendments must be mechanically mixed in the soil and require a long time (3–5 yr) to react. Liquid products are generally faster reacting (3–6 months) and do not require mechanical operations unless a strong surface crust, hardpan, or other near-surface cemented layers restrict infiltration and penetration of treatment into the soil (Deuel, 1999).

4.2. Loss of Plant Nutrients

Soil erosion results in a net removal of nutrients from the remaining soil. In addition, the continued use of compound, inorganic fertilizers in most countries is causing marked deficiencies of many major, secondary and micro-nutrients. In a number of cases, imbalanced use of fertilizers dominated by nitrogen has caused multiple nutrient deficiencies. For example, in Bangladesh, more than 50% of the total area has suffered from soil fertility decline (Table 7).

Sulfur deficiency is estimated to affect about 25 million ha of cultivated land across the region and zinc deficiency is also extensive. For example, in Bangladesh, approximately 4 million ha of cultivated land is estimated to suffer from sulfur deficiency. Sulfur and zinc deficiency is widespread in flooded rice soils (Saheed, 1992).

In China, the nutrient status of agricultural soils in 1990 showed a deficiency of phosphorus (in 53% of the surveyed soils), potassium (30%), boron (26%), molybdenum (35%), manganese (16%), zinc (39%) and copper (5%) (Zitong, 1994).

In Nepal, over-exploitation of land resources and accelerated soil erosion have led to a major loss of plant nutrients and organic matter, particularly in hilly areas with steep slopes, high rainfall intensity and shifting cultivation. The mountain soils suffer from deficiency of calcium, phosphorus and other elements. In Pakistan practically all soils are low in organic matter (less than 1%) and nitrogen content, and about 80 to 90% of agricultural soils, especially calcareous and alkaline soils, have low to medium availability of phosphorus.

We can refer also to the above-mentioned description of biogeochemical cycling in the main Asian ecosystems for understanding the peculiarities of nutrient declining in agricultural soils.

Remediation of macro- and micro-nutrients in agricultural soils is connected with the application of mineral complex fertilizers like $N-P-K-S-Zn$ for rice, $N-P-K-Cu$ for wheat, and various other combinations of fertilizers depending on the monitored deficiency of nutrients in soils. The application of organic fertilizers is also important since this amendment will improve not only mineral nutrients, but also the content of organic matter in soil will be increased.

4.3. Soil Acidification

Soil acidification affects extensive areas in many countries of the region, caused by both natural factors and human interventions. Assessment of natural aspects of soil acidification was shown in Section 2. The anthropogenic problems of soil degradation due to acidification are related to acid rains (mainly forest soils, see more details in Chapter 4) and application of mineral fertilizers, most of which are physiologically acid and increase the soil acidity during perennial application, especially application of nitrogen and phosphorus fertilizers.

In addition, the causes of soil acidity are: strong leaching of base cations under high rainfall intensity; negative balance between addition and removal of basis like calcium and magnesium; liberation of acids from organic matter decomposition; continued waterlogging; and low buffering capacity of soils.

Several countries have a significant proportion of their land area with acid sulfate limitations, notably Brunei (2.3%), Cambodia (1.2%), Malaysia (2.0%), Thailand (2.0%) and Vietnam (4.6%). Such soils are commonly developed on estuarine and marine alluvium (mangrove swamps) and contain considerable amounts of sulfides.

The total area of human-induced soil acidification is uncertain in many Asian countries. However, it has been noted in India that an area of 93.7 million ha suffers from soil acidity (both natural and anthropogenic) in varying degree.

The only rehabilitation measure against soil acidification is liming. The most used materials are carbonate and dolomitic powder or granulated lime applied in rates of 2–5 ton/ha in every 3–5 years.

4.4. Soil Pollution

Unregulated dumping of untreated industrial, municipal, agricultural wastes and mining give rise to land degradation. Heavy metals like mercury, lead, cadmium, nickel and arsenic cause serious land pollution problems. For example, in Japan, waste from the mines and factories located in agricultural areas have been found to have contaminated the soil with heavy metals (see Chapter 12 for more information on

HM pollution). Oil and other fuel pollution are of environmental concern in many countries of Asia at present.

In China as much as 10 million ha of farm land have been reported to be polluted to a varying degree: 3.3 million ha by sewage irrigation; 5.3 million ha by acid pollution (mainly as acid rain, see Chapter 4); and 0.9 million ha by leaching from solid wastes and garbage (NEPA, 1994). We should also mention the soil pollution due to mining, for instance, the extensive areas of tin tailings from former tin mining operations in Malaysia.

Soil pollution owing to high use of agrochemicals such as fertilizers, pesticides, insecticides, fungicides and herbicides, has also been reported in recent years. The harmful influence of nitrogen fertilizers is connected with accumulation of nitrate in soil solution, ground waters and vegetables (see Chapter 13). The extensive use of pesticides, particularly with the advent of high yield varieties, poses environmental problems resulting from the harmful accumulation of non-biodegradable residues. This results in deterioration of soil quality, due to the effects on the microfauna, bacteria, fungi, etc (see also Chapter 14). Biological soil degradation is also associated with lowering or depletion of soil organic matter.

Remediation of soil chemical pollution
A challenge facing remediation engineers is to choose, among effective treatment processes, the best (or the best grouping of treatment steps) for cleanup of soil at a particular site. More detailed discussion on the remediation of soils polluted by heavy metals (HM) and persistent organic pollutants (POP) will be presented in Chapters 12 (HM) and 14 (POP). Here we will pay more attention to treatment of soils polluted by oil and oil fuel products.

Oils and fuels generally enter the environment as a result of leaking storage tanks and associated piping (both above and below ground). Prompt removal of accidental spills is seldom done, and leaking storage tanks and piping is hardly ever detected promptly. Hence the spilled and leaked hydrocarbons leach underground and pollute soil body and ground waters.

Petroleum can be classified into various products: those with boiling points in the 30 °C–100 °C range are called *light gasolines*, those with boiling points in the 100 °C– 200 °C range are called *heavy gasolines*, those with boiling points in the 200 °C–300 °C range are called *diesel* and *kerosene*. Each type of petroleum is a mixture of hydrocarbons. Alkanes (paraffins and cycloalkanes), aromatic hydrocarbons, and polycyclic hydrocarbons are common components.

Hydrocarbons migrate through the high organic soils at slower rates than through low organic soils. This is because adsorption of petroleum hydrocarbons (PHC) into soil organics can reduce PHC migration significantly. In low organic content subsoil and ground stratas, adsorption levels of the various hydrocarbons are not high. A weak van der Waals force dominates the interaction of nonpolar PHC molecules with clay surfaces. If the concentration of hydrocarbons in soil solution is more than what is soluble in water, micellar forms of hydrocarbons will exist, and these hydrocarbons absorb on clay surfaces. The accomodation concentration of the hydrocarbon is approximately inversely proportional to the percent adsorbed by the clay surfaces. This means that the very toxic aromatic compounds have very low probabilities of being adsorbed on clay surface. Water inhibits the adsorption of nonpolar PHC molecules by soil functional groups, since relatively nonpolar organic molecules cannot effectively compete with high polar water for adsorption sites on the mineral surface with a hydrophilic nature. If water does not exist there, PHC can adsorb in the large surface area of a clay (Wilkins, 1999).

The physical and chemical removal of oil spills can often be successful, but mostly expensive. We can mention soil incineration, soil vapor extraction and thermal desorption methods, all of which are very expensive and destructive (Hyman, 1999). Besides that the primary goal of remediation is the preservation of public health and safety, it is necessary to carry out the remediation in a cost-effective manner to keep the owner/operator of the site in business and out of bankruptcy court. In this case, biodegradation by natural population of microorganisms or bioremediation, is often considered as the primary mechanism because of its low cost and effectiveness in ambient conditions. The scientific basis is biochemical conversion of organic chemicals by bacteria and fungus in natural processes.

The ease of biodegradation will depend on the type of hydrocarbon. Moderate to lower molecular weight hydrocarbons (C_{10} to C_{24} alkanes, single ring aromatics) appear to be the most easily degradable hydrocarbons. As the molecular weight increases, so does resistance to biodegradation. The vast majority of gasoline components can be readily degraded by a mixed microbial population.

A hydrocarbon biodegradation process is an oxidation-reduction reaction where the hydrocarbon is oxidized (donates electrons) and an electron acceptor (e.g., oxygen) is reduced (accepts electrons). There are a number of different compounds that can act as electron acceptors, such as oxygen, nitrate (NO_3^-), iron oxides [e.g., $Fe(OH)_3$, sulfate (SO_4^{2-})], water and carbon dioxide. Anaerobic bacteria use molecular oxygen as electron acceptors. Oxygen is the most preferred electron acceptor because microorganisms gain more energy from aerobic reduction. Water and carbon dioxide are the least preferred because micro-organisms gain least energy from these reactions.

The growth and metabolism of micro-organisms in natural environments is always limited by the availability of electron donors, electron acceptors or other essential nutrients. In petroleum spills, petroleum hydrocarbons are electron donors, so microbial metabolism is generally limited by the availability of electron acceptors (such as oxygen, NO_3^-, $Fe(OH)_3$ and SO_4^{2-}) or by availability of essential nutrients (such as nitrogen, phosphate or potassium). The basic principle of engineered bioremediation systems is to relieve the lack of electron acceptors and nutrients and thereby increase rates of PHC degradation. There are many fertilizer mixtures for applying such as a C : N : P ratio of 100 : 10 : 0.2 or 100 : 10 : 2 or 75 : 5 : 1. US EPA recommends ranges for C : N ratios of 10 : 1–100 : 1 for stimulating hydrocarbon degradation in soil (U.S. EPA, 1995). Using nitrate as an oxidant has been widely applied.

At present, bioremediation oil treatment technology can be similar to that shown in Figure 4.

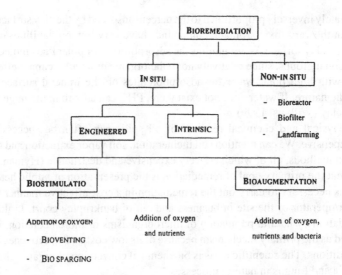

Figure 4. Generalized scheme of bioremediation for oil polluted soil.

The application of bioremediation treatment for oil polluted soil and groundwater in Thailand is shown in Box 1.

Box 1. Bioremediation of hydrocarbon in contaminated site of Bangkok metropolitan area (Mahatnirunkul et al, 2002)

The contaminated site is at the Rayong province in the East part of Thailand. On November 7, 1996, an accidental explosion of an illegal oil storage occurred in this area. The oil from eight underground tanks (60,000 liter/tank) burned for about 5 hours. During this accident about 200,000 liters of oil flowed down into soil and unconfined aquifer at the southern area of the explosion site. Soil and groundwater contamination damaged fruit plants and soil-shell turtle farm and fish production farm nearby this site. Eventually the contaminated groundwater may directly flow to Nong Pla Lai Reservoir that served as the main water source supply to the Rayong and Chon Buri Provinces for municipal and industrial water consumption.

For this contaminated site the research framework was set as follows:

- Develop the bioremediation process from initial site assessment to *in situ* treatment;

- Verify the mathematical model of major components for spill screening pattern;

- Develop the strategic environmental guideline for soil and groundwater *in situ* bioremediation treatment.

FURTHER READING

1. ESCAP, 1995. *State of the Environment in Asia and Pacific.* United Nations, p. 638.

2. Hyman M. H., 1999. Groundwater and soil remediation. In: Meyers R. A. (Ed.), *Encyclopedia of Environmental Pollution and Cleanup,* John Wiley, NY, 684–714

3. Deuel L. E., 1999. Salt waste in oil and gas production, remediation. In: Meyers R. A. (Ed.), *Encyclopedia of Environmental Pollution and Cleanup,* John Wiley, NY, 1475–1477.

4. Radojevic M. and Bashkin V., 1999. *Practical Environmental Analysis.* RSC, UK, 274–376.

5. ESCAP, 2000. *State of the Environment in Asia and Pacific.* United Nations, p. 905.

WEBSITE OF INTEREST

1. http//:www.escap.un.org

QUESTIONS AND PROBLEMS

1. Discuss the applicability of soil and ecosystems mapping for characterizing the Asian regional biogeochemical fluxes.

2. Present the definitions of biogeochemical uptake coefficient and active temperature coefficient and give examples of these coefficients for various ecosystems.

3. Characterize biogeochemical cycling in Asia. Analyze the alteration of biogeochemical parameters from north to south and from west to east.

4. Estimate the role of climate in transformation of biogeochemical features of ecosystems in Asia. Select the ecosystem and consider the relevant alterations.

5. Discuss the similarities and differences in natural and anthropogenic processes of soil acidification in Asia. Explain the possible reasons.

6. What are the characteristic features of biogeochemical fluxes in Mountain Asian ecosystems? Describe the role of relief in manifestation of biogeochemical cycles.

7. Describe the role of soil chemical pollution and land degradation in alteration of biogeochemical cycles in Asia.

8. Discuss the role of various factors in land degradation in Asia. Consider the modern trends in the whole region and in your country.

9. Estimate the degree of salinization and alkalization in various parts of the Asian region. Give the determination and discuss the possible sources of these degradation processes.

10. How does one measure the degree of salinization and alkalization processes in soil? Give an example of determination methods.

11. Present a general description of remediation processes for saline and sodic soils. Explain a possible modification of these processes for polluted soils in your country.

12. What nutrient losses are predominant in soil degradation in your country? Give examples and explain the reasons for soil nutrient losses.

13. Explain the natural and anthropogenic causes of soil acidification in the whole Asian region and in your country. What types of remediation can be applied for acid soils?

14. List the general sources of soil pollution and rank them in accordance with your regional peculiarities.

15. Describe the oil pollution processes that occur in soil after accidental spill or leakage. Present the explanation of soil physico-chemistry for oil adsorption.

16. Characterize the application of bioremediation treatment for oil-polluted soils. Present the scientific basic of this process.

CHAPTER 8

FRESHWATER POLLUTION

1. INTRODUCTION

Freshwater, an essential element for all forms of life, is a crucial resource for Asia. The withdrawal of freshwater from rivers, lakes, and underground reservoirs for human consumption has grown tremendously since the later part of the 19th century. Increasing population, urbanization and rapid growth of economic activities is imposing severe demand on the limited freshwater supply. The growing imbalance between supply and demand has already led to shortages owing to competition, which is likely to become more critical with time. Unfortunately, the scarcity of water is also being accompanied by deteriorating water quality owed to pollution and environmental degradation. This carries serious consequences to human health, terrestrial and aquatic ecosystems.

We will consider in this chapter the sources of freshwater supply in the Asian region and the main sources of water pollution, principal water chemical processes and accumulation of different pollutants in surface and ground waters with degradation of water quality. Finally, we review briefly the policies and programs that have been undertaken in the region to promote sustainable development of water resources at national and international levels.

2. WATER RESOURCES IN THE ASIAN REGION

Renewed through the global water cycle, freshwater is unevenly distributed among and within the Asian countries (Table 1).

Water resources availability and use in the Asian region vary widely between countries. The variation depends on the country's geographical conditions such as relief, climate and catchment size, accessibility of water resources and the level of economic development.

In the Asian region water resources are correlated with precipitation, which has extremely uneven distribution. For example, the maximum average annual precipitation is recorded in the western slope of India (more than 10,000 mm) and vast areas in Indo-China and the Indonesian islands receive annually from 1,500 mm to excess of 3,000 mm. On the other hand, almost all the northwestern part of the Region is extremely dry, with the annual precipitation less than 200 mm. Precipitation also varies

Table 1. Water resources and their use in selected countries of Asia (ESCAP, 2000).

| Country | Population, million, mid-1999 | Total area, km^2 | Water resource use | | |
			Total resources, km^3/yr	Annual use, km^3/yr	% from total
Afghanistan	21.92	652,090	60	26	43
Bangladesh	18.90	7,682,640	398	24	6
Bhutan	2.06	47,000	95	< 1	1
Cambodia	11.93	181,035	88	1	1
China	1266.83	9,600,000	2,812	500	18
North Korea	23.70	120,410	67	14	21
India	998.05	3,287,260	1,142	552	48
Indonesia	209.25	1,811,570	2,986	49	2
Iran	64.96	1,636,000	130	75	58
Japan	126.69	377,800	435	90	21
Laos	5.29	236,800	270	1	< 1
Malaysia	22.70	328,500	556	12	2
Mongolia	2.62	1,566,500	25	< 1	4
Nepal	22.40	147,181	207	12	6
Pakistan	138.72	796,000	247	180	73
Philippines	74.45	298,170	356	105	30
South Korea	46.85	99,290	70	30	42
Thailand	61.80	511,000	210	33	16
Vietnam	78.70	330,000	318	65	20

considerably during different periods of the year. Monsoon rainfall is the dominant pattern, with distinctive dry and a rainy season in wide parts of Asia. During the long dry season, temporary water shortage is experienced in many river basins, while during the rainy season, severe floods may cause tremendous damage.

The total annual runoff in the Asian region is about 12,260 km^3, which is about 30% of the global total. However, Asia has the lowest per capita freshwater resources among the World's regions. Several developing countries are already facing severe water shortages, especially during the dry season.

In Southeast Asia, annual per capita internal renewable water resources range from 172 m^3 a year in Singapore to more than 21,000 m^3 in Malaysia. In South Asia, India, Pakistan and Iran, freshwater supplies are between 1,400 and 1,900 m^3 per capita per year. On the other hand, Bhutan and Laos have around 50,000 m^3 per capita per year.

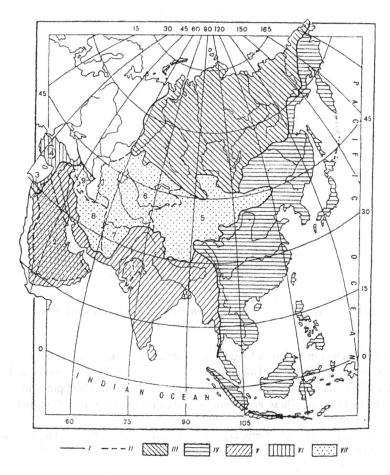

Figure 1. Ocean drainage basins of Asia (UNESCO, 1978) I—boundaries of drainage basins, II—Boundaries of internal runoff regions; III—Arctic Ocean; IV—Pacific Ocean; V—Indian Ocean; VI—Atlantic Ocean. 1—Thar desert (India); 2—Arabian Peninsula; 3—Dead Sea basin; 4—Inland Anatolia (Turkey); 5—Central Asia; 6—Kazakhstan and Middle Asia; 7—Seistan depression and adjacent regions; 8—Iranian highlands; 9—Pre-Caspian area.

2.1. Surface Water

Rivers, lakes and man-made reservoirs are the main sources of surface water. The Asian region has several of the World's important river systems, including Chang Jiang (Yangtze), Huang He, Mekong, Ayeyarwaddy, Brahmaputra, Ganges and Indus. These rivers have a total drainage area of more than 6 million km^2. International rivers in the region include Mekong, which flows through Vietnam, Laos, Cambodia, Myanmar and Thailand; and the Ganges, Brahmaputra, and Meghna River Systems, which are shared by India, China, Nepal, Bangladesh and Bhutan (Figure 1).

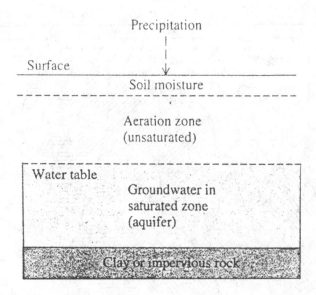

Figure 2. Groundwater in relation to the regions in the soil (Baird, 1999).

The Asian region is also endowed with a substantial number of lakes. Among the largest and most utilized are Dongting-hu in China, Tonge Sap in Cambodia, Lake Toba in Indonesia, Kasumigaura in Japan, Laguna de Bay in the Philippines and Lake Songkha in Thailand. These are extensively used for drinking, irrigation, fisheries, and recreational purposes.

Reservoir construction is primary for irrigation, flood control, and hydropower. There are over 800 major reservoirs with total volume of about $2,000 \, km^3$.

2.2. Groundwater

The great majority of the available fresh water on the Earth lies underground, half of it at depths exceeding one kilometer. As one digs into ground below the initial belt of soil moisture, the aeration or unsaturated zone, where the particles of soil are covered with a film of water but in which air is present between the particles, is next encountered. At lower depths is the saturated zone, in which water has displaced all the air. Groundwater is the name given to the freshwater in the saturated zone (Figure 2).

Groundwater in Asia occurs in many different type of rocks. These range from ancient crystalline basement rocks, which store minor quantities of water in their shallow weathered and jointed layers, to alluvial plain sediments, which may extend to depths of several hundred meters and contain enormous volumes of water (ESCAP, 2000). The Asian region has vast groundwater reservoirs in some areas, which receive extensive amounts of water from abundant recharge available during rainy seasons.

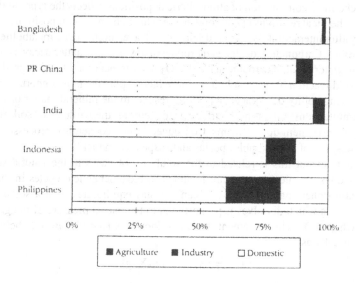

Figure 3. Freshwater withdrawals by sectors in selected Asian countries (ESCAP, 1995).

Bangladesh, India, Indonesia, Nepal and Myanmar particularly have large and deep aquifers. Many Asian countries are dependent on groundwater exploitation to supplement scarce surface water resources. This dependency reaches 30 to 35% of the total supply in Bangladesh, India and Pakistan (ADB, 1998).

2.3. Sectoral Water Use in Asia

The biggest users of water are the agriculture, industry and domestic sectors, whose demands are continuously increasing, resulting in increasing competition for supply (Figure 3).

Agricultural water usage is normally accorded a lower governmental priority than domestic and/or industrial supplies and at times of water shortages, agricultural water supplies are reduced to ensure adequate supplies to the other sectors.

3. HYDROCHEMISTRY OF NATURAL WATERS

The interactions between living matter and the hydrosphere is one of the global processes occurring in the biosphere. Living matter of the Earth is inalienably linked to liquid water. Water accounts for 60% of the total mass of terrestrial living organisms. All biochemical reactions and physiological processes proceed in aqueous media. Enormous amounts of water undergo decomposition during the activity of photosystem-II of the photosynthetic process.

The chemical composition of atmospheric depositions reflects the types and rates of biogeochemical reactions in the troposphere. These chemical compositions change seriously after interactions with humid acids of soil layer, higher plant metabolites, and soil microbes. Carbon dioxide, the end product of any organic matter decay, is readily soluble in water to yield carbonic acid, H_2CO_3, the dissociation of which produces a slightly acid reaction of water solution. These factors play the most important role in the dissolving ability of surface waters with respect to the mineral matter of Earth's crust. Simultaneously the runoff surface waters entrain the particles of soil mineral matter and carry them up to streams. Consequently biogeochemical processes affect the composition of both soluble species and suspended matter.

To some extent this is related to the solubility of gases in the natural waters. Many gaseous chemical species are included in biogeochemical cycles in the system of atmosphere-natural waters. When any gas equilibrates with a solvent, the amount of gas which dissolves is proportional to the partial pressure of the gas (see Box 1, Chapter 3). This statement, which is known as Henry's law, can be written mathematically as

$$K_H = [X, solv]/p(X, g),$$

where X, solv is the mass of chemical species X, dissolved in water and X, g is the partial pressure of the gas in the atmosphere in equilibrium with water. The proportionality constant is the equilibrium constant or Henry's Law constant, usually expressed in mol L^{-1}. Table 2 gives the K_H values for some biogeochemically active gases at 25 °C.

The thermodynamic principles of gas solubility in water are very important. An example of these principles is discussed briefly in Box 1.

Box 1. Thermodynamic principles of gas solubility in water (After Bunce, 1994)

Gases become less soluble with increasing temperature. From this variation of K_H with temperature, one can deduce the underlying thermodynamic principles governing the dissolution of gases, namely, that change from the gaseous state to solution is a process for which $\Delta H°$ and $\Delta S°$ are both negative. Enthalpic stabilization accompanies dissolution, but the dissolved state is more ordered than the gas. A rise in temperature thus favors the gaseous state ($-T\Delta S°$ for dissolution becomes more positive).

The low solubility of non-polar gases, such as methane, in water can be also understood using thermodynamic principles. The process $CH_4(g) \rightarrow CH_4(aq)$ is exothermic ($\Delta H°$ negative) and the low solubility is due to the enthropic factor. The drawback of this process is the intrinsically greater order of a condensed phase compared with the gas phase, and the property of water in ordering itself around the non-polar solute molecule. The latter phenomenon has been likened to forming a miniature iceberg around the solute, thereby greatly reducing the entropy of the water. Please remember that we must consider the whole system, not just the methane or other non-polar gases.

Table 2. Examples of Henry's Law constants
at 25°C and 1 atm pressure.

Gas	K_H, mol/L/atm
H_2	7.8×10^{-4}
N_2	6.5×10^{-4}
CO_2	3.4×10^{-2}
CO	9×10^{-4}
O_2	1.3×10^{-3}
O_3	1.3×10^{-2}
NO_2	6.4×10^{-3}
HO_2	2.0×10^3
SO_2	1.2
CH_2O	6.3×10^3
NO	1.9×10^{-3}
NO_3	15.0
H_2O_2	7.4×10^4

Stream waters can be considered as the composite solutions containing soluble ions and particulate solid matter. The main forms of the occurrence of chemical elements in stream water are:

(i)   simple and complex ions;

(ii) neutral molecules, mostly existing as a ligand with an inorganic complexing ion;

(iii) colloidal particles from 0.001 to 0.1 μm, with ions adsorbed at their surface;

(iv) finely dispersed, mainly clay mineral particles of 0.5–2 μm; and

(v)   large suspended particles from clastic minerals of 2–3 to 10 μm.

In analytical practice, the suspended particles are separated by means of membrane filters with various pore diameters. The centrifugation is also of use. Methods for determining the chemical species in waters are described in detail in our previous textbook (Radojevic and Bashkin, 1999).

In the river waters, the most common soluble species are HCO_3^-, SO_4^{2-} and Cl^- anions accounting, respectively, for 48.8%, 10.0% and 5.3% of the sum of total soluble compounds. Among the cations, calcium (10.8%), magnesium (2.7%) and potassium (1.2%) are typical. The other elements are present in various trace amounts.

Table 3. Typical concentrations of ions in rivers and sea water.

Ion	C(river), mol L^{-1}	C(ocean), mol L^{-1}
HCO_3^-	9.5	0.0023
Ca^{2+}	3.8	0.010
Cl^-	2.2	0.55
Na^+	2.7	0.46
Mg^{2+}	1.7	0.054
SO_4^{2-}	1.2	0.028
K^+	0.59	0.010

The amount of solid dissolved in natural water varies widely. The values in Table 3 are typical of river and ocean water, although we shall see that river water is quite variable in its mineral content. Groundwater is at least as high in dissolved solids as lake and river water; sometimes it is much higher, with total dissolved solids exceeding 1000 ppm. With the exceptions of Ca^{2+} and HCO_3^-, there is a parallel between the average concentrations of ions in fresh and in ocean water. There is relatively more Ca^{2+} and HCO_3^- in river water because some hydrochemical and biogeochemical principles: rivers dissolve ancient rocks containing $CaCO_3$, whereas the oceans precipitate $CaCO_3$ in the form of marine organisms' exoskeletons.

Some chemical species play the most important roles in water chemistry. We will consider them in more detail.

3.1. Dissolved Oxygen

By far the most important oxidizing agent in natural waters is dissolved molecular oxygen. Upon reaction, each of its oxygen atoms is reduced from zero oxidation state to the -2 state in H_2O or OH^-.

Reaction (1) occurs in acidic water solutions

$$O_2 + 4H^+ + 4e^- \longrightarrow 2H_2O, \tag{1}$$

and reaction (2) in a basic aqueous solution

$$O_2 + 2H_2O \ H^+ + 4e^- \longrightarrow 4OH^-. \tag{2}$$

The concentration of dissolved oxygen in water is small and therefore precarious from the ecological point of view. For the reaction (3)

$$O_2(g) \longleftrightarrow O_2(aq). \tag{3}$$

The appropriate equilibrium constant is the Henry's Law constant K_H, which for oxygen at 25 °C has the value $1.3 \times 10^{-3}\,mol\,L^{-1}\,atm^{-1}$:

$$K_H = [O_2(aq)]/P_{O_2} = 1.3 \times 10^{-3}\,mol\,L^{-1}\,atm^{-1}. \qquad (4)$$

Since in dry air the partial pressure, P_{O_2}, of oxygen is 0.21 atm, it follows that the solubility of O_2 is $8.7\,mg\,L^{-1}$ of water. This value can also be stated as 8.7 ppm since for water phase, ppm (part per million) concentrations are based on mass rather than moles. Because the solubilities of gases increase with decreasing temperature, the amount of dissolved oxygen at 0 °C (14.7 ppm) is greater than is the amount that dissolves at 35 °C (7.0 ppm). This means that during hot summer conditions, most Asian surface waters are depleted of dissolved oxygen.

A river or lake which has been artificially warmed can be considered to have undergone thermal pollution in the sense that it will contain less oxygen than colder water because of the decrease in gas solubility with increasing temperature. To sustain their lives, fish require water containing at least 5 ppm of dissolved O_2.

Thermal pollution often occurs as a result of the operation of electric power plants, since they draw cold water from a river or lake, use it for cooling purposes, and then return the warmed water to its source.

Oxygen Demand

The most common substance oxidized by dissolved oxygen in water is organic matter having a biological origin, such as dead plant matter and animal wastes. If, for the sake of simplicity, the organic matter is assumed to be entirely polymerized with an approximate empirical formula of CH_2O, the oxidation reaction will be as follows (reaction (5))

$$O_2 + CH_2O(aq) \longrightarrow CO_2(g) + H_2O(aq). \qquad (5)$$

Similarly, dissolved oxygen in water is consumed by the oxidation of dissolved ammonium ion (NH_4^+), a substance that like organic matter is present in water as a result of biological activity, to nitrate ion (NO_3^-), reaction (6)

$$2.5\,O_2 + NH_4^+(aq) \longrightarrow NO_3^-(aq) + 2\,H_2O(aq). \qquad (6)$$

Water that is aerated by flowing in shallow streams and rivers is constantly replenished with oxygen. However, stagnant water or that near the bottom of a deep lake is usually almost completely depleted of oxygen because of its reaction with organic matter and the lack of any mechanism to replenish it quickly, diffusion being a slow process due to thermal stratification (Figure 4).

Water's unique temperature-density relationship results in the formation of distinct layers within non-flowing bodies of water, as shown in Figure 4. During the summer in moderate Asian regions or during most of the year in tropical and subtropical regions, a surface layer (*epilimnion*) is heated by solar radiation. It flows owing to its lower

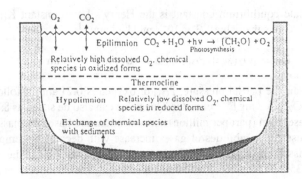

Figure 4. Stratification of lake (Manahan, 2000).

density upon the bottom layer, or *hypolimnion*. This phenomenon is called *thermal stratification*. When a significant temperature difference exists between the two layers, they do not mix but behave independently and have very different chemical and biological properties. The shear-plane, or layer between epilimnion and hypolimnion, is called the *thermocline*.

The capacity of the organic matter in a sample of natural water to consume oxygen is called *biochemical oxygen demand, BOD*. It is usually evaluated experimentally by determining the concentration of dissolved O_2 at the beginning and at the end of the five-day period, in which the sealed water sample is maintained in the dark at a constant temperature 25 °C. The BOD equals the amount of oxygen consumed as a result of the oxidation of dissolved organic matter in the sample and noted as BOD_5 (see this method in Radojevic and Bashkin, 1999).

The oxidation reactions are catalyzed in the sample by the action of microorganisms already present in natural water. If you suspect that the sample will have a high BOD, you have to dilute the water sample with pure oxygen-saturated water. This is necessary to be sure that oxygen is non-limiting to oxidize all the organic matter. The results are corrected for this dilution.

A faster determination of oxygen demand in a water sample can be made by evaluating the *chemical oxygen demand, COD*. Dichromate ion, $Cr_2O_7^{2-}$, can be dissolved as one of its salts, such as $K_2Cr_2O_7$ or $Na_2Cr_2O_7$ in sulfuric acid. This reactant is a very strong oxidizing agent. COD is measured by reacting the water sample with a fixed amount of $Na_2Cr_2O_7/H_2SO_4$ under saturation conditions of time and temperature, and then titrating the unreact $K_2Cr_2O_7$ against a standardized Fe^{2+} solution. Each mole of $Cr_2O_7^{2-}$ consumed is equivalent, in acidic solution, to 1.5 moles of O_2. In other words, one mole of $Cr_2O_7^{2-}$ can oxidize as much organic matter as 1.5 moles of O_2 (reaction (7))

$$Cr_2O_7^{2-} + 14\,H^+(aq) + 6\,e^- \longrightarrow 2\,Cr^{3+}(aq) + 7\,H_2O(aq). \qquad (7)$$

In practice, excess dichromate is added to the sample and the resulting solution is back-titrated with Fe^{2+} to the end-point.

The difficulties with the COD method as a measure of oxygen demand is that acidified dichromate is such a strong oxidizer that it oxidizes substances that are very slow to consume oxygen in natural waters and that therefore pose no real threat to their oxygen content. In other words, dichromate oxidizes substances that would not be oxidized by O_2 in the determination of the BOD. Because of this excess oxidation, namely of stable organic matter such as cellulose to CO_2, and of Cl^- to Cl_2, the COD value for a water sample as a rule is slightly higher than its BOD.

There are two other methods for measuring oxygen demand in natural waters. *Total organic carbon, TOC*, is measured by oxidizing all the organic matter to CO_2, and then analyzing CO_2 formed gas chromatographically. TOC is usually reported in ppm of carbon. The parameter *dissolved organic carbon, DOC*, is used to characterize only organic material that is actually dissolved.

Each of these parameters to measure the water oxygen demand is arbitrary because not all organics oxidize at equal rates. For example, carbohydrates, which are polyalcohols, oxidize rapidly, whereas alkanes have no functional group and oxidize very slowly. Total organic carbon is an arbitrary measure of oxygen demand because all carbon compounds are included, even through they oxidize at different rates. Chemical oxygen demand goes some way to compensating for this, because acidic dichromate readily oxidizes functionalities such as alcohols and alkenes, but it gives too rapid rates (see above) to mimic natural conditions. Biochemical oxygen demand would seem at first to be the ideal approach, since oxidation is accomplished biologically. However, both the choice of time (five days or some other period) and temperature (20 or 25 °C) for the test and degree of dilution, are arbitrary parameters. All these measures thus have their uses, in terms of convenience of analysis, but none can reflect accurately what happens in a real environmental system.

Anaerobic decomposition of organic matter in natural waters
Dissolved organic matter will decompose in water under anaerobic (oxygen-free) conditions if appropriate bacteria are present. Anaerobic conditions occur naturally in stagnant water such as swamps and at the bottom of deep lakes. The bacteria operate on carbon so as to disproportionate it. In other words, some carbon is oxidized to carbon dioxide and the rest is reduced to methane (reaction (8))

$$2\,CH_2O(aq) \xrightarrow{\text{bacteria}} CO_2(g) + CH_4(g). \qquad (8)$$

This is an example of *fermentation* reaction, which is defined as one in which both oxidizing and reducing agents are organic materials. Since the methane produced in this process is almost insoluble in water (see Box 1), it forms bubbles that can be seen rising to the surface in swamps. We recall that methane was originally called marsh or swamp gas.

Here we can also refer to the same chemical reaction that occurs in digester units used by rural inhabitants in Asian countries, such as India, to convert animal wastes into methane gas than can be used as fuel.

Since anaerobic conditions are reducing conditions in the chemical meaning, insoluble Fe^{3+} species that are present in sediments at the bottom of lakes are converted into soluble Fe^{2+} compounds, which then dissolve into the lake water (reaction (9))

$$Fe^{3+}(s) + e^- \longrightarrow Fe^{2+}(aq). \tag{9}$$

It is not uncommon to determine both aerobic and anaerobic conditions in the same lake, but at different depths, especially in summertime (see Figure 4).

3.2. Dissolved Carbon Dioxide

The solubility behavior of CO_2 in water is inherently more complex than that of oxygen. We should remember that carbon dioxide may react chemically with water molecules in addition to the solubility of gaseous species (reaction (10))

$$CO_2(g) \longleftrightarrow CO_2(aq) \overset{H_2O}{\longleftrightarrow} H_2CO_2(aq) \overset{-H^+}{\longleftrightarrow} HCO_3^-(aq) \overset{-H^+}{\longleftrightarrow} CO_3^{2-}(aq). \tag{10}$$

We can make some simplifications for calculating the solubility of carbon dioxide in water and for estimating the final pH value:

- The total system $CO_2(g) \leftrightarrow H_2CO_2(aq)$ is considered together;

- The $CO_2(aq)$ and undissociated $H_2CO_2(aq)$ are interchangeable.

On this basis, the value of K_H for dissolution of $CO_2(g)$ in water at 25 °C is 3.4×10^{-2} mol L^{-1} atm^{-1}, and the concentration of dissolved $CO_2(aq)$ in equilibrium with the air [$pCO_2(g)$ equal 3.0×10^{-4} atm] is 1.0×10^{-5} mol L^{-1} or 0.44 ppm of CO_2.

To estimate the resulting pH, we should consider the consequence part of reaction (10), namely:

$$H_2CO_3(aq) \longleftrightarrow HCO_3^-(aq) + H^+(aq). \tag{11}$$

We rewrite this reaction scheme as follows:

$$H_2CO_3(aq) \longleftrightarrow HCO_3^-(aq) + H^+(aq)$$

Equilibrium concentration = $\quad 1.0 \times 10^{-5} \qquad x \qquad\qquad x$

The concentration of $H_2CO_3(aq)$ remains constant, since the total system {$CO_2(g) \leftrightarrow H_2CO_2(aq)$} is considering together in the equilibrium to the atmospheric reservoir of $CO_2(g)$. Thus for $H_2CO_3(aq)$,

$$K_a = \{[H^+][HCO_3^-]\}/[H_2CO_3] = [x^2]/[1.0 \times 10^{-5}] = 4.2 \times 10^{-7} \text{ mol } L^{-1}.$$

Hence:

$$[x] = \{[4.2 \times 10^{-7}] \cdot [1.0 \times 10^{-5}]\}^{1/2} = 2.1 \times 10^{-6} \quad \text{and} \quad pH = 5.67.$$

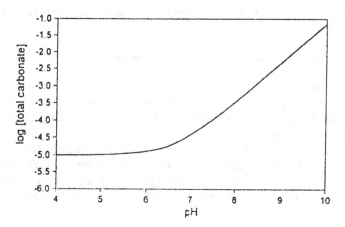

Figure 5. Calculated concentration of dissolved carbonate {[H₂CO₃(aq), carbonic acid] + [HCO₃⁻(aq), bicarbonate] + [CO₃²⁻(aq), carbonate]} in equilibrium with 350 ppmv of [CO₂(g)] as a function of pH (Bunce, 1994).

This gives the important result that pure water in equilibrium with air is not at pH 7; it is slightly acidic owing to the presence of dissolved carbon dioxide. Remember from Chapter 4, that the pH of even unpolluted rainwater is 5.6 rather than 7.0.

From the calculations above, we can see also that the total carbonate is

$$[HCO_3^-] + [H_2CO_3] = 1.2 \times 10^{-5} \, mol \, L^{-1}.$$

As pH rises, the total amount of dissolved carbonate as {[CO₂(aq), carbon dioxide] + [H₂CO₃(aq), carbonic acid] + [HCO₃⁻(aq), bicarbonate] + [CO₃²⁻(aq), carbonate]} in equilibrium with the atmosphere increases. As before, the atmosphere is an inexhaustible reservoir of $CO_2(g)$, maintaining its pressure at 3.0×10^{-4} atm (Figure 5).

3.3. Acid-Base Chemistry of Natural Waters

We will consider in this section two main hydrochemical parameters, alkalinity and hardness.

Alkalinity

The alkalinity of a water sample is a measure of its capacity to neutralize acids. In other words, it is the sum of all the titratable bases. Alkalinity is an operationally defined concept, i.e., it is defined in terms of the amount of acid required to react with the sample to a designated pH. Hence, experimentally determined values vary considerably with the end-point pH used in the titrimetric analysis. Alkalinity is commonly expressed in mg $CaCO_3 \, L^{-1}$.

Alkalinity of waters is mainly due to salts of strong bases and salts of weak acids. Alkalinity of many surface waters is primarily a function of hydroxide, carbonate and bicarbonate concentrations. Bicarbonate is the major contributor to alkalinity and it arises from the action of CO_2 in percolating water and basic minerals in soils and rocks (reaction (12))

$$CO_2(aq) + CaCO_3(s) + H_2O(aq) \longrightarrow 2\,HCO_3^-(aq) + Ca^{2+}(aq). \tag{12}$$

Other compounds present in natural water may also make a minor contribution to alkalinity. For example, ammonia and salts of weak inorganic acids such as borates, silicates and phosphates may contribute to alkalinity, as may salts of organic acids (e.g., humic, acetic, propionic).

Alkalinity is determined by titrating a measured volume of water with sulfuric acid, H_2SO_4. If the pH of the sample is greater than 8.3, the titration is carried out in two stages. In the first stage, the titration is carried out to a pH of 8.3, the phenolphthalein end point. By the time, the pH has been reduced to 10, all the hydroxides would have been neutralized (reaction (13))

$$2\,OH^- + H_2SO_4 \longrightarrow 2\,SO_4^{2-} + 2\,H_2O. \tag{13}$$

The phenolphtalein end point corresponds to the equivalence point for reaction (14)

$$2\,CO_3^{2-} + H_2SO_4 \longrightarrow SO_4^{2-} + 2\,HCO_3^-. \tag{14}$$

This is known as *phenolphthalein alkalinity*. Subsequently, the titration is carried out to a pH of about 4.5, the methyl orange end point, which corresponds approximately to the equivalence point for reaction (15)

$$2\,HCO_3^- + H_2SO_4 \longrightarrow SO_4^{2-} + 2\,H_2CO_3. \tag{15}$$

This is referred to as *total alkalinity*. If the sample has a pH less then 8.3, a singe titration to the methyl orange end point is carried out. Alternatively, a potentiometric titration can be carried out, measuring the pH with an electrode and plotting the titration curve. A typical potentiometric titration curve is shown in Figure 6, illustrating the location of the stoichiometric end points and the contribution of various species to alkalinity.

The technical description of this method for alkalinity determination is shown in reference (Radojevic & Bashkin, 1999, 180–189).

The three principal contributors to alkalinity are hydroxide (OH^-), carbonate (CO_3^{2-}) and bicarbonate (HCO_3^-) ions. It is possible to calculate the contribution of each of these to the alkalinity on the basis of the alkalinity determination, if it is assumed that other species do not contribute to the alkalinity and if is assumed that OH^- and HCO_3^- cannot coexist. The following alkalinity conditions are then possible in the sample: (a) hydroxide alone, (b) hydroxide and carbonate, (c) carbonate alone, (d) carbonate and bicarbonate and (e) bicarbonate alone. The results of the alkalinity determination can be used to work out the alkalinity relationships from Table 4.

We can also determine the contributing anions from pH and alkalinity measurements. Given that $[OH^-] = K_w[H^+]$, the hydroxide alkalinity can be calculated from

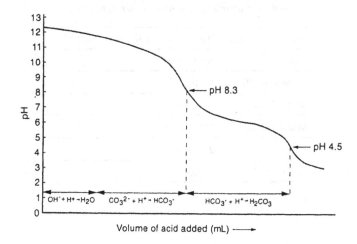

Figure 6. Alkalinity titration curve (Radojevic & Bashkin, 1999).

Table 4. Alkalinity relationships.

Condition	Titration result	Alkalinity as CaCO₃		
		Hydroxide	Carbonate	Bicarbonate
A	PA = TA	TA	0	0
B	PA > 0.5TA	2PA − TA	2(TA − PA)	0
C	PA = 0.5TA	0	2PA	0
D	PA < 0.5TA	0	2PA	TA − 2PA
E	PA = 0	0	0	TA

Note: TA—total alkalinity; PA—phenolphtelein alkalinity.

the sample pH. Given that $1 \, mol \, L^{-1}$ of OH^- is equivalent to $50{,}000 \, mg \, L^{-1}$ of alkalinity as $CaCO_3$, the hydroxide alkalinity (HA) can be determined from:

$$HA \, (mg \, CaCO_3 \, L^{-1}) = 50000 \times 10^{(pH-pKw)}. \tag{16}$$

Carbonate alkalinity (CA) and bicarbonate (BA) alkalinity (in mg $CaCO_3 \, L^{-1}$) can then be determined from the following expressions:

$$CA = 2(PA - HA), \tag{17}$$

$$(BA = TA - (CA + HA). \tag{18}$$

Table 5. Classification used for water hardness.

Degree of hardness	Hardness (mg equivalent $CaCO_3\ L^{-1}$)
Soft	< 50
Moderately soft	50–100
Slightly hard	100–150
Moderately hard	150–200
Hard	200–300
Very hard	> 300

Hardness

The term *hardness* refers to the ability of water to precipitate soap. Hard waters are undesirable for two reasons:

- They require considerable amount of soap to produce a lather;

- They produce scale in industrial boilers, heaters and hot water pipes.

The former is of concern to domestic users of water, whereas the latter is a problem for engineers. On the other hand, soft waters have been linked with an increased incidence of cardiovascular disease. Other advantages of hard water include the neutralization of acid deposition and the reduction of the solubility of toxic metals (see Chapter 4 for more details).

The major contributors to water hardness are dissolved calcium and magnesium ions. These ions combine with soap to form insoluble precipitates. Other polyvalent cations such as those of iron, zinc, manganese, aluminum and strontium may also contribute to hardness, but their contribution is usually insignificant due to low concentrations of these metals in water. The hardness of water derives largely from the weathering of minerals, such as limestone ($CaCO_3$), dolomite ($CaCO_3 \cdot MgCO_3$) and gypsum ($CaSO_4 \cdot 2H_2O$) and it varies considerably from place to place, depending on the nature of geological formations. Groundwater is generally harder than surface waters. *Total hardness* is defined as the sum of Ca and Mg concentrations expressed as calcium carbonate in mg L^{-1} or ppm. Waters are classified according to the hardness scale as shown in Table 5.

Scaling problems tend to occur with moderately hard and very hard waters. Calcium is associated with bicarbonate ions in solution and upon heating these are converted to calcium carbonate, which forms a thick scale on the surface of domestic and industrial boilers, water heating appliance, kettles, pipes, etc. (reaction (19))

$$Ca^{2+} + 2HCO_3^{-} \longrightarrow CaCO_3(s) + CO_2 + H_2O. \tag{19}$$

This hardness, which can be removed by heating, is called *temporary* or *carbonate hardness*. Temporary hardness is derived from contact with carbonate (limestone and dolomite). Hardness which cannot be removed by boiling is called *permanent* or *non-carbonate hardness* and it is due to anions, such as chloride, nitrate, sulfate and silicate. This hardness does not contribute to scale formation. Contact with gypsum would result in permanent hardness. *Calcium hardness* is that due to Ca only, while *magnesium hardness* is due to Mg only. Magnesium hardness can be calculated from a determination of total and calcium hardness:

Magnesium hardness = Total hardness − Calcium hardness.

A simple, rapid and inexpensive method commonly used in the water industry for hardness determination is the direct complexation titration with ethylenediaminete-traacetic acid (ADTA), $(HOOCCH_2)_2NCH_2CH_2N(CH_2COOH)_2$.

EDTA forms $1:1$ complexes with divalent metals such as calcium (reaction (20))

$$Ca^{2+} + EDTA \longrightarrow \{Ca \cdot EDTA\}_{complex}. \qquad (20)$$

Erichrome Black T or Calmagite can be used as indicators. If a small quantity of indicator is added to a water sample containing Ca and Ng ions at pH 10, the solution becomes wine-red. The indicator forms complexes with Ca and Mg ions, which give the solution a wine-red color (reaction (21))

$$Ca^{2+} + Eriochrome\ Black\ T \longrightarrow \{Ca \cdot Eriochrome\ Black\ T\}_{complex}. \qquad (21)$$

As EDTA is added it displaces the cations from the cation-indicator complex by forming more stable complexes with the cations. When all of the Ca and Mg is complexed with EDTA, (at the end point), the solution turns from wine-red to blue due to the free Eriochrome Black T indicator. In order to obtain a sharp end point, a small amount of magnesium ions must be present. This is generally not a problem with natural water samples, which tend to contain some Mg, but it is a problem when standardizing EDTA solutions with pure $CaCO_3$. A small quantity of $MgCl_2$ is added to the standard EDTA solution to ensure the presence of Mg ions.

The technical description of method for hardness determination is shown in (Radojevic & Bashkin, 1999, 175–180).

4. POLLUTION LOADING AND CONSEQUENCES

The threats to freshwater (surface and ground waters) pollution in the Asian region come from such pollutants as organic matter, nutrients, heavy metals and toxic chemicals, sediments or suspended solids, silts, and salts. In many cases the pathogens are also of great environmental concern (Figure 7).

The type and extent of water pollution vary depending on factors such as location, ecosystem characteristics, land-use, and degree of economic development. The relative severity of water quality problems in the Asian sub-regions is summarized in Table 6.

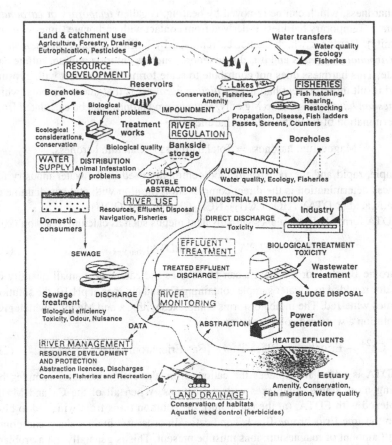

Figure 7. Sources of water pollution in the Asian region.

Table 6. Relative severity of water pollution in the Asian region (after ESCAP, 2000).

Pollutant	Asian sub-regions				
	Northeast	Southeast	South	China	India
BOD	∗∗	∗∗∗	∗∗	∗∗∗	∗∗∗
Nutrients	∗∗	∗	∗∗∗	∗∗	∗∗∗
HM		∗∗∗	∗	∗	∗
Solids		∗∗	∗∗	∗∗∗	∗∗
Pathogens		∗∗∗	∗∗	∗∗	∗∗∗

Note: Degree of severity: ∗—moderate; ∗∗—severe; ∗∗∗—very severe.

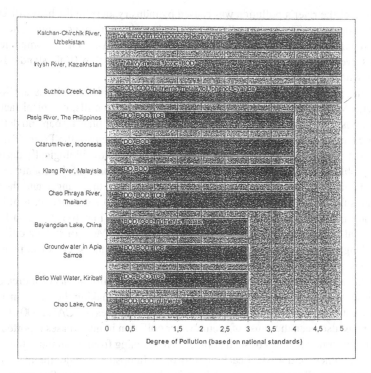

Figure 8. Selected cases of water pollution in the Asian rivers (ADB, 1999; ESCAP, 2000).

Among Asian rivers, the Yellow River (China), Ganges (India), Amu and Syr Darya (Central Asia) top the list of the World's most polluted rivers according to a report of the World Commission on Water (The Independent, 1999). Selected cases of water pollution for different rivers are highlighted in Figure 8.

4.1. Organic Matter and BOD

Organic matter is the most common pollutant discharged into surface waters. The primary source is domestic sewage, but industrial effluent, such as that from tanneries, paper mills and textile factories, are also significant sources. The oxygen required for breaking down or decomposing organic matter is taken from the surrounding water, thus diminishing its total dissolved oxygen content. The amount of microbial decomposition can be measured as biochemical oxygen demand, BOD. Another common variable used to estimate amounts of organic matter present in the water is the chemical oxygen demand, COD, which indicates through chemical reactions the oxygen required to oxidize organic compounds (see Section 3).

In the Asian region, the waters in China, India, and Southeast Asian countries are most severely polluted by organic matter from sewage and processing industries, such as pulp and paper and food. A detailed study of the Sarabaya River in Indonesia found that 80% of the river's pollution was caused by industry, despite the heavy discharge of domestic waste. The major polluters were pulp and paper, monosodium glutamate, dyes, sugar, tiles, coconut oil and metal fabrication industries. Of 28 companies and firms surveyed in detail, only four complied with applicable BOD standards, and eleven with COD standards.

In the Philippines, about 69% of the total 15,000 industrial firms are located in Metro Manila. These include industries such as textile mills, chemical factories, paper mills, distilleries, food manufacturing plants, plastic plants, soap and detergent factories and tanneries. As a result, four major river systems—the Pasig, Tullahan-Tenejeros, San Juan and Paranaque-Zapote Rivers—are now biologically dead. The same problems exist in various riverine systems in other parts of the country.

In India, 114 towns and cities dump raw sewage into the Ganges and the rivers of peninsular India. The Vrishabavathi River near Bangalore is now a mass of human and industrial waste and sewage, which eventually flows into Byramangala Lake, a traditional feeding ground for thousands of waterfowls. Most of India's cities situated alongside rivers have similar problems where paper mills, chemical plants and tanneries dump untreated waste into rivers.

In China, the Huaihe River is facing a disaster as 7 million tons of untreated domestic and industrial waste are dumped every day into it making the water unsuitable for domestic consumption and even for agricultural purposes (ESCAP, 2000). Organic matter has also been the cause of groundwater pollution in many cases reported in the region. Important sources of such pollution are leaching from unsanitary dumping of refuse and other solid waste. Leaching from open dumpsites is major threat to groundwater in almost all the major cities of the Asian developing countries. Sewage pollution in groundwater has been observed in Jakarta and Manila, which have 900,000 and 600,000 septic tanks, respectively and also in several cities of Pakistan.

4.2. Nutrients

Small doses of nutrients are essential to the metabolism and growth of all aquatic organisms. However, man-made sources of nutrients, such as organic matter in municipal wastewater and runoff from fields fertilized with chemicals and manure, can upset the natural balance of aquatic ecosystems. The enrichment of surface waters with nutrients, especially phosphorus and nitrogen, often leads to enhanced plant growth (algal blooms) and depleted oxygen levels as the plant material decays. This phenomenon is called *eutrophication*. This process is natural in its characteristics, but anthropogenic activities enhanced greatly its development in all Asian countries.

The first signs of eutrophication came to light only about 30 years ago. Today 54% of lakes and reservoirs in the Asian region are eutrophic, and the figure is rising (ESCAP, 2000). In China, the growth of algae stemming from human, agricultural and industrial wastes has reduced oxygen levels in rivers to the point that recent sampling near urban areas indicated that only 5 of out 15 rivers reached would support fish.

Fertilizer use in the region has increased substantially in recent years, especially in developing countries where intensive irrigation allows for double or triple cropping.

Use of nitrogen fertilizers increased two or fourfold in many countries during the last three decades. As a result, there has been a marked increase in surface and groundwater pollution caused by direct leaching of fertilizer residues from agricultural lands. For instance, China, Indonesia, India, Malaysia, South Korean, the Philippines and Thailand have experienced increasing contamination. Pollution caused by the leaching of nitrogen fertilizers has also been detected in the groundwater in many areas of India. In Harayana, for instance, some well water is reported to have nitrate concentrations ranging from $114\,mg\,L^{-1}$ to $1,800\,mg\,L^{-1}$, far above the national standard of $45\,mg\,L^{-1}$ (World Resources Institute, 1994). In Sri Lanka, latrines contributed to the increasing nitrate pollution in shallow groundwater (Hiscock, 1997). Nitrate pollution also exits in groundwater in Japan, Laos, Malaysia, South Korea and Indonesia.

Pollution of freshwater lakes by nitrogen and phosphorus is also common in China. Lake Tai-hu has nutrient loading with agrochemical discharge estimated at more than 40,000 tons of nitrogen and 3,000 tons of phosphorus annually via smaller rivers and streams (see more details in Chapter 13).

In Asia, surface water eutrophication also is greatly enhanced by phosphorus from detergents and this source of pollution is still of importance in many countries.

4.3. Heavy Metals and Toxic Chemicals

The growing concentrations of heavy metals and toxic chemicals such as arsenic, cadmium, mercury and lead exceeding basic water quality standards in the Asian region is disturbing. Chemical substances, such as DDT, PCB, and industrial solvents, which originate primary in industries such as coal mining, petrol refining and in textile, wood pulp, and pesticide factories, are also being found increasingly in the rivers.

Heavy metals including lead, cadmium, chromium, copper, nickel and zinc are released in water due to both direct and indirect causes of human activities, such as processing of ores and metals, the industrial use of metal compounds (chromium in tanneries and zinc in petrol, for example), leaching from domestic and industrial waste dumps and mining operations, and the use of lead pipes. Industrial and mining activities are major contributors to the increasing trend of heavy metal contamination of water bodies. Moreover, there is a clear risk of increased heavy metal pollution as the region's industrial sector expands to include modern, highly polluting manufacturing industries, such as synthetic chemicals, electronics and electroplating. Already the median level of dissolved mercury from a range of sampling places in Asian rivers ($0.2\,\mu g\,L^{-1}$) far exceeds the recommended standard of $0.001\,\mu g\,L^{-1}$ (ESCAP, 1995). More specifically, in Indonesia, high concentrations of chromium, cadmium, mercury and selenium have been detected along the river adjacent to the Tangerang Industrial Zone (IIED, 1994).

In China, the daily discharge of wastewater and solid wastes by large and medium enterprises has reached 80 million tons and 1 million tons, respectively. In Beijing alone, there are 5,700 industrial enterprises, many of which release toxic wastes. Altogether, they produce about 1 million m^3 per day of contaminated wastewater,

contributing approximately 45% of the volume and 60% of the total water pollution
load in the municipality. Furthermore, the Songhua and Liaohe River Basins, the
Yangtze River within major cities, branches of the Pearl River, the Huaihe River systems and the Haihe River Basin are all seriously polluted. Among the main pollutants
identified are arsenic and mercury (ESCAP, 1995).

The water bodies in Southeast Asia are most polluted in terms of heavy metals
and toxic chemicals. Sixteen rivers in Johor, Malaysia, including Sungai Skudai,
were found to be loaded with mercury exceeding local water quality standards. Lead
was also relatively high in 34 rivers, particularly in the State of Sabah, with Sungai
Brantian registering the highest concentration. In certain rivers, the mercury, lead,
cadmium, zinc and copper levels exceed the nationally set standards. Heavy metal
pollution in Malaysian rivers is caused mainly by industrial discharge and mining.
The Damit/Tuaran River in Sabah is particularly polluted, exceeding standards for all
heavy metals (Encyclopedia of Malaysia, 1997).

A survey in Thailand revealed lead contamination in several rivers such as Pattani
and Colok in the south, the Moon river in the northeast, the Pa Sak river in the north,
and the Mae Klong river in the central region. Mercury contaminations are reported
in the lower central region's Pranburi River, in the Mae Long, Chao Phraya, the
Petchburi's rivers of the central region, as well as the Wang River in the northern
region. High levels of arsenic poisoning were also reported in Tambon Ron Phibun,
in the Ron Phibun District, which is located close to tin and wolfram mines. Arsenic
in groundwater was also found in several villages in southern Thailand situated near
the mines. A similar, but much more severe arsenic contamination was recorded in
the Mae Moh valley due to lignite mining and acid tail drainage.

In Bangladesh in 1998, a case of chronic poisoning from an arsenic-contaminated
water supply gained worldwide attention (see Box 2).

Box 2. Millions in Bangladesh face slow arsenic poisoning (after Mantell, 1998)

Millions of people in rural areas of Bangladesh are being slowly poisoned as they drink
water contaminated with small but potentially fatal quantities of arsenic. Estimates
by World Bank and other experts claim that from 18 to 50 million people out of
a total population of about 120 million in the country are at risk. Thousands are
already showing symptoms of poisoning. Nineteen rural districts covering an area of
500 km^2 near the border of Bangladesh and India have arsenic-contaminated wells.
Many villages adjacent to the capital city, Dhaka, are also affected. In the neighboring
Indian state of West Bengal, an estimated 6 million Indians are drinking contaminated
water and 300,000 are showing signs of poisoning. Many victims are children (45%)
who have been consuming the contaminated water since birth. The contaminated
water comes from underground tube wells introduced widely over the last 20 years
as a cheap alternative water supply to prevent outbreaks of diseases such as diarrhea
and cholera. Tube wells are steel cylinders sunk into the ground to varying depths to
provide underground water for irrigation and drinking.

The United Nations International Children's Emergency Fund (INICEF) initiated well drilling as a means of providing clean water in rural areas in Bangladesh. When the program began, no water or soil tests were carried out. It was estimated that there are now 5 million tube wells, providing 96% of all water to over 50 million people. Testing is meant to be carried out on new installations but mainly takes place at government installed wells. Of the 20,000 tube wells tested so far, 25% have dangerous levels of arsenic, 40% have unsafe levels and only 35% were safer or below $0.01 \, \text{mg L}^{-1}$ of arsenic. The World Health Organization recommends a level of $0.01 \, \text{mg L}^{-1}$ of arsenic but the governments of Bangladesh and India regard $0.05 \, \text{mg L}^{-1}$, a level five times higher than the WHO standard, as acceptable.

Various theories for the contamination have been advanced. According to one theory, overuse of the water supply has increased oxygen levels in underground waterways, resulting in higher rates of leaching of minerals containing arsenic. Other scientists say that biological oxidation processes are involved. In 1997, the Bangladesh Center for Advanced Studies hypothesized that only the upper 150 meters of grounds contained high levels of arsenic.

We can add here that these high arsenic-containing geological rocks occupy the vast area between the Indostan and Indo-China peninsulas, and similar problems are shown in Thailand and Malaysia. This can be called *arsenic biogeochemical sub-region of the biosphere* in the context of a biogeochemical mapping (Bashkin and Howarth, 2002).

Other toxic species are also of environmental concern. A study in some japanese cities showed that even there, 30% of all groundwater supplies are contaminated by chlorinated solvents from industry. In some cases, the solvents from spills traveled as far as 10 km from the source of pollution (UNEP, 1996). Along with nutrients, various herbicides, pesticides and other chemicals in agricultural runoff are also increasingly polluting water bodies in the Asian region. For example, in Central Asia, herbicides, pesticides and defoliants used in agriculture are contaminating the Amu Darya and Syr Darya Rivers and causing health hazards (Galiulin and Bashkin, 1996). In Tashauz Velayat and other districts of Turkmenistan, chemicals are accumulated in storage lakes and water reservoirs. For example, water of the Sarykamysh Lake is contaminated by pesticides and other chemicals brought by drainage water (ESCAP, 2000). A special study conducted by the World Bank in Central Asia revealed that cancer diseases are the highest in the developing countries. The corresponding data are shown in Tables 7–9.

4.4. Solids and Irrigation

Many rivers, lakes and reservoirs in the Asian region are suffering from *siltation*, which is related to the accumulation and transport of suspended solids in the river waters. This is caused by accelerated soil erosion through the overuse or misuse of arable, grazing and forest lands within their catchment areas. With the increasing

Table 7. Alteration of chemical composition of the Syr Darya river water used for drinking by aborigines.

Chemical species	Content in river water, ppm	
	1956	1990
Nitrate	1.1	66.0
Nitrite	0.0	5.6
Chloride	40.0	480.0
Sulfate	141.0	900.0

Table 8. Concentration of chlourorganic pesticides in the Syr Darya river water used for drinking by aborigines, μg/L.

Sampling point	DDT	DDE	HCH
Middle flow	0.026–0.095	0.001–0.012	0.068–0.14
Kzyl-Orda city	0.04–0.09	0.009–0.06	0.065–0.14
Kazalinsk city	0.03–0.11	0.00–0.012	0.05–0.09
Delta, Aral Sea	0.068	0.116	0.140

Table 9. Dynamics of cancer diseases in the Kzyl-Orda administrative district, Kazakhstan (Syr Darya drainage area).

Malignant tumors	1955	1960	1965	1970	1975	1980	1985	1995
	Cancer case per 100,000 individuals							
All tumors	80.1	91.7	186.8	117.6	137.7	140.0	130.2	213.8
Stomach cancer	—	12.4	32.6	23.6	19.7	12.1	19.1	32.7
Esophageal cancer	—	28.1	86.1	51.8	56.3	52.9	49.6	69.5
Lung cancer	—	4.1	9.5	8.1	9.1	7.9	13.0	21.4

deforestation and land degradation, soil erosion is exacerbating the natural processes of sedimentation and siltation of water bodies in China, India and Pakistan. In the Ganges, Brahmaputra, and Yellow River basins, erosion is responsible for an annual yield of over 1,000 tons of sediment per km^2 of land.

In fact, the concentration of suspended solids in rivers and lake waters is often closely correlated with the percentage of land in the catchment that is devoted to agriculture. The suspended load per km^2 in Asian rivers is 3 to 8 times the world average. The China, South and Southeast Asian sub-regions are the worst cases. In the catchment areas of many rivers, the erosion rates are 20–40 tons per hectare (ESCAP, 1995; 2000). Poor catchment management has often led to high levels of sedimentation in reservoirs and water bodies, reducing the efficiency of irrigation systems and shortening the lifespan of many dams and reservoirs. In the first 10 years (1975–1985) after construction of Tarbela Dam on the Indus in Pakistan, the biggest earth fill dam in the World, the reservoir lost a full 14% of its storage capacity. Improvement measures have to date proved unsuccessful and the siltation of the reservoir continues. It is claimed the Indus goes through such difficult country, and that it carries such a high sediment load, that little can be done to reduce the load by improved watershed management.

A recent survey of lakes and reservoirs showed that 14% of those in the Asian countries have serious siltation problems. In Cambodia, heavy siltation of Tonle Sap Lake resulting from deforestation in the upper catchment has been reported to have significantly reduced its depth and decreased its fish yield. The rapid advance of sedimentation and land reclamation works have also resulted in the shrinkage of Lake Dongting in Central China, which receives flood water from the Chiang-Jiang (Yangtze River) every summer. Some 50 years ago it stood at $6200\,km^2$, whereas now it only covers $2740\,km^2$. The lake is currently accumulating new sediments at a rate of 5–6 cm per year, as a result of extensive cultivation in the steep slopes of the upper catchment. If this continues, the lake, with an average depth of 6.7 m, will be completely silted in some 100 years.

Irrigation and salinity of natural water

Salt content in surface and underground water is also reported to be increasing in some countries of the region. One cause is that salts dissolve in irrigation water as it percolates through the soil; another is that excessive abstraction, e.g., for irrigation, as in some places of Bangladesh, has led to saline intrusion from adjacent seas as piezometric heads in freshwater aquifers fall below sea levels. Altogether twenty-five locations in Bangladesh have been identified as vulnerable to saltwater intrusion. Also, in Thailand's Chao Phraya Basin, dry-season river flows are carefully managed to maintain the saline/freshwater balance in the delta.

Increasing salinity in the Asian region is occurring not only as a consequence of excessive irrigation but also due to rapid deforestation and in some places mining, that mobilizes salts naturally present in groundwater and rocks and contributes to their concentrations in the surface and drainage water. Moreover, the effluents from power stations and industrial cooling systems, paper mills and other industrial processes are also sources of salt accumulation in the receiving water bodies. Other causes of increasing salinity are rising water tables due to excessive seepage of water from canal systems, and capillary rise and evaporation of saline groundwater, a situation commonly found in basins and abandoned channels of relatively recent river plains. Inadequate availability of water from rains or excessive irrigation without adequate drainage is also one of the important causes of salinity. This occurs when the leaching requirements of soil are not met, resulting in the net upward movement or concentration within the soil profile of salts released by the soil material or left as residues by irrigation water.

According to (ESCAP, 1995) the Asian region has more salt effected soils than the rest of the World. Cambodia, Thailand, Malaysia, Afghanistan, Bangladesh, Mongolia, Pakistan, Indonesia, China, India, Iran, Kazakhstan, Uzbekistan, and Turkmenistan are facing problems of salinity at varying degrees (see Chapter 7). Increasing salinity of ground and surface waters renders land unsuitable for agriculture and leaves water unsuitable for domestic and industrial use. This is a major issue of water pollution in many countries of Asia.

4.5. Pathogens

Pathogens come primarily from domestic sewage that is discharged directly into water sources, although they can also enter water supplies from stormwater runoff, or as a result of soil percolation from landfills or from agricultural areas where untreated wastewater is used on crops. Since water borne pathogens are difficult to detect in the laboratory, water is instead tested for more easily measured indicators, the presence of which indicate that water is contaminated with fecal matter. The most commonly used indicator is the microorganism *Echerisias coliform*, measured in count per 100 ml of water. WHO clean water standards require that 98% of water samples from any one area be completely free of coliform bacteria, a standard which is rarely met in developing countries of the region. This is primarily because sewage treatment is very limited or non-existent in many places where domestic sewage is frequently discharged untreated into surface waters. Urban rivers in particular have been observed to have a high level of fecal contamination. The levels of coliforms typically found in many of the region's rivers (500 per 100 ml) far exceed the maximum recommended for potable water supplies (10 per 100 ml), and even that which defines the surface water as polluted (100 per 100 ml). In some developing countries, river pollution from raw sewage can even reach levels thousands of times higher than the recommended safe limits for drinking and bathing. With no sewage disposal system, Bangkok alone discharges an estimated 10,000 tons of raw municipal waste into its rivers and canals each day.

Even in South Korea about 40% of the total pollution load of the rivers comes from sewage. At the end of the 1990s, sewage treatment plants treated less then 70% of municipal wastewater. In Vietnam, Hanoi city releases several thousand cubic meters of untreated sewage containing organic toxins, bacteria and parasites directly into inland waters daily, whilst Ho Chi Minh (Saigon) City discharges hundreds of thousands of cubic meters of sewage into the Mekong River.

In the Philippines, water pollution in Metro Manila is caused by increased domestic and industrial discharges. Many urban areas lack an efficient sewage system and only 12% of Metro Manila's population are served by a sewage treatment system. As a results, the Laguna de Bay is already in a hyper-eutrophic state owing to pollution loads from domestic and industrial waste (ESCAP, 1995).

Similar situations with pathogens have been registered in Sri Lanka and India, especially in the Ganges River.

5. POLICIES AND STRATEGIES FOR ABATEMENT OF WATER POLLUTION IN ASIA

We have seen from previous discussion that pollution of surface and groundwater in the Asian region depends on both national and international decisions and agreements. To a major degree this is related to the solution of transboundary problems in water resources use and water protection of such great Asian rivers as the Mekong and the Indus.

5.1. National Level

Traditionally governments' policies and strategies on water management were aimed at supply expansion in order to meet the ever increasing demands for water in agriculture, domestic and industrial sectors. However, policy frameworks are increasingly focused on an integrated approach to water resources management by placing emphasis on demand management, addressing issues related to water-use efficiency, conservation and protection, institutional arrangements, legal regulatory and economic instruments, public information, and interagency cooperation. A sustainable development concept implies the use of a dynamic, interactive, iterative and multi-sectoral approach to water resources management. The main purpose of this management is rational water use and protection from chemical and biological pollution.

Based on this concept, many countries of the Asian region have reviewed or revised their national policies on water resources development and management (ESCAP, 2000). At present, common elements include:

- Integration of water resources development and management into national socio-economic development;

- Assessment and monitoring of water resources and water chemical and biological indices;

- Protection of water and associated resources, such as forest and mountain ecosystems;

- Provision of safe drinking supply and sanitation;

- Conservation and sustainable use of water for food production and other economic activities;

- Institutional and legislative development, and public participation.

We will consider a few examples, which provide the explanation of these elements' use in different Asian countries.

The concept of management of water resources within a river basin or sub-basin context, facilitating integration of land and water related issues, has been applied in Japan. Similarly, in India and China, the national water policy asserts that water resources planning be undertaken for a hydrological unit, such as a drainage basin or sub-basin.

The problem of assessment and monitoring of water resources and water quality is aggravated because of an extremely uneven density of hydrometeorological networks. It varies from a dense network in Japan to only a few gauges in the Tibetan high mountainous regions. This makes a proper assessment of water resources and water quality a very difficult task. The development of comprehensive data and information system is of crucial importance for many Asian countries. Some examples of development of such a system are shown now in India, Thailand and Sri Lanka.

Several countries of the Asian region are undertaking steps which include large-scale and ambitious programs and action plans to rehabilitate degraded streams and depleted aquifers. These are Thailand, the Philippines, Malaysia, Bangladesh, China, India, South Korea and Singapore. The success of river rehabilitation in Singapore is a good example of the implementation of such a program.

The main problem in provision of safe water in developing Asian countries is backlog. The population growth in Asia has slowed down but with a heavy base, it continues to increase in large numbers, which means not only providing safe water to the existing population with no connections but also to incrementally increasing population requiring huge investments. The challenge is how to use limited resources in the most efficient ways. Another challenge integral to the provision is the need to invest in efficient water service delivery infrastructure with increased community involvement in operation, maintenance and resource conservation. Lack of technical assistance to build the infrastructures and service facilities for safe water supply and sanitation is also a constraint for many countries of the region (see also Chapter 10).

5.2. Regional/International Level

Although managing or resolving transboundary water resources problems and conflicts in the Asian region is a slow and cumbersome process, some headway has been made in the region, particularly in Southeast and Northeast Asia in adopting a basin-wide approach (ESCAP, 2000). This approach is beneficial to the management of a large number of transboundary water systems when the riparians agree to cooperate in the formulation and implementation of development plans. For example, the Mekong River Commission is identifying and implementing various projects from Indicative Plan for the development of land, water and related resources in the Lower Mekong. Since the establishment of the Commission as a primary transnational organization, many dialogues between riparian states have been initiated and various projects have been realized. A key aspect of the Commission's new legal arrangement is related to dam development. Currently, proposals and programs are lined up with regard to mainstream dam development.

In Northeast Asia an example of a basin-wide approach is the Tumen River Area Development Program participated in by China, North Korea, Mongolia, South Korea and Russia.

The Indus Basin water sharing accord between India and Pakistan, the acclaimed "Water Sharing Treaty" between India and Bangladesh, The Indian-Bhutan cooperation in hydropower development, and India-Nepal cooperation in harnessing transboundary rivers are examples of transboundary cooperation on water management in South Asia.

In Central Asia the agreed joint management of transboundary waters in the Aral Sea Basin by Kazakhstan, Kyrgizia, Tajikistan, Turkmenistan and Uzbekistan is also being strengthened and enhanced through agreed consultation and coordination. The multi-sectoral Aral Sea Basin Program approved in 1994 by the five Heads of State is being coordinated by the Interstate council on the Aral Sea Rehabilitation Fund.

International organizations are playing diverse roles in water resources management within the Asian region. We should note the remarkable activity of the Economic and Social Commission for Asia and the Pacific (ESCAP) of United Nations, United Nations Environmental Program (UNEP), Asian Development Bank and World Bank and many other international organizations. They promote activities related to irrigation and drainage flood control, fisheries, hydropower, water supply, sanitation, urban drainage, inland navigation and port development.

FURTHER READING

1. Bunce N., 1994. *Environmental Chemistry*. second edition, Wuerz Publishing Ltd, Winnipeg, Chapter 5.

2. ESCAP, 1995. *State of the Environment in Asia and Pacific*. United Nations, NY, Chapter 4.

3. Radojevic M. and Bashkin V., 1999. *Practical Environmental Analysis*. Royal Society of Chemistry, UK, Chapter 4.

4. Baird C., 1999. *Environmental Chemistry*. second edition, W. H. Freeman and Company, NY, Chapter 8.

5. ESCAP, 2000. *State of the Environment in Asia and Pacific*. United Nations, NY, Chapter 4.

WEBSITE OF INTEREST

1. Surface and ground water pollution in Asia: Indian Subcontinent, http://wsws.org/news/1998/dec1998/bang-d02.shtml

2. Groundwater Foundation, http://www.grounwater.org

QUESTIONS AND PROBLEMS

1. Characterize the importance of surface and groundwater pollution in the Asian countries.

2. Present a description of surface water resources in Asia. Compare the Asian water resources with the World's.

3. Discuss the formation of unsaturated and saturated zones and the development of underground water aquifers in the Asian region.

4. Discuss the sharing of water resources in various sectors and indicate the most important water users in different countries of Asia.

5. Characterize the biogeochemical conditions and factors that affect the formation of chemical composition of natural waters.

6. Describe Henry's Law and its application to solubility of gases in water. Present the Henry's Law constant for the most important gases.

7. Discuss the thermodynamic principles of solubility of gases in water and give some examples.

8. Consider the solubility of oxygen in surface water and discuss the role of this process in aquatic ecosystem life.

9. Consider thermal pollution and its role under Asian conditions. Discuss the causes of this phenomenon.

10. Describe the methods that apply to the assessment of oxygen demand in surface waters. Pay attention to the advantages and disadvantages of each method.

11. What are the principal differences between dissolution of oxygen and carbon dioxide in surface and ground waters?

12. Present an example of a calculation of surface water pH under equilibrium with atmospheric carbon dioxide.

13. Describe the speciation of carbonates in surface waters in accordance with their pH and CO_2 content.

14. Give the definition of alkalinity and present the methods for measuring this hydrochemical index.

15. Characterize water hardness and its origin. How can you determine this parameter in water and what type of hardness can be measured?

16. What are the principal sources of water pollution in the Asian countries? Consider examples from various sub-regions.

17. Discuss organic pollution in various Asian countries. Use the BOD and COD indices to characterize this type of pollution.

18. Describe the causes which enhance the development of eutrophication process in surface waters. Compare the contributions from agriculture and industry.

19. What are the main sources of heavy metals in surface and ground waters in various Asian countries? Discuss the natural and anthropogenic causes.

20. Discuss the microbiological pollution in the Asian rivers. How can you measure the level of pathogens in natural water bodies or wastewater?

21. What measures must be implemented in the Asian countries to reduce the pollution of natural waters?

22. Discuss the national and international initiatives for water pollution abatement strategy in Asia.

CHAPTER 9

POLLUTION OF MARINE WATERS

1. INTRODUCTION

The first evidence of marine water pollution on a global and local scale was shown in the mid-20th century both from waste dumping and oil spills. In recent times a number of ship accidents in various places affected seriously the quality of marine water and presented physical and chemical damage to key ocean habitats. This, in turn, weakens the ability of marine plants and animals to survive the second danger, which is related to highly efficient industrial fishing. Regulating fishing is difficult, but possible. Saving marine water from the chemical impact of civilization is one of the greatest challenges in the Asian region in the 21st century.

This chapter summarizes the major environmental sources of marine water pollution in the Asian region, marine water chemistry and its changes under an increasing pollution load. We will discuss also the consequences of seawater pollution to various coastal and marine ecosystems in Asia.

2. SOLUBLE AND SOLID DISCHARGE OF POLLUTANTS INTO THE PACIFIC AND INDIAN OCEANS

It is notable that over 75% of the pollution entering the Pacific and Indian oceans word-wide comes from human activities on land (Figure 1).

Most nutrients, sediments, pathogens, persistent toxicants and thermal pollution come from land based sources, through rivers, direct discharge or airborne emissions. Even oil pollution, which is typically associated with tanker operations and accidents at sea, actually comes as much from land as from the sea. Coastal waters in particular, which are less mixed than areas farther from the shore, are under increasing pressure from environmental pollution.

The river waters in the Asian region are often heavily contaminated by municipal sewage, industrial effluent and sediments, which form the most important sources of pollution. Of the estimated total of 13.5×10^9 tons per year of sediment transported by the World's rivers, Asian rivers accounted for nearly 50%, although they constitute only 17% of the World's total drainage area. Basically, this is the result of lack of sanitation, discharge of untreated industrial effluents into the rivers, and sediment

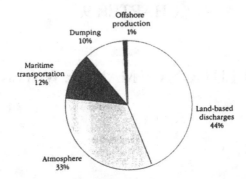

Figure 1. Sources of marine pollution in Asia (ESCAP, 1995).

loads added through catchment erosion (see Chapter 8). Other major sources of the sediment load include the tillage from mining operations and earth from construction works. Mining operations in most developing countries in the Asian region still use conventional methods, which leave large amounts of tillage to be washed away by rain water, which is finely deposited in the coastal waters (ESCAP, 1995).

Municipal and industrial waters are the most serious land-based pollutants entering the seawater through rivers or direct runoff. Most of the coastal cities in the region discharge their domestic and industrial wastes directly into the sea without any treatment. It has been conservatively estimated that over 90% of human sewage in both the Asian Pacific and Indian Ocean regions finds its way into coastal waters without prior treatment.

River waters also feed coastal zones with nutrients and chemical contaminants contributed by fertilizers and pesticides leached or washed off from agricultural land. Fertilizer consumption in the Asian region doubled during the 1990s and at present represents the biggest part of the World's fertilizer value. Use of pesticides appears to be increasing even more, especially in the developing countries. In South Asia, for example, pesticide use is considerable, with India alone using 55,000 tons a year (ESCAP, 1995). About 25% of these fertilizers and pesticides are thought to end up in the sea. This increases the level of coastal water pollution.

The marine pollution from sea-based activities in the Asian region is largely associated with marine transportation and offshore mineral exploration and production activities. The Malacca and Lombok-Makassar Straits are the main shipping routes from the Indian to Pacific Ocean and in the opposite direction. Incidence of accidental oil spills have been frequently reported along these routes. In the Strait of Malacca alone, for example, 490 shipping accidents had been reported over a five year period from 1988–1992 (Figure 2).

This has resulted in a considerable amount of oil spillage at sea. A similar accidental history is being recorded at present. Approximately 5×10^6 tons of oil enter the Arabian Sea each year from marine shipping, while the Bay of Bengal receives some 4×10^5 tons from similar sources.

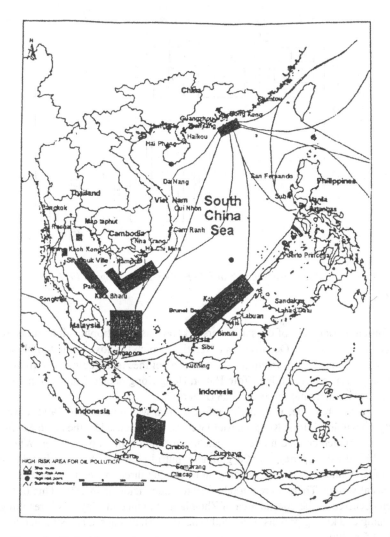

Figure 2. High risk areas for oil pollution in the South China Sea (ESCAP, 2000).

We should point out here that the Asian region has also a number of natural seeps of oil, such as off the coast of Vietnam, in the central Philippines' seas, off the northwest coast of Borneo and to the west of Sulawesi.

Some observations indicate that marine oil pollution from land-based sources still seems to be relatively high compared to that from sea-based sources. In the port of Chittagong in Bangladesh, the estimate of crude oil spillage is about 6,000 tons annually, while crude oil residue and wastewater effluent from land based refineries amount to 50,000 tons per year (Khan, 1993).

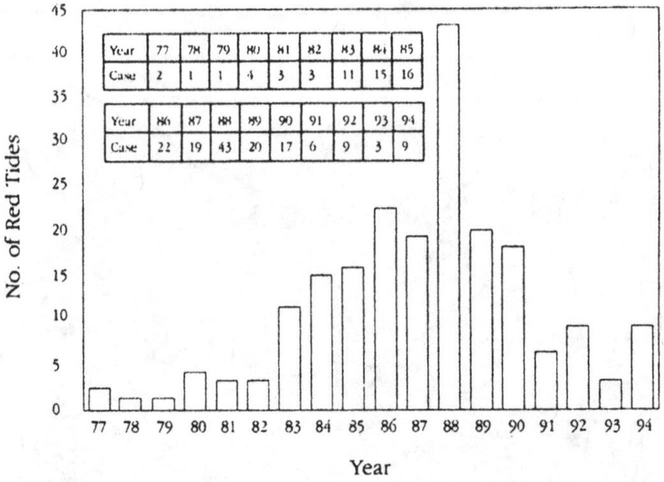
Year 77 78 79 80 81 82 83 84 85
Case 2 1 1 4 3 3 11 15 16
Year 86 87 88 89 90 91 92 93 94
Case 22 19 43 20 17 6 9 3 9
No. of Red Tides
0 5 10 15 20 25 30 35 40 45
77 78 79 80 81 82 83 84 85 86 87 88 89 90 91 92 93 94
Year

Figure 3. Number of reported red tides in Tolo Harbor, Hong Kong, in 1977–1994 (ESCAP, 1995).

Nutrient enrichment of coastal waters is known as an eutrophication and is manifested in algal blooms. Occurrences of red tide, which is a toxic plankton bloom, is a major environmental problem in coastal waters of the Asian region. The frequency of the appearance of red tide in Tolo Habor, Hong Kong ranged from two in 1977 to nine in 1994 with a maximum in 1988 (Figure 3).

The largest red tide, in terms of infested area, was recorded in 1992, covering nearly the whole of Hong Kong's eastern waters. It was caused by the algae *Noctiluca scintillans*, which created numerous patches of bright salmon pink water. When the algae drifted ashore, they created glutinous putrid slimes that fouled beaches and killed fish. China is also experiencing an increasing threat of red tide occurrence in its coastal waters. There were a total of 19 red tide incidents in 1993. One of these, which occurred in the coastal area of Zhejniang Province in May 1993, resulting in poisoning and killing prawns over a 5,000 hectare extent, caused an estimated loss of US$ 34.5 millions.

The water and bed sediment of the coastal areas of the region are also suffering from heavy metal contamination such as lead, mercury, cadmium and zinc. Major sources of this contamination include industrial effluent and dumping of land-based solid waste into the sea. In India, for instance, exceptionally high concentrations of Pb, 820 μg L^{-1} and Cd, 336 μg L^{-1}, respectively were observed in Thane creek on the Bombay coast, while Hg concentration was 778 μg L^{-1}. Sediment along the creeks and offshore stations was reported to contain significant concentrations of lead. In Pakistan, heavy metal pollution has been detected in water and sediment from the coastal area within the north of the Indus River and there is increasing evidence of these toxic substances getting into the biogeochemical food webs (ESCAP, 1995).

3. MARINE WATER CHEMISTRY

3.1. Solubility of Calcium Species in Natural Waters

Calcium carbonate has intrinsically low solubility in water, $K_{sp} = 6 \times 10^{-9}$ (mol/L)^2 at 25 °C. However, because the carbonate anion is basic, $CaCO_3$ becomes increasingly soluble as the pH drops. It is known that even unpolluted rain water has pH 5.6 and the rainwater is slightly acidic. The reaction may be summarized as follows:

$$CaCO_3(s) + H_2CO_3(aq) \longleftrightarrow Ca(HCO_3)_2(aq).$$

This reaction is responsible for formation, over thousands of years, of caves and gorges in limestone areas, as CO_2-laden rainwater very slowly dissolves the rock. The process is slow because the equilibrium constant is small. This constant can be estimated as the following:

$$CaCO_3(s) \longleftrightarrow Ca^{2+}(aq) + CO_3^{2-}(aq) \qquad K_1 = K_{sp} \text{ for } CaCO_3$$

$$H^+(aq) + CO_3^{2-}(aq) \longleftrightarrow HCO_3^-(aq) \qquad K_2 = 1/K_a \text{ for } HCO_3^-$$

$$H_2CO_3(aq) \longleftrightarrow H^+(aq) + HCO_3^-(aq) \qquad K_3 = K_a \text{ for } H_2CO_3$$

$$CaCO_3(s) + H_2CO_3(aq) \longleftrightarrow Ca(HCO_3)_2(aq) \qquad K_4 = K_1 \times K_2 \times K_3$$

where $K_4 = (6.0 \times 10^{-9})(1/4.8 \times 10^{-11})(4.2 \times 10^{-11}) = 5.3 \times 10^{-5}$ (mol/L)^2.

Owing to the dynamic equilibrium between atmospheric carbon dioxide and oceanic bicarbonate and carbonate anions, a great amount of soluble calcium cations is contained in the ocean. This mass is four orders of magnitude higher than the total mass of bound calcium in living and dead matter of both terrestrial and aquatic organisms. The average calcium content in the seawater is 408 mg L^{-1}, and the overall pool is 559×10^{12} tons.

The actual calcium concentration in marine waters is about 30 times higher than in riverine waters. This is owing to limited solubility of calcium carbonate and, especially, related to the extensive calcium uptake by planktonic organisms followed by calcium deposition as pellets. These processes have facilitated the vast accumulation of calcium as a component of massive layers of limestone, dolomite, marl, calcareous clay, and other Ca-containing rocks (see Box 1).

Box 1. Why calcium carbonate does not spontaneously precipitate? (After Bunce, 1994)

An interesting paradox about seawater is that calcium carbonate does not spontaneously precipitate, but neither do sea shells on the beach dissolve. This suggests that the oceans are not far from equilibrium with respect to the system $CaCO_3(s)$/$Ca^{2+}(aq)$/$CO_3^{2-}(aq)$. However, this precipitation is not confirmed by calculations unless considerable care is taken.

At 15 °C, K_{sp} for $CaCO_3$ is 6.0×10^{-9} (mol/L)^2. From the ionic concentration (Table 1) we can write:

Table 1. Ions in seawater.

Ion	Concentration, mol L^{-1}	Input from rivers, 10^{10} mol yr^{-1}	Residence time, 10^7 yr
Ca^{2+}	0.010	1220	0.1
Mg^{2+}	0.054	550	1
Na$^+$	0.46	900	7
K$^+$	0.010	190	0.7
Cl$^-$	0.55	720	10
SO$_4{}^{2-}$	0.028	380	1
HCO$_3{}^-$	0.0023	3200	0.01

$$\begin{aligned} Q_{sp} &= c\left(Ca^{2+}, aq\right) \times c\left(CO_3{}^{2-}, aq\right) \\ &= (0.010) \times \left(2.7 \times 10^{-4}\right) \\ &= 2.7 \times 10^{-6} \, (mol/L)^2. \end{aligned}$$

Hence $Q_{sp} \gg K_{sp}$, and according to this calculation, the oceans are vastly supersaturated with respect to calcium carbonate. We would therefore expect CaCO$_3$ to precipitate spontaneously.

It turns out that there are two factors which have been overlooked in this simple calculation, namely *ionic strength* and *ion complexation*.

We will consider these processes.

Ionic strength. According to chemical laws, equilibrium constants have to be written in terms of activities rather than concentrations. Activity is the effective concentration, by definition. For low concentrations, we can usually approximate by writing concentration where we really mean activity, since

Activity = activity coefficient × concentration.

The approximation is equivalent to saying that the activity coefficient is unity. The approximation fails in a solution of high ionic strength such as seawater. Activity coefficients appropriate for seawater are 0.26 for Ca^{2+} and 0.20 for CO$_3{}^{2-}$. Now the reaction quotient looks like this:

$$\begin{aligned} Q_{sp} &= a\left(Ca^{2+}, aq\right) \times a\left(CO_3{}^{2-}, aq\right) \\ &= (0.010 \times 0.27) \times \left(2.7 \times 10^{-4} \times 0.20\right) \\ &= 1.4 \times 10^{-7} \quad \text{(no units, activities are dimensionless).} \end{aligned}$$

This represents an improvement of nearly an order of magnitude, but we would still predict that the solution is supersaturated with respect to CaCO$_3$.

Table 2. Complexation of species in seawater.

Species	K_{assoc}, L mol^{-1}	% of total
Calcium, total concentration: 0.010 mol/L		
Ca^{2+} (aq)	—	91
$(CaSO_4)°$	2×10^2	8
$(CaHCO_3)^+$	2×10^1	1
Hydrocarbonate, total concentration: 0.0023 mol L^{-1}		
HCO_3^- (aq)	—	75
$CaHCO_3^+$	2×10^1	3
$MgHCO_3^+$	1×10^1	12
$(NaHCO_3)°$	2	10
Carbonate, total concentration: 0.00030 mol L^{-1}		
CO_3^{2-} (aq)	—	10
$(MgCO_3)°$	3×10^3	64
$(CaCO_3)°$	3×10^3	7
$(NaCO_3)°$	3×10^1	19

Complexation. The ionic strength/activity coefficient effect arises because at high ionic strength a particular ion (say, Ca^{2+}) is not completely free in solution; it will be surrounded by ions of opposite charge. This is a generalized effect in that the identities of these counter ions are not important. Besides this general effect, the ion in question can associate with specific counter ions to form recognizable chemical species whose concentrations can be measured. For Ca^{2+}, these include $(CaSO_4)°$ and $(CaHCO_3)^+$. These tight ion pairs are observable chemical species. Note that $(CaSO_4)°$ for example is an aqueous entity, different from $CaSO_4(s)$. As an aside, the existence of these complexes is the reason that simple K_{sp} calculations frequently underestimate the true solubility of ionic complexes. Calculations have shown that in seawater, Ca^{2+} and CO_3^{2-} are speciated as shown in Table 2.

We can now put the finishing touches to our Q_{sp} calculations

$$Q_{sp} = a(Ca^{2+}, aq) \times a(CO_3^{2-}, aq)$$

$$= (0.010 \times 0.27 \times 0.91) \times (2.7 \times 10^{-4} \times 0.20 \times 0.10)$$

$$= 1.3 \times 10^{-8} \quad \text{(no units, activities are dimensionless).}$$

We are now within a factor of two to the value of K_{sp}, and have shown $CaCO_3$ in seawater to be close to saturation, as found experimentally.

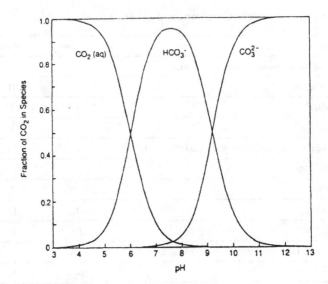

Figure 4. Distribution of dissolved carbon species in seawater as a function of pH at 15°C and a salinity of 35‰. Average oceanic pH is about 8.2 (Butcher et al, 1992).

For the global calcium fluxes, the biological cycle and the aqueous migration of ions with the land-ocean system are of primary importance. About $1.5–3.1 \times 10^9$ tons of Ca per year is involved in the biological cycle in terrestrial ecosystems. In the oceanic ecosystems, this value is about 1.1×10^9 tons/yr.

3.2. Carbon Dioxide Interactions in Air-Sea Water System

The interaction between carbon dioxide in the atmosphere and the hydrosphere is the principal for understanding the large carbon biogeochemical cycle. As mentioned above, the gases of the troposphere and the surface layer of ocean persist in a state of kinetic equilibrium.

In a contrast to the atmosphere, where most carbon is presented by CO_2, oceanic carbon is mainly present in four forms: dissolved inorganic carbon (DIC), dissolved organic carbon (DOC), particulate organic carbon (POC), and the marine biota itself.

DIC concentrations have been monitored extensively since the appearance of precise analytical techniques. When CO_2 dissolves in water, it may hydrate to form $H_2CO_3(aq)$, which, in turn, dissociates to HCO_3^- and CO_3^{2-}. This process depends on pH and specification is shown in Figure 4.

The conjugate pairs responsible for most of the pH buffer capacity in marine water are HCO_3^-/CO_3^{2-} and $B(OH)_3/B(OH)_4^-$. Although the predominance of HCO_3^- at the oceanic pH of 8.2 actually places the carbonate system close to a pH buffer minimum, its importance is maintained by the high DIC concentration ($\sim 2\,mM$). Ocean water in contact with the atmosphere will, if the air-sea gas exchange rate is

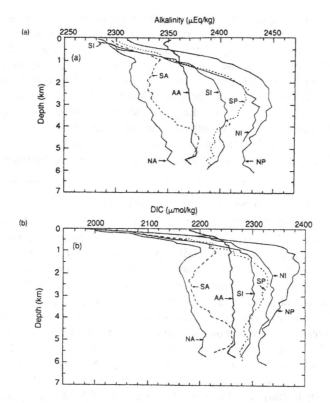

Figure 5. The vertical distribution of alkalinity (a) and dissolved inorganic carbon in the World's Oceans. Ocean regions are shown as NA (North Atlantic), SA (South Atlantic), AA (Antarctic), SI (South Indian), NI (North Indian), SP (South Pacific) and NP (North Pacific) (Butcher et al, 1992).

short compared to the mixing time with deeper water, reach the equilibrium according to Henry's Law. At the pH of oceanic water around 8.2, most of the DIC is in the form of HCO_3^- and CO_3^{2-} with a very small proportion of H_2CO_3. Although H_2CO_3 changes in proportion to $CO_2(g)$, the ionic form changes little as a result of various acid-base equilibria.

From chemical aqueous carbon specification, the alkalinity, representing the acid-neutralizing capacity of the solution, is given by the following equation:

$$Alk = [OH^-] - [OH^+] + [B(OH)_4^-] + [B(OH)_3] + 2[CO_3^{2-}].$$

Average DIC and Alk concentrations for the World's Ocean are shown in Figure 5

With an average DIC of 2.35 mmol/kg sea water and the world oceanic volume of 1370×10^6 km^3, the DIC carbon reservoir is estimated to be 37900×10^9 tons C. The surface waters of the World's ocean contain a minor part of DIC, $\sim 700 \times 10^9$ tons C. However, these waters play an important role in air-deep water exchange (see above).

Oceanic surface water is everywhere supersaturated with respect to the two solid calcium carbonate species calcite and aragonite. Nevertheless, calcium precipitation is exclusively controlled by biological processes, specifically the formation of hard parts (shells, skeletal parts, etc.). The very few existing amounts of spontaneous inorganic precipitation of $CaCO_3(s)$ come from the Bahamas region of the Caribbean.

The detrital rain of carbon-containing particles can be divided into two groups: the hard parts comprised of calcite and aragonite and the soft tissue containing organic carbon. The composition of the soft tissue shows the average ratio of biophils as $P : N : C : Ca : S = 115 : 131 : 26 : 50$, with $C_c : C_o$ ratio as $1 : 4$.

More details of carbon transformation in bottom sediments are presented in Box 2.

Box 2. Heterogenity of carbon cycle in bottom sediments (after T.Fenchel et al, 1998)

The carbon cycle, in both lake and sea sediments, is predominantly heterotrophic, but there are some aspects of autotrophy. The DOM produced from hydrolysis of fine particulate organic matter (FPOM) is processed by a number of oxidative and fermentative processes in aqueous sediments. The oxidants are O_2, NO_3^-, NH_4^+, Fe^{3+}, SO_4^{2-}, and CO_2 in sequence from the top of the sediment downward. The DOC component of DOM has limited possibilities: it can be oxidized by one of the listed oxidants or it can leave the sediment unoxidized. This is an obvious conclusion, but it has some interesting connotations, for example, the proportion of C oxidized by O_2, etc. and the determination of the factors influencing this proportion. Clearly, the quantity of DOC will determine the depth of O_2 penetration and the extent to which O_2 can participate in C oxidation. An equally important factor is the depth at which DOC is produced by hydrolysis of POM: the deeper the site of POM hydrolysis, the more likely will be the anoxic processing of the soluble products. Another very important factor is the degree to which HS^- is free to diffuse in marine sediments. If HS^- can diffuse to the sediment surface and react with O_2, the depth of O_2 penetration will be greatly reduced.

Simulation modeling of the fate of DOC under different conditions emphasizes these simple relationships in marine sediments. Figure 6 shows the rates, at which DOC is stipulated to be produced by hydrolysis of POM at an proximately 6, 36, and 60 mmol/m^2/day either at the surface of the sediment (TOP), or as a linear gradient from the surface down (LINEAR), or equally to all sediment layers down to 5 cm (MIX). The matrix in Figure 6 represents sediments receiving increasing quantities of POM (arrow down) at increasing frequencies (arrow across).

Both processes, with limits, will increase benthic faunal populations and bio-turbation, but sediments receiving excessive organic loading will tend to become anoxic and devoid of macrofauna. This matrix represents the variety of sediment types that might be found in various parts of the World and under various water depth and productivity conditions. These hypothetical types are used as a proxy for actual sediment variety, but if the World's sediments could be categorized in terms of quantity, quality and frequency of POM input, simulation could predict all the important biogeochemical rates and nutrient profiles. More continuous productivity

Frequency of POM arrival

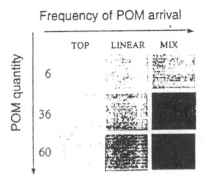

Figure 6. A representation of the addition of particulate organic matter (POM) to sediments. The amount of POM added increases from top to bottom. The frequency of POM addition increases from left to right. The amount of POM hydrolysis per day is 6, 36 or 60 mmol/m²/day. Infrequent addition of POM will not support macrofaunal development and will result in POM hydrolysis at the sediment surface (TOP distribution), whereas more frequent additions will result in increasing macrofaunal populations (LINEAR and MIX distributions) and greater sediment mixing. Dark shading represents the degree of bioturbation (Fenchel et al, 1998).

(longer seasons) would result in a more continuous input than that found in ice covered seas. The TOP distribution would represent the sediment subject to very little disturbance either by water movement or by bioturbation. At the lower DOC addition rates, this would be equivalent to the shelf sediment with no macrofauna. The absence of macrofauna could be attributed to a single pulse of POM per year, as might occur in the Arctic Ocean. The highest DOC inputs to the sediment surface may be unrealistic, as macrofauna might be expected to arise in response to this amount of organic input, unless the input occurs as a single pulse. The LINEAR distribution might be found in sediments receiving multiple POM pulses, supporting a moderately active macrofaunal population, which transports a small amount of the fresh POM to 5 cm depth. The MIX distribution is equivalent to a very active macrofaunal population or very efficient mixing by waves or tides.

The estimation of C_c and C_o mass annually eliminated from the biogeochemical cycles in oceans is a very uncertain task. The carbonate-hydrocarbonate system includes the precipitation of calcium carbonate as a deposit:

Atmosphere $\quad CO_2$

$\uparrow\downarrow$

Surface ocean layer $\quad H_2O \leftrightarrow H_2CO_3 \leftrightarrow H^+ + HCO_3^- \leftrightarrow H^+ + CO_3^{2-} + Ca^{2+}$

$\uparrow\downarrow$

Deep ocean water $\qquad\qquad\qquad\qquad\qquad\qquad\qquad\qquad\qquad\qquad CaCO_3$

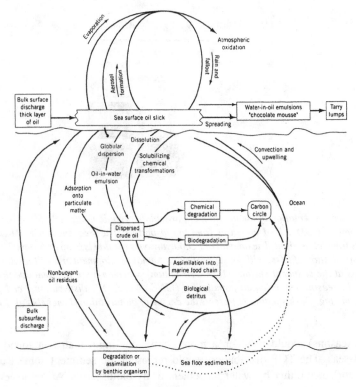

Figure 7. Rate and weathering of polluting oil in estuarine and marine waters showing various abiotic and biotic processes (Kennish, 1999).

The binding of carbon into carbonates is related to the activity of living organisms. However, the surface runoff of Ca^{2+} ions from the land determines to a significant degree the formation of carbonate deposits. The Ca^{2+} ion stream is roughly 0.53×10^9 tons/year, which can provide for a $CaCO_3$ precipitation rate of 1.33×10^9 tons per year. This would correspond to the loss of 0.57×10^9 tons CO_2, or 0.16×10^9 tons C from the carbonate-hydrocarbonate system.

3.3. Biogeochemical Transformation of Oil Compounds in Marine Water

Various biogeochemical and physical-chemical processes act within hours of an oil spill on the sea surface to alter its composition and toxicity, most importantly evaporation, dissolution, photochemical oxidation, advection and dispersion, emulsification, and sedimentation (Figure 7).

After a spill, oil spreads across the sea surface as a slick, varying in thickness from micrometers to a centimeter or more. A slick's behavior is a function of the composition of the oil and the prevailing abiotic factors in the area, water temperature, wind, wave action, tides, and currents (Figure 8).

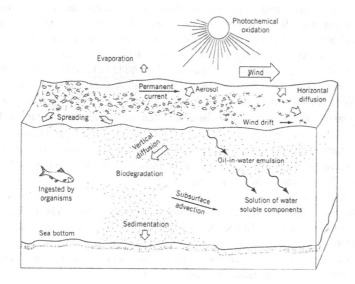

Figure 8. Effects of wind and other factors on the movement of polluting oil at sea (Kennish, 1999).

Currents transport, or *advect*, the oil from a spill site. Oil slicks travel downwind at 2.5–4.0% of the wind velocity (Kennish, 1999). Light oils spread faster than heavy oils. The spreading rate is more rapid at higher than lower temperatures and depends on the volume and density of the oil. As wind and wind waves develop, the slick breaks up into distinct patches of oil that drift slowly apart by horizontal eddy diffusion. Subsurface *advection* mixes the oil in subsurface waters to depths of about 10 m; vertical diffusion plays a less integral role in subsurface mixing of the oil.

During the first 24–48 hours of an oil spill, evaporation and dissolution produce the greatest change in composition of the oil by causing a rapid loss of the lighter, more toxic, volatile components. This process is of especial concern in tropical and subtropical marine waters of the Asian region. The low molecular weight aromatics, such as toluene and xylene, present at peak concentrations of $10–100\,\mu gL^{-1}$ are lost relatively quickly. The water soluble components gradually dissolve in seawater and are effectively isolated from the sea-surface slick and incorporated into subsurface waters. Dissolved constituents attain highest concentrations below the oil slick within the first 24 hours of a spill.

Photochemical oxidation acts on oil at or near sea surface to convert high molecular weight aromatic hydrocarbons to polar oxidized compounds. Some of the oxidation products formed by this process include alcohols, ethers, dialkyl peroxides, and aliphatic acids. The photochemical oxidation process, which operates most efficiently on thin oil slicks ($\approx 10^{-8}$ cm thick), is more important after the volatile oil fractions have evaporated.

Agitation of waves and currents mixes the oil and seawater leading to turbulence and dispersion of oil slicks. In very rough seas, aerosol formation ensues at sea surface. Intense agitation and mixing cause the formation of oil-in-water emulsions or water-in-oil emulsions. As the heavier oil fractions mix with seawater, viscosity increases and a water-in-oil emulsion forms. Viscid pancake-like masses, called chocolate mousse, which develop from 50–80% water-in-oil emulsions, may persist for months at sea. The heaviest residues of crude oil-tar balls, measuring 1 mm to 25 cm in diameter, are often advected, like chocolate mousses, to remote impact sites (Kennish, 1999). These products degrade very slowly, being extremely stable and persistent in marine waters and, consequently, a major problem for shorelines subsequent to stranding. As we have mentioned already, vast areas of the Indian shores both in the Arabian Sea and Bay of Bengal are polluted by these tar balls.

Evaporation, dissolution, photooxidation, and the formation of viscous water-in-oil emulsions increase the density of oil on the sea surface. The change in oil density through time is facilitated by the incorporation of particulate matter and the agglomeration of oil particulate mixtures. The sorption of hydrocarbons to particulate matter creates a high specific gravity mixture more than twice that of seawater alone (1.025 g cm^{-1}), which promotes its sedimentation.

Oil is subject to biodegradation by many species of bacteria, fungi, and yeast. Bacteria are the most important biogeochemical agents in the breakdown process. Microbes collectively degrade as much as 40–80% of a crude oil spill with structurally simple hydrocarbon and nonhydrocarbon compounds degrading most readily, especially at higher temperatures. Alkanes and cycloalkanes, for example, break down rapidly, whereas the high molecular weight species, for instance polycyclic aromatic hydrocarbons, degrade slowly. Asphaltenes, the highest molecular weight components of crude oil, persist with little alteration to form tarry residues that ultimately settle to the seafloor or become stranded along coastlines. Biodegradation rates of oil are influenced most greatly by water temperature, nutrient availability, dissolved oxygen levels and salinity. Thus, we can estimate that in tropical and subtropical coastal waters of South and Southeast Asia, the water temperature, high nutrients and relatively low salinity will stimulate the biodegradation process, however, low oxygen content (the higher temperature, the lower oxygen content, see Chapter 7) will slow this process. You can estimate the rate of the biodegradation process in open marine waters far away from the coast.

Microbial breakdown of oil is active in coastal marine zones and slows down in high latitudes because of low temperatures.

Marine water bioremediation

Four main approaches are followed to clean up oil spills from marine environments:

1. Chemical application;

2. Mechanical cleanup;

3. Shoreline cleanup; and

4. No spill control.

The most appropriate treatment hinges on conditions existing at the cleanup site. Oil treatments on sandy beaches usually differ from those on rocky shorelines or in salt marshes. For oil spills at sea, the best approach may be no treatment at all, especially when prevailing winds are offshore.

Chemical treatment of a slick usually involves spraying dispersants from ships or aircraft to accelerate the emulsification of the oil. Solvents and agents to reduce surface tension are also utilized to remove oil slicks from the surface of pools and enclosed inshore areas. However, dispersants are not effective on heavy or weathered oils, and some of them are toxic to marine life. Despite these disadvantages, chemical treatment of oil slicks has been instrumental in preserving many threatened coastal habitats.

Devices that physically contain floating oil are of great value in cleaning efforts in harbors and inshore regions. Included here are floating booms, which corral oil in a confined zone from which it can be pumped or diverted and removed by other means, and slick lickers with rotating belts of absorbent, which extract oil and pass it into barges. These devices are most effective in dealing with small oil spills in sheltered waters (Kennish, 1999).

When oil is stranded on beaches or rocky intertidal areas, cleanup crews often use physical removal. Rocky surfaces are cleaned by high pressure hoses, high pressure stream, hand scrubbing, or dispersants. These cleanup methods may cause additional harm to benthic communities. They are ineffective on sandy beaches because the oil simply drains onto the beach and endemic flora and fauna are threatened. The oiled surface layers of affected beaches commonly are stripped via bulldozers or manually.

The cleanup of oiled beaches is labor-intensive and costly. This is one of the reasons for still polluted beaches in South Asia. Hence the best strategy is to deal with an oil spill expeditiously before it becomes stranded along the coastline.

4. CONSEQUENCES OF COASTAL AND MARINE WATER POLLUTION

4.1. Damage to Coastal Zone

In terms of pollution, only 18.7% of China's coastal waters met Grade I water quality of the Pearl River Estuary. Of China's four major sea areas, the East China was the most polluted, followed by the Bohai Sea, the Yellow Sea and the South China Sea (State of the Environment, China, 1997).

As the pollution intensifies, the destabilized coastal systems undergo wild gyrations in population densities—some species dying off while others bloom in huge numbers and then die again. The least desirable creatures, small, tough, and often poisonous, bloom in polluted environments while the most important creatures die. Biodiversity plummets and the ability of the ecosystem to maintain its high value for the production of food, recreation and tourism drops with it.

4.2. Mangroves Ecosystems

Of the 16.6 million hectares of mangroves that are estimated to exist in the World, over 7.5 million hectares occur in Asia. Of these, more than 4 million hectares are found in Indonesia, which has the largest area of mangroves that exist in any single country of the World.

In addition to the traditional uses of mangroves, which, by and large, were fairly sustainable, recent population and economic pressures have led to an over-exploitation of the trees themselves as well as the conversion of the wetland they occupy.

The greatest threat to mangroves in the Asian region is the development of aquaculture. More than one million hectares of mangrove forest ecosystems have been converted to aquaculture ponds, which causes changes in drainage patterns, nutrient availability and the frequency of tidal inundation.

In the 20th century, the coastal ecosystems in the Asian region were seriously damaged by pollution or destruction. People in just four Asian countries—Malaysia, the Philippines, Thailand and Vietnam—cleared 750,000 hectares of mangroves, an estimated 10% of all remaining mangrove forests in South and South East Asia. The Philippines alone lost 70% of its mangroves (ESCAP, 2000).

4.3. Coral Reefs

Coral reefs are among the World's most diverse natural ecosystems—the marine equivalent of tropical rain forests. Globally they are about $600,000 \, km^2$ of coral reefs. Of these, more than half are in the Indian Ocean with the rest eventually distributed between the Caribbean and Pacific (ESCAP, 1995).

Coral organisms are extremely sensitive to physical and chemical conditions of marine water, with narrow tolerance ranges: water temperature between 25 °C and 28 °C, high salinity, preferably 35 parts per thousand (35‰), water clarity, high content of dissolved oxygen and an ample supply of sunlight. Any departure from these conditions results in stressful conditions for coral, hence their inability to survive near river mouths (low salinity and high turbidity), and at depths in excess of 50 m (lack of sunlight penetration). Many urban and industrial developments pose a threat to coral systems through: discharges of domestic and industrial effluents; agricultural/aquacultural runoff contaminated with fertilizers and pesticides; dredging and reclamation activities with their high suspended sediment loads; and other stresses, which arise from high concentrations of population.

ESCAP (1995), chemical pollution and increased sedimentation have been observed to cause "bleaching" in coral's symbiotic algae, which gives coral its color, causing it to abandon the coral due to environmental stress. Without these algae, the coral will eventually die. It is also expected that should sea temperature rise due to global warming, this will also exacerbate this condition.

4.4. Seagrass Meadows

Seagrass beds are composed of rooted, seed-bearing marine plants (halaphytes). They occur in shallow, near-shore waters, which are sheltered from high wave energy, and in estuaries and lagoons. They are often associated, both physically and ecologically, with mangrove forests and coral reefs, often forming the spatial link between these two ecosystems. Like mangroves and coral reefs, seagrass meadows with their halophytic plants, epiphytes and abundant detritus, comprise a highly productive habitat that supports a large quantity of commercially important species. They exhibit very high rates of primary productivity, with 3–5 kg m^{-2} of standing stock. It has been estimated that the economic return obtained from seagrass beds can be up to US$ 215,000 per hectare (ESCAP, 1995).

Thirteen species of seagrass have been reported from Indonesian waters alone and extensive beds of seagrass are also reported from southern India and from around Sri Lanka.

In Southeast Asia, seagrasses are under threat from loss of mangroves, which act as a filter for sediment from land, coastal development, urban expansion and bucket dredging for tin. Other impacts include substrate disturbance, industrial and agricultural runoff, industrial wastes and sewage discharges.

4.5. Species at Risk

Over 90% of the World's living biomass is contained in the oceans, and new marine and coastal phenomena, communities and species are constantly being discovered. Marine and coastal species include many that are rare, threatened and endangered, often as a result of habitat loss. In the Asian region, habitats are affected both directly, such as by oil spills and ocean dumping, and indirectly, through activities such as waste disposal and abusive land use practices that send sediments and pollutants into rivers, and therein to the oceans.

Of all marine habitats, those in coastal waters are under the most pressure and coastal species in particular are at risk from pollution, reclamation, dredging, dumping and other coastal alterations. Habitat loss often has an irreversible impact on a threatened species since it affects whole populations at once. The major threats to Asian wetland habitats, which include the coastal margins, are hunting, human settlements, agriculture, pollution, fishing, logging, wood fuel collection, land degradation, water diversion, aquaculture development and foreshore development.

In addition to habitat destruction, several vulnerable marine species are affected by direct exploitation in the Asian region: whales, dugongs, birds and turtles are threatened by pollution, poaching and tourism encroachment with relevant marine water pollution.

FURTHER READING

1. Bunce N., 1994. *Environmental Chemistry*. 2nd edition, Wuerz Publishing Ltd, Winnipeg, Canada, Chapter 5.

2. ESCAP, 1995. *State of the Environment in Asia and Pacific*. United Nations, Chapter 5.

3. Kennish M. J., 1999. Marine pollution. In: Meyers R. A. (Ed.), *Encyclopedia of Environmental Pollution and Cleanup*, John Wiley, NY, 909–943.

4. ESCAP, 2000. *State of the Environment in Asia and Pacific*. United Nations, Chapter 5.

WEBSITES OF INTEREST

1. Indian Ocean chemical and physical characteristics,
   http://www.indoex.ucsd.edu/whatlearned/wahtlearned.html#11

2. Climate hotspots,
   http://psbsgi.nesdis.noaa.gov/8080/PSB/EPS/SST/climohot.html

3. Monitoring,
   http://www.nova.edu/ocean/ncri/conf99.html

4. Management,
   http://www.nova.edu/ocean/ncri/conf99.html

5. Marine biodiversity,
   http://www.wri.org/wri/wr-96-97/bi_txt4.html

QUESTIONS AND PROBLEMS

1. What are the two main global threats to marine waters and how pronounced are they in Asia?

2. Discuss the sources of coastal water pollution in the Asian region; focus your attention on untreated sewage discharge.

3. Characterize the role of land-based and marine-based sources in oil pollution of seawaters in the Asian oceanic waters.

4. Describe the modern state of coastal water pollution by heavy metals in South Asia. Show the typical concentrations of different heavy metals.

5. Discuss calcium equilibrium in marine waters. Estimate the role of ion complexation in this process.

6. What does ionic activity mean and how does this phenomenon effect the behavior of various chemical species in seawater? Present an example of such a calculation.

7. Describe the chemical processes involving the dissolution of carbon dioxide in seawater.

8. Discuss biogeochemical processes of carbon transformation in bottom sediments and point out the role of meso- and micro-fauna.

9. Characterize the biogeochemical processes in marine waters after oil spill. Stress the role of abiotic and biotic factors.

10. Highlight the role of micro-organisms in biodegradation of oil slicks in seawater. Show the physical-chemical parameters that will accelerate or slow these processes in the Asian marine environments.

11. Discuss the existing approaches to remediation of seawaters from oil pollution and indicate the characteristic details of remediation for shorelines.

12. Discuss the consequences of coastal water pollution for mangrove ecosystems in the Asian region.

13. What anthropogenic factors are of great environmental concern for coral reefs in the Asian marine waters?

14. What sources of environmental pollution are the most dangerous for seagrasses ecosystems?

15. Characterize species at risk in seawaters of the Asian region and rank the anthropogenic factors that influence this risk.

CHAPTER 10

DRINKING WATERS

1. INTRODUCTION

During the last decades of the 20th century, the global water supply and sanitation coverage just managed to keep pace with population growth. The percentage of people served in the rural areas increased significantly as a result of special international programs, such as International Drinking Water Supply and Sanitation Decade, ISWSSD, in the 1980s, although less improvement was made in the urban areas. Such an assessment also holds true for the Asian region, where the proportion of rural population with access to an adequate water supply rose from 28% in 1980 to 67% in 1990 and 95–98% in 2000. Similarly, the share of rural population with adequate sanitation was increased from 42% to 65% in 2000. The above improvements in the region were to a large extent due to substantial increases in coverage reported by China. Central Asia in particular is still poorly served, for example only 23% of farmers in Uzbekistan, 20% in Kyrgyzia, 14% in Tajikistan and 12% in Turkmenistan currently have a piped water supply. The progress in sanitation was less impressive, particularly in South India, where less than 5% of the rural population in Bangladesh, Bhutan, India and Nepal had access to adequate sanitation in mid-1990s. The urban water supply and sanitation is still far away from the 100% value (Figure 1).

This means that millions of people—mostly women—spend many hours every day carrying water, often of dubious quality, from a distant well to their houses. Between 15 to 20 million babies die every year as a result of water-borne diarrheal diseases, such as typhoid fever, ameboid dysentery, and cholera; the largest part of this number is in the Asian region.

In this chapter we discuss chemical and physical methods of drinking water purification, with special attention to water disinfection methods.

2. PURIFICATION OF FRESHWATER IN ASIA

We have considered potential sources of freshwater supply in Chapter 8 for various sub-regions in Asia. Generally we can say that groundwater tends to be less contaminated than surface water because organic matter in the water has had time to be decomposed by soil bacteria. The ground itself acts as a filtering device so that less suspended matter is present.

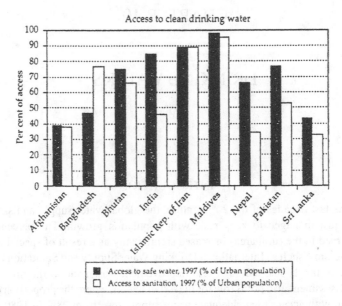

Figure 1. Access to safe water supply and sanitation (% of urban population) in the Asian region (ESCAP, 1995).

2.1. Water Supply in the Asian Cities

Efforts during the IDWSSD 1980–1990 managed to increase the number of persons with access to adequate water supply in the urban areas by 46% and sanitation by 39%. However, population growth meant that the proportion of urban population served remained unchanged. In fact, the absolute number of people unserved increased by 18% for water supply and 39% for sanitation.

Across the region, the most significant progress was made in East Asia. With the exception of China, the East Asia countries achieved 100% coverage of safe drinking water supply and sanitation by 1990 and China, 87%. Other sub-regions have also seen great improvements in urban water supply since 1980, particularly in South Asia. However, progress in sanitation has lagged behind, most notably in urban India and Bangladesh. Whilst urban areas throughout the region are now better served with water and water treatment, service (water disinfection) is still poor, especially in slums and squatter settlements.

Water supply status also varies even within different urban areas of the same country. For example, in Pakistan availability of a piped water supply system varies 35% in Faislabad to 92% in Karachi. In Uzbekistan, there is not enough water for everyone especially in the cities of Samarkand and Andijan. In Thailand, it varies from 78% in Bangkok to only 8.4% in Nakhon Si Thammarat. In China, water scarcity is particularly severe in and around Beijing. Among the megacities of the region, only Seoul and Shanghai have 100% water service coverage and, along with Bangkok, have 24 hours water supply daily (ESCAP, 2000). Most megacities have 65–83% water

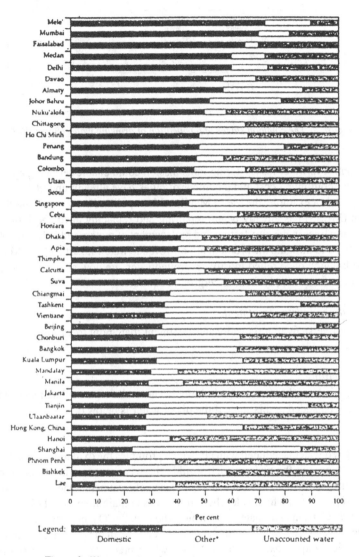

Figure 2. Water use in urban areas in Asia (ESCAP, 2000).

service coverage. Water supply in these cities operates from 10 to 19 hours daily except in Dhaka and Karachi where water supply lasts for 6 and 4 hours, respectively.

A major issue in urban water supply is the high rates of unaccounted for or non-revenue water owed to leakage and illegal connections. Rates of unaccounted for water are particularly very high in cities, such as Dhaka, Hanoi, Mandalai, Metro Manila, Phnom Pehn, Lae, Calcutta and Apia, among others (Figure 2).

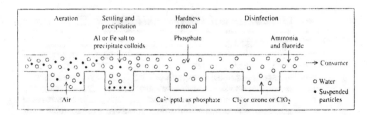

Figure 3. The common stages of purification of drinking water (Baird, 1999).

A control in water losses during transmission could enable servicing a large population without much cost and a number of cities have already undertaken action in this regard. For example, the Metro Manila's Metropolitan Waterworks and Sewerage System (MESS) has already begun repairing and replacing aged and broken water distribution lines in the city, testing and replacement of meters and reduction of illegal connections to mitigate the problem of non-revenue water. Jakarta, Metro Manila and Bangkok which once relied mostly on groundwater, are now also drawing less than 5% of their water supply from underground aquifers.

2.2. Outline of Water Treatment

There are five steps in a typical water treatment that apply to various extents in the Asian region (Figure 3).

Primary settling
The water is brought into a large holding basin to allow particulate matter to settle. Lime may be added at this stage if the pH of water is below 6.5.

Aeration
The clarified water is agitated with air, which promotes the oxidation of easily oxidizable substances in the water, which would otherwise consume the chlorine or other disinfecting material to be added at a later stage of treatment. One impurity removed in this step is Fe^{2+}, which is leached from rocks containing pyrite, FeS_2, in granitic areas, or $FeCO_3$, in carbonate areas. Iron carbonate, $FeCO_3$, in particular, is modestly soluble in water, $K_{sp} = 4.0 \times 10^{-5}$ mol L^{-1} at 25 °C. In the presence of atmospheric oxygen, Fe^{2+}(aq) is oxidized to Fe^{3+}(aq), which precipitates as $Fe(OH)_3$(s) at any pH > 3.5. Such iron staining, a brown deposit on washbasins and toilets is quite common with water from domestic wells; besides the staining, which fortunately can be removed easily with acid cleansing agents, the dissolved iron gives the water an unpleasant taste.

In a municipal treatment plant, aeration circumvents this problem by oxidizing any Fe^{2+} to Fe^{3+}. The rate of this reaction (1) increases with increasing pH, as follows

$$Fe^{2+} + O_2 + H^+ \longrightarrow Fe^{3+} + H_2O,$$

$$-d[Fe^{2+}]/dt = 8 \times 10^{13}[Fe^{2+}][OH^-]^2 \cdot p(O_2)L^2 \, mol^{-2} atm^{-1} min^{-1}. \qquad (1)$$

Since aeration is not practical for individual homeowners, especially in the rural areas of the Asian region, severe cases of iron contamination can be remediated through the use of special filter cartridges, which contain both an oxidant and a filter. They are replaced at intervals when the oxidant is all consumed or the clogging of the filter creates excessive back pressure.

Please remember that this reaction (1) will be again discussed in Chapter 12 when we will discuss the acid mine drainage.

Coagulation

Primary settling of raw water is not sufficient to remove the finest particles, such as colloidal minerals, bacteria, pollen, and spores. Their removal is necessary to give the finished water a clear sparkling appearance. The commonest filter aid used in water treatment is *filter alum* $Al_2(SO_4)_3 \cdot 18H_2O$. In the pH range 6–8, $Al(OH)_3$ is formed in accordance with reaction (2)

$$Al^{3+}(aq) + 3\,HCO_3{}^-(aq) \longleftrightarrow Al(OH)_3(s) + 3\,CO_2(aq). \tag{2}$$

At these pH values, $Al(OH)_3$ is close to its minimum solubility (see Chapter 8) and at equilibrium very little aluminum is left dissolved in the water. The acidic cation $Al^{3+}(aq)$ reacts with $HCO_3{}^-(aq)$ and hence slightly reduces the alkalinity of the water. The amount of alum added must be controlled carefully at the water treatment plant to avoid an excessive concentration of toxic aluminum ions in the treated water.

Aluminum hydroxide forms a very gelatinous precipitate, as is evident if a dilute solution of alum in a test tube is neutralized by NaOH: a milligram or so of precipitated $Al(OH)_3$ will fill the whole tube. Because of this property, the precipitate settles very slowly, and as it does so, it travels down with the fine particles in the treated water. Secondary settling is thus an integral part of the coagulation treatment.

Other useful coagulating agents are ferric sulfate and activated silica. Ferric hydroxide is gelatinous like aluminum hydroxide, and the chemistry is analogous. The coagulating effect of activated silica is due to the formation of gelatinous alkali metal silicates (Bunce, 1994).

Hardness removal

If the water comes from wells in areas having limestone bedrock, it will contain significant amounts of Ca^{2+} and Mg^{2+} ions (see Chapter 8), which are usually removed during the processing. Calcium can be removed from water by addition of phosphate ions. More commonly, calcium ions are removed by precipitation and filtering of the insoluble salt $CaCO_3$; the carbonate ions are either added as sodium carbonate, Na_2CO_3, or if sufficient $HCO_3{}^-$ is naturally present in the water, hydroxide ions can be added in order to convert bicarbonate ions to carbonate (reactions (3) and (4))

$$OH^- + HCO_3{}^- \longrightarrow CO_3{}^{2-} + H_2O, \tag{3}$$

$$Ca^{2+} + CO_3{}^{2-} \longrightarrow CaCO_3(s). \tag{4}$$

Magnesium ions precipitate as magnesium hydroxide, $Mg(OH)_2$, when the water is made sufficiently alkaline, that is, when the OH^- ion content is increased. After removal of the solid $CaCO_3$ and $Mg(OH)_2$, the pH of the water is readjusted to near-neutrality by injecting carbon dioxide into it.

Disinfection

Disinfection is the most important part of water treatment. Filtration and coagulation afford a material that is pleasant to look at, but it is disinfection that makes the water safe to drink. Disinfection kills any bacteria and viruses which have escaped filtration but, at least as important, it prevents recontamination during the time the water is in the distribution system. In the suburbs of a large city, the water may remain in the distribution system for five days or more before it is drunk. Five days is plenty of time for any missed microorganisms to multiply; furthermore, leaks and breaks in the water mains permit recontamination, especially at the extremities of the distribution systems where the water pressure is low.

Recontamination is a serious problem in the Asian urban slums. These slums usually grow faster than the city can extend its distribution system to them. The people must depend either on contaminated ground or surface waters, or on municipally-treated water of sub-optimum quality. We can explain this as follows:

- low pressure at the outer edges of the distribution system, and many breaks because of inadequate installation, allow contamination from the ground. The high pressure in a properly installed system means that the flow through any leak is always from inside to outside the pipe. Low pressure at the fringe of the pipeline water system occurs when cities in the Asian region grow so rapidly that the demand on the system exceeds its capacity;

- individual homes lack their own faucets. Water is drawn from a common, often dirty, communal outlet;

- the first two problems both exist when the city water supply is broken illegally at the point close to the newly developing slums (see above).

The common disinfecting agents are shown in Table 1.

Chlorine is the most commonly used disinfecting agent in the Asian cities. Other disinfectants are ozone, chlorine dioxide, and ultraviolet radiation. Some new disinfecting agents are now in advanced water treatment systems, like bactericides, surfactants, antibiotics, irradiation, sonification, and electric shock.

Chlorine is unique in that it is the only one of the group to possess the residual disinfectant activity; in other words, it maintains its protection of the drinking water throughout the distribution pipeline system. All the other disinfectants, with exception of chlorine dioxide, must be followed with a low dose of chlorine in order to preserve the protection in marginal sites of the water distributing systems.

Table 1. Methods for disinfection of drinking water.

Disinfecting action			
	Chemical		Physical
	Chlorine	Heat	Microwave
	Chlorine water		IR
	Hypochlorite		
Oxidant	Chloramine	Irradiation	UV
	Chlorine dioxide		X-ray
	Fluorine		
	Ozone	Sonification	UV
	Hydrogen peroxide		
Heavy metals	Copper		
	Silver		
Surfactant	Various species	Electric shock	High voltage
Antibiotic	Various species		

3. CHEMICAL, TOXICOLOGICAL AND ECONOMIC ASPECTS OF VARIOUS DISINFECTANTS IN ASIAN CITIES

Disinfection, using chemical and physical methods (see Table 1) is the final step in drinking water purification.

3.1. Chlorine

Before explaining the disinfecting effects of chlorine, we should consider the chemistry of chlorine dissolution in water. When chlorine dissolves in water, the equilibria (reactions (5)–(7)) are rapidly established.

$$Cl_2(g) \longleftrightarrow Cl_2(aq), \quad K_H = 8.0 \times 10^{-3} \text{ mol L}^{-1} \text{ atm}^{-1}, \tag{5}$$

$$Cl_2(aq) + H_2O \longleftrightarrow H^+(aq) + Cl^-(aq) + HOCl(aq), \quad K_c = 4.5 \times 10^{-4} \text{ mol}^2 \text{ L}^{-2}, \tag{6}$$

$$HOCl(aq) \longleftrightarrow H^+(aq) + OCl^-(aq), \quad K_a = 3.0 \times 10^{-8} \text{ mol L}^{-1}. \tag{7}$$

Hydrochloric acid, HCl, which is completely dissociated into $H^+(aq)$ and $Cl^-(aq)$, is a product of the reaction (6), and hence chlorination is a process which reduces the total alkalinity of water. From these equilibrium constants, we can calculate the

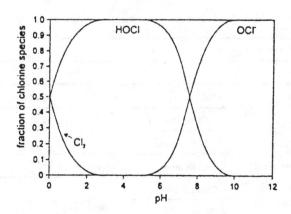

Figure 4. Speciation of active chlorine as a function of pH (Bunce, 1994).

speciation of chlorine species or active chlorine, by which is meant Cl_2, HOCl, and OCl^-. All these species are oxidizing agents; Cl^- is not. The speciation of chlorine is shown in Figure 4, which shows that significant concentrations of free chlorine molecules will only be present in solution below about pH 1, and hence are almost absent at the pH of drinking water.

At the pH of drinking water, the speciation between hypochlorous acid, HOCl, and hypochlorite ion, OCl^-, is of crucial importance. Hypochlorous acid has pK_a 7.5, so that HOCl will predominant below pH 7.5 and OCl^- will predominate above pH 7.5. This is significant, because HOCl is approximately 100 times more effective a disinfectant than OCl^-; probably because the neutral HOCl molecule can penetrate the cell membranes of microorganisms more easily than the ionic OCl^-. HOCl is therefore more destructive. Consequently water at pH $>$ 7.5 requires a higher dose of chlorine to achieve a specified level of disinfection. In other cases, the longer time is needed to achieve disinfection for a fixed dose of chlorine.

Since some of the chlorine will have been used up in destroying microorganisms, as well as in some purely chemical reactions, which will be considered shortly below, the amount of chlorine in the finished water is less than the total amount that was used originally. The following terms are used by water engineers.

- *Chlorine dose*: the amount of chlorine originally used;

- *Chlorine residual*: the amount remaining at the time of analysis;

- *Chlorine demand*: the amount used up; i.e., the difference between the chlorine dose and the chlorine residual;

- *Free available chlorine*: the total amount of HOCl and OCl^- in solution.

Figure 5 shows the relationship between chlorine dose, chlorine demand and chlorine residual.

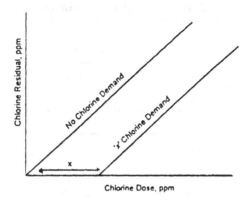

Figure 5. Chlorine dose, chlorine demand and chlorine residual (Bunce, 1994).

If the chlorine demand is zero, the graph is a straight line of unit slope, passing through the origin. When there is a chlorine demand, the residual stays as zero until the demand has been met, and then increases in direct proportion to the additional dose.

In practical terms, the chlorine is supplied as a bulk liquid under pressure since Cl_2 has a normal boiling point of $-35\,°C$ at 1 atm. It is injected at a controlled rate in a large tank of water, such that the residence time of the finished water in the tank is about 20–60 minutes. A typical concentration of chlorine in the purified water is about 1 ppm.

Drawbacks to the chlorine disinfection

Although elemental chlorine is a cheap and effective disinfectant, there can be problems with its use. These include so-called taste and odor problems and the problem of possible toxicity of chlorine itself and chlorinated by-products.

Taste and odor

The commonest of these occur in industrialized and urban areas where discharges from industry cause the surface water (and groundwater to a lesser extent) to be contaminated with phenol or its derivatives, like cresol. Examples include the manufacture of certain herbicides or their precursors, the pulp and paper industry, the manufacture of phenol for use in plastics, such as phenol-formaldehyde, the manufacture and use of pentachlorophenol as a wood preservative, use of phenol derivatives (cresol, etc.) in hospitals, etc. The runoff of these compounds to surface waters and seepage to the groundwater are enlarged in rainy seasons in many Asian cities.

Phenols are a source of trouble because they are chlorinated very rapidly to *chlorinated phenols*, which have a penetrating antiseptic odor. Odor thresholds for these compounds are in ppb ($\mu g\,L^{-1}$) range; at ppm ($mg\,L^{-1}$) level, they make the water completely unusable for drinking or cooking. Some threshold values for Cl-phenol species are shown in Table 2.

*Table 2. Odor threshold values for chlorophe-
nol species in drinking water (Bunce, 1994).*

Cl substituents	Odor thresholds, ppb
None	> 1000
2-	2
4-	250
2,4-	3
2,6-	3
2,4,6-	> 1000

Chlorination of phenol is an example of electrophilic aromatic substitution; phenol is so much more reactive than, let us say, benzene that no Lewis acid catalyst is required, and the reaction takes place rapidly even in aqueous solution. The odor-causing 2-, 2,4-, and 2,6-chlorinated phenols are among the by-products because phenol is chlorinated preferentially in the ortho- and para-positions. In cases where phenols are present in the water supply, the options are to use other disinfectants than chlorine, or to remove the contaminants by the use of activated charcoal as a water filter either in water treatment plant or in home outlets (see below).

We strongly recommend the use of these filters for people staying in various Asian cities, like Bangkok downtown, especially during the rainy season.

Trihalomethanes

Production of trihalomethanes (THMs) is another drawback of using chlorine as a disinfectant. The general formula of THMs is CHX_3, where the three X atoms can be chlorine, bromine or a combination of the two. The compound of principal concern is chloroform, $CHCl_3$, which is produced when hypochlorous acid reacts with organic matter that is dissolved in the water. HOCl reacts with the acetyl $\{-C(=O)CH_3\}$ groups in humic acids, which are breakdown products of plant materials such as lignin or soil humus. This reaction (8) is similar to the well-known iodoform reaction in organic chemistry

$$HOCl + -C(=O)CH_3 \longrightarrow -CO_2H + CHCl_3 + 2\,H_2O. \qquad (8)$$

Chloroform is a suspected liver carcinogen in humans, and it may also have negative reproductive and development effects. Its presence, even at very low levels of ≈ 30 ppb, raises the specter that chlorinated drinking water may pose a health hazard, although the benefits that it confers in the elimination of fatal waterborne diseases are far more significant for the Asian region. Currently, the limit of THMs has been proposed at 100 ppb by the World Health Organization (WHO) and the same limit has been set in the USA. The 100 ppb level is set not only to regulate the THM chemicals themselves but also as an indicator that the production of other chlorinated organic by-products is not excessive.

Before leaving this sub-section on chlorine as a disinfectant, we should note that the majority of the Asian cities still do use chlorine to disinfect drinking water. The benefits in term of protection from water-borne diseases in Asia far overweigh the possible hazard from, for example, 10 or 20 ppb of trihalomethanes. Disinfection with chlorine may possibly pose a cancer risk in old age; no disinfection in the Asian region does mean a substantial chance of dying from typhoid fever or cholera as a child or young adult. However, as we have seen in Table 1, there are different methods of disinfection and we consider some of these alternatives to chlorination.

3.2. Ozone

The use of ozone to disinfect drinking water started at the end of the 19th century in the Netherlands and in the beginning of the 20th century in Germany. At present, ozonation is used in many municipalities worldwide, including some Japanese cities.

Ozone is a slightly more powerful disinfectant than chlorine; it cannot be stored or transported. Ozone is prepared by passing a high voltage electric discharge ($\approx 15,000$ V) through dry air, and then absorbing the ozone in water ($K_H = 1.3 \times 10^{-2}$ mol L^{-1} atm^{-1}). Under optimum conditions, up to 6% of the air oxygen can be transformed to ozone, although in practice, the ozonized air used to disinfect water contains about 1% ozone (reaction (9))

$$3\,O_2(g) \xrightarrow{\;15,000\,V\;} 2\,O_3(g). \tag{9}$$

The chemistry of ozone is different from chlorine since ozone is a powerful oxidant. For example, hydroxylation is the characteristic reaction with phenol. Ozone is of course free of any complications from formation of chlorinated derivatives. It decomposes quite rapidly in water, the kinetics of the reaction (10) being pH dependent

$$\text{rate} = 2.2 \times 10^5 [OH^-]^{0.55}[O_3]^2 \text{ mol } L^{-1} \text{ s}^{-1}. \tag{10}$$

The rate of ozone decomposition reaction and instantaneous ozone demand also depend on the content of dissolved organic matter. An example of different treated waters with varying content of total organic carbon (TOC) is shown in Figure 6.

Since ozone leaves no residual, post-treatment chlorination of finished water with a light dose of chlorine is needed to maintain its protection in the distribution system. The equipment needed to generate ozone is expensive, and offers economy with large-scale operation in the Asian communities.

3.3. Chlorine Dioxide

Chlorine dioxide ClO_2 is a very effective disinfectant, about twice as powerful as HOCl, but also more expensive. Typical dose rates are 0.1–5 ppm in raw water.

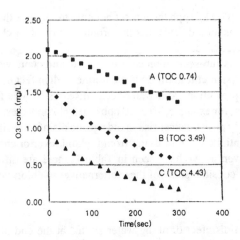

Figure 6. Ozone decomposition in water with different content of total organic matter as TOC (Park et al, 1998).

An application of this disinfectant has been started in the mid-20th century in the USA and now thousands of municipalities in North America and Europe are using this species for chlorination of the water to be treated.

A major drawback to the use of chlorine dioxide compared with chlorine is that chlorine dioxide cannot be stored. This species has to be prepared and used on site, whereas chlorine can be delivered in tank cars. Chlorine dioxide is chemically unstable. It is an endothermic compound and decomposes explosively.

Two reactions are known for generation of chlorine dioxide (reactions (11) and (12))

$$10\,NaClO_2 + 5\,H_2SO_4 \longrightarrow 8\,ClO_2 + 5\,Na_2SO_4 + 2\,HCl + 4\,H_2O, \quad (11)$$

$$12\,NaClO_2 + Cl_2 \xrightarrow{pH<3.5} 2\,ClO_2 + 2\,NaCl. \quad (12)$$

Reaction (11), the disproportionation of the ClO_2^- ion to ClO_2 and Cl^-, affords chlorine-free chlorine dioxide. Reaction (12) is less useful when ClO_2 is being used to combat taste and odor, since the product is inevitably contaminated with hypochlorous acid, thus defeating the objective of using a chlorine substitute. The precursor in either case is sodium chlorite, $NaClO_2$, which is a powerful oxidizer, and has to be stored carefully.

Like ozone, chlorine dioxide is also beneficial for controlling taste and odor problems, for oxidizing iron and manganese, for oxidizing pollutants in raw water, and for bleaching color in water. From a practical perspective, chlorine dioxide is also five times more soluble in water than chlorine and is effective over a wide pH range. Chlorite dioxide also produces fewer chlorinated by-products than chlorine, as evidenced by lower organic chlorite levels produced.

Table 3. Lowest observed toxic effects of chlorinated species (Bunce, 1994).

Toxicant	Observed effect	Lowest concentration, ppm
$NaClO_2$	Red blood cell damage	20–100
$NaClO_2$	Lower meaning rates	100
ClO_2	Red blood cell damage	> 100
ClO_2	Reduced growth rates	100

Toxicological studies of the chlorite ion and of ClO_2 have shown that sodium chlorite causes hemolysis at 50 ppm. As a precaution, it is recommended that finished water should contain no more than 1 ppm of ClO_2, which can therefore be reduced to no more than 1 ppm of ClO_2^- ion. Animal studies have shown the lowest levels of chlorine dioxide and sodium chlorite, which cause observable toxic effects to be substantially greater than 1 ppm or so, which is used for water treatment (Table 3).

Chlorine dioxide, in contrast to ozone, may be generated using relatively inexpensive equipment, which is why it finds application as a replacement for chlorine on occasions when taste and odor problems are experienced, like during the rainy season in South or Southeast Asia. By contrast, ozone will be used all the time or not at all.

3.4. Ultraviolet Radiation

Ultraviolet light can be also used to disinfect and purify water. Powerful lamps containing mercury vapor, the excited atoms of which emit UV-C light centered at 254 nm, are immersed in the water flow; about ten seconds of irradiation is usually sufficient to eliminate the toxic micro-organisms. An advantage of UV disinfection technology is that small units can be employed to serve small population bases whether in the developed or developing countries of the Asian region, so that continuous monitoring of chemical systems is made unnecessary. The germination action of the light is owed to its disrupting the micro-organism's DNA so severely that its subsequent replication is impossible and the cell is inactivated.

The use of UV light to purify water is complicated by the presence of dissolved iron and of humic substances, both of which absorb the light and thus reduce the amount available for disinfection. Small particles of solid suspended in the water also inhibit the action of the UV since they can shade or absorb bacteria and also scatter or absorb the light.

3.5. Combination Treatments of Disinfectants

Drinking water treatment plants sometimes use combinations of disinfectants for treating raw drinking water. A combination of ozone and chlorine or chloramine is an important example of combining disinfectants. In this case, ozone is the primary disinfectant (at the beginning of the treatment process) and chlorine or chloramine is a secondary disinfectant at the end of the process (just before the water leaves the treatment plant to enter the distribution system). The primary reason that a secondary disinfectant is often used with ozone is because ozone does not persist long in water (see above), and a secondary disinfectant is necessary to maintain disinfection in the distribution system from the treatment plant and the consumer tap. Chlorine, chloramine, and chlorine dioxide are all used for this purpose because they persist long enough to provide disinfection throughout the distribution system. If a secondary disinfectant is not used with ozone, bacteria and other micro-organisms regrow in the distribution system, causing illness of people who drink the water.

Although chlorine is commonly a secondary disinfectant with alternative disinfectants, its use does not usually defeat the purpose of using alternative disinfectants to minimize THM levels. This is because lower levels of chlorine (or another secondary disinfectant) are required as a secondary disinfectant, as opposed to the primary disinfectant, which must kill the higher number of microorganisms that enter with the raw water at the beginning of the treatment process. And, when lower levels of chlorine are applied, THMs and other chlorinated by-products are usually found at lower levels than those with chlorine as the primary disinfectant. Although chlorine is widely used as a secondary disinfectant, chloramine and chloride dioxide are better choices when THM control is difficult because these disinfectants do not produce significant levels of THNs. When chloramine is a secondary disinfectant, however, care must be exercised to ensure that microorganisms are killed effectively, because chloramine is a weaker oxidant. Many drinking water treatment plants in Europe and the USA, however, successfully use chloramine as a secondary disinfectant. Some studies in the Asian region showed the successful application of other oxidants, like hydrogen peroxide (see Box 1).

Box 1. Disinfection of filtered water using ozone and hydrogen peroxide
(S. Muttamara & C. Sales, 2000)

Laboratory scale experiments were carried out to investigate the bactericidal effectiveness of ozone-hydrogen peroxide and to ascertain the extent of its oxidation. It included the investigation of the effects of both ozone and ozone-hydrogen peroxide combination on THM formation. Filtered unchlorinated water from Banklen Water Treatment Pant in Bangkok was used as source water.

The results revealed that ozone-hydrogen peroxide combination is a more active bactericidal agent compared to that of ozone alone and chlorine. Nearly 99.99% of coliform inactivation was attained at 0.34 mg L^{-1} of ozone and ®Object ratio of 0.5. The application of ozone-hydrogen peroxide also increased the THM removal by ozone alone to 20–30%. The addition of hydrogen peroxide also increased the ozone residuals and therefore improved the oxidizing capacity of ozone in the removal of organic materials in the natural water. The ozone-hydrogen peroxide combination, aside from its effectiveness in coliform inactivation, is a good agent for the control of THM formation through maximizing total organic matter removal.

Table 4. Cost effectiveness of various disinfection methods (after Bunce, 1994).

Disinfectant	Plant capacity, L $\times 10^6$ per day	Cost, US cent per 1000 L
Cl_2	0.19	6.8
	1.9	0.8
	3.8	0.4
ClO_2	0.19	13.8
	1.9	2.9
	3.8	2.2
O_3	0.19	19.1
	1.9	2.5
	3.8	1.7
UV	0.19	5.7
	1.9	1.9
	3.8	1.4

3.6. Cost Effectiveness of Various Disinfection Methods

Table 4 presents a cost comparison between the various disinfectants. We can note the cost effectiveness of UV for the smallest plants, its second place cost relative to chlorination for larger plants, and the extremely high cost of ozonation, associated with high capital equipment cost for a small installation.

3.7. Analysis of Residuals

Iodometric titration is a simple and reliable method. The relevant reactions (13)–(15) between the disinfectant and iodine ion are shown below

$$HOCl + 2\,I^- + H^+ \longrightarrow I_2 + Cl^- + H_2O, \tag{13}$$

$$ClO_2 + 5\,I^- + 4\,H^+ \longrightarrow 2/5\,I_2 + Cl^- + 2\,H_2O, \tag{14}$$

$$O_3 + 2\,I^- + 2\,H^+ \longrightarrow I_2 + O_2 + H_2O. \tag{15}$$

We can note that although ClO_2 functions as a one-electron oxidant in its reactions with organics, it is a five-electron oxidant with I^-, with the chlorine changing from the $+4$ oxidation state in ClO_2 to the -1 oxidation state in the reaction product Cl^-. In each case, the iodine liberated is titrated against standard sodium thiosulfate solution (see Radojevic and Bashkin, 1999).

4. POLLUTANTS IN DRINKING WATER

In the Asian region, prospective drinking water supplies may become contaminated with various organics, heavy metals, and nitrates in many ways. Some of these are natural, as in the decay of biological materials, weathering of rocks containing different trace metals, and nitrification; others are the result of human activities, which cause contamination by natural substances (runoff fertilizers from agricultural fields; food processing and meat packing plants; manure from feedlots on farms; leaking of heavy metals from landfills) or by synthetic compounds (insecticides and herbicides used in agriculture; seepage from unsecured municipal waste dumps or industrial waste dumps; sewage outflows; water used for cooling purposes in industry; and even equilibrium of organic air pollutants between lakes and the atmosphere).

We will consider several of these issues. For some pollutants, such as the most dangerous heavy metals, nitrates, and organochlorine compounds, more details will be also presented in Chapters 12, 13 and 14.

4.1. Organics in Drinking Water

The presence of organics in drinking water came to prominence in the 1960s with the development of gc/ms (gas chromatography combined with mass spectrometry), which made possible the detection and quantitation of organic compounds in water in sub-ppm level. In a now classic study carried out in the late 1960s, the lower Mississippi River, from which the city of New Orleans, USA, takes its drinking water, was found to be contaminated with literally hundreds of organic compounds, and these contaminants were carried through the water treatment process into the finished drinking water. Public complaints about the poor taste of the water were instrumental in initiating this study.

The substances detected included herbicides, such as Alachlor, butachlor, atrazine, cyanazine, propazine, and simazine; insecticides, such as chlordane, heptachlor, dieldrin, eldrin, and DDE as a DDT metabolite; and industrial organics, which included alkylbenzenes such as toluene, xylenes, ethylbenzene; alkanes such as decane through pentadecane; naphtalene and methylnaphtalenes; alkyl phthalates, used as plasticizers; chlorinated methanes and ethanes, especially $CHCl_3$; chlorinated benzenes and chlorinates phenols; benzaldehyde; dicyclopentadiene, etc. Of these compounds, the hydrocarbons are typical of oil refinery operations, and the chlorinated benzenes and phenols are associated with herbicide and insecticide production.

Later studies of the 1970s–1980s carried out on water from other major waterways such as the lower Great Lakes (Lake Erie and Lake Ontario in Canada and the USA), the Rhine, Danube and Volga Rivers in Europe, many Asian rivers from Japan, China, Thailand, India, the Philippines, Indonesia and other countries, have shown similar findings (see Chapter 8).

4.2. Heavy Metals

Iron has already been cited as a potential nuisance in water. It stains bathroom fixtures and gives an unpleasant metallic taste to the water, but it is rarely present in amounts that might be toxic since, upon oxidation, it precipitates from solution. Other metals which are of special concern in the Asian region for their accumulation in water sources and toxicity due to the mobility in biogeochemical food webs are aluminum, cadmium, lead and mercury. Sources of these metals in raw water are both natural and anthropogenic. In Asia, aluminum is abundant in many tropical and subtropical ancient soils as well as volcanic ash soils (see Chapter 7) and additionally its migration and accumulation in drinking water supplies has been accelerating due to acid rains (see Chapter 4). Migration and accumulation of other heavy metals is also enhanced in drinking water and food chains (see Chapters 4 and 12).

High sodium concentrations are suspected of being responsible for promoting cardiovascular disease, and can render the water unpalatable.

Aluminum

Until the 1980s, aluminum was generally regarded as innocuous in drinking water and in the diet. However, the current situation regarding the toxicology of Al is extremely confused (Bunce, 1994). Although aluminum is very abundant in the Earth's crust and especially in many Asian soils, it appears to have no natural biochemical function in living organisms. Aluminum is kept out of the body by three barriers: the gastrointestinal tract, serum binding to transferrin (this can interfere with iron transport, however), and the blood-brain barrier.

Allegations have been made that patients suffering from renal failure are susceptible to aluminum toxicity. Anemia (interference with iron transport), softening of the bones (replacement of calcium phosphate by aluminum phosphate), and a form of senile dementia can result. As well, a link to Alzheimer's disease (also a form of senile dementia) has been proposed, with the brains of dementia patients showing characteristic tangles of nervousness and hot-spots of unusually high Al concentration.

Some epidemiological studies showed an increase in the incidents of Alzheimer's disease in parts of the United Kingdom, where the water naturally contains elevated levels of Al species (Bunce, 1994).

Aluminum cookware is another potential source of Al species in the diet. High concentrations (up to 100 ppm) of Al species can be recorded in fruits and fruit juices cooked in aluminum cookware. The combination of high $[H^+]$ and high [citrate] is responsible, since aluminum forms strong complexes with citrate ions, thereby dissolving the protective oxide from the metal surface and releasing toxic free Al^{3+} ions in solution.

Sodium

Sodium ions may naturally be present in drinking water if the supply is brackish, like many water sources in Central Asia and the Middle East. Above about 300–400 ppm of sodium chloride, the water is unpalatable on account of its chloride content, and special purification measures, such as reverse osmosis (see below Section 6) are required. Naturally soft water frequently contains sodium rather than the "hardness" cations calcium and magnesium, while water softening artificially in domestic water softeners has the hardness cations replaced by sodium (see below Section 5). Even though drinking water is only one source of dietary sodium, it is recommended that those who use domestic water softeners retain one outlet of unsoftened water for drinking. The use of sodium salts by industry and the practice of using sodium chloride as road salt de-icer in the moderate and cold climate contribute substantially to the problem of surface and ground water by sodium.

Excessive intake of sodium has been statistically associated with hypertension (high blood pressure) and cardiovascular disease. For this reason, the standard for sodium has been set as 100–200 ppm for drinking water (Table 5).

5. BOTTLED DRINKING WATER

In many Asian cities, the content of various dangerous chemicals (both mineral and organic species) exceeds the relevant guidelines and standards shown in Table 5. Tap water is unacceptable for drinking in Bangkok, Seoul, Bangalore, New Delhi, Karachi and many other municipalities in the Asian region. People have to drink so-called bottled drinking water that receives inadequate special treatment after raw water purification in the municipal treatment plants. This is caused by both natural and, to a greater extent, by anthropogenic pollution of water supplies.

There are different methods for removal of dangerous chemicals for bottled drinking water and we will consider some of them.

5.1. Activated Carbon

Activated carbon is used for the removal of both natural and anthropogenic organic compounds responsible for causing taste and odor problems in drinking water. This is an expensive procedure; however it is widely used in the Asian cities. The finished water is passed over a bed of carbon, which has been activated by partly burning wood chips in a limited supply of air, to produce a very large surface area. Absorption of organics by charcoal may be familiar from an organic chemical laboratory, where charcoal is often used to remove colored impurities from organic compounds. Activated coal has only a finite capacity to absorb impurities from drinking water; when it is spent, it may be reactivated, although with some loss, by further partial burning. This drives off and burns away the absorbed organics.

5.2. Ion Exchange

This method is suitable for the removal of harness cations and heavy metals from drinking water. A cation exchanger is an insoluble inorganic or organic polymer, which carries multiple negative charges on its backbone. An original polymer, for example, might be carboxylate or sulfonate substituents on its carbon backbone. These negative charges are balanced by cations, but the cations are free to move within the structure of the polymer. When the ion exchanger is ready for use, the associated cations are normally Na^+. When drinking water passes through ion exchanger resin, the cations in the water can exchange with cations associated with the resin.

We consider the example for Ca^{2+} as a water cation; however similar physical-chemical reactions can be shown for any cation, including heavy metals.

Table 5. Guidelines and standards for drinking water quality, ppm.

Parameter	WHO	US EPA	Russia	Japan
Al	0.2	0.2	0.5	0.2
As	0.01	0.05	0.01	0.01
Ba	0.7	2.0	0.7	—
B	0.3	—	0.3	0.2
Cd	0.003	0.005	0.003	0.01
Cl^-	250	250	350	200
Cr(III)	0.05	0.1	0.05	0.05
Cu	1–2	1.0	2.0	1.0
CN^-	0.07	0.2	0.07	0.01
F	1.5	2–4	1.5	0.8
Fe(II)	0.3	0.3	0.3	0.3
Hg	0.001	0.002	0.001	0.0005
Mn	0.1	0.005	0.5	0.01
Mo	0.07	—	0.07	0.07
Na	200	100	100	200
Ni	0.02	0.1	0.02	0.01
NH_4^+	1.5	—	2	—
NO_3^-	50	45	45	45
NO_2^-	3	3	3	3
Pb	0.01	0.015	0.01	0.05
Sb	0.005	0.006	0.005	0.002
Se	0.01	0.05	0.01	0.01
SO_4^{2-}	250	250	500	—
Zn	3	5	1	1
TDS	1000	500	1000	500
PH	6.5–8.5	6.5–8.5	6.0–9.0	5.8–8.6

Designating the anionic sites on the polymer as (A^-) we have reaction (16) as follows

$$Ca^{2+}(aq) + 2\,Na^+(A^-) \longrightarrow 2\,Na^+(aq) + Ca^{2+}(A^-)_2. \tag{16}$$

Calcium (or any other cation) from the water becomes associated with the ion exchanger, and sodium leaves the resin and enters the water.

When most of the sodium ions have been exchanged for calcium and other water cations, the ion exchanger loses its effectiveness. It can be regenerated by passing a concentrated salt solution through the resin, whereupon the reverse reaction (17) occurs

$$Ca^{2+}(A^-)_2 + 2\,Na^+(aq) \longrightarrow 2\,Na^+(aq) + Ca^{2+}(aq). \tag{17}$$

The cations released are discharged to the wastewater, together with the excessive NaCl.

Since the reaction by which the resin is regenerated is the reverse of the reaction by which the exchanger operates, the chemical requirements for the design of a successful water exchanger can be specified. The equilibrium constant for the exchange reaction (16) must be large enough that the low concentrations of the pollutants in the finished water will successfully displace Na^+ from the resin sites. However, it must be not too large, otherwise it will not be possible to get the calcium (and other cations) ions off the resin again; this would give a one-time-only water filter. A moderately large equilibrium constant for reaction (16) means that the reverse reaction (17) can be made to proceed if the regenerating solution of NaCl is strong enough, and that is how the resin is regenerated in practice.

Common ion exchangers are suffonated polystyrene, cross-linked for rigidity with divinylbenzene, and zeolites, which are inorganic structures, based on linked tetrahedral arrangements of SiO_4 and AlO_4 units. Each oxygen bridges two other atoms, $Si - O - Si$ or $Al - O - Al$, etc. to form a rigid network, similar to quartz, SiO_2. Quartz itself has no charge, but one negative charge on the framework is created for every aluminum atom, which replaces silicon. This charge must balanced by one mobile cation in the channels in the zeolite.

Deionized water

Excessive dangerous anions must also be removed from the bottled drinking water. This is carried out by the use of two ion exchangers in series. The cation exchanger is similar to that just described, except that H^+ is the counter ion in the active form of the resin

$$Ca^{2+}(aq) + 2\,H^+(A^-) \longrightarrow 2\,H^+(aq) + Ca^{2+}(A^-)_2. \tag{18}$$

The resin is regenerated with HCl rather than NaCl, reaction (19)

$$Ca^{2+}(A^-)_2 + 2\,H^+(aq) \longrightarrow 2\,H^+(A^-) + Ca^{2+}(aq). \tag{19}$$

The second exchanger is an anion exchanger. The polymer backbone this times carries trialkylammonium $(-NR_3{}^+)$ substituents, and the counter ions are OH^-. Writing (C^+) as a cationic site on the resin, and Cl^- as an example of the anion to be exchanged, the reaction (20) for anion exchange is as follows

$$Cl^-(aq) + (C^+)OH^- \longrightarrow OH^-(aq) + (C^+)Cl^-. \tag{20}$$

This resin is regenerated with an NaOH solution.

The cation exchanger replaces unwanted cations by $H^+(aq)$. Charge balance requires that the anion exchanger replaces unwanted anions by an equal number of OH^- ions. These react with the $H^+(aq)$ to form water, thus removing all the unwanted ions from drinking water, reaction (21).

$$H^+(aq) + OH^-(aq) \longrightarrow H_2O(l). \tag{21}$$

6. DESALINATION

Many parts of the Asian region lack sufficient sources of potable water. The Middle East is one such region, where a growing population lives in an area which is largely desert. Such wells and water holes as exist are often brackish, that is, rather high in salt and hence unpalatable. Therefore there is much interest in methods for the desalination of both these brackish waters and seawater.

An obvious method of purifying salty water is by distillation. Because of the large energy requirement (ΔH° of vaporization for water $= 44 \, kJ \, mol^{-1}$ at 25 °C), the cost is high, about US\$1 per 1000 L. The technology is practical in Middle East countries, where energy costs are low.

Reverse osmosis, by definition, is the reverse of conventional osmosis, which is familiar from biology. If a dilute solution and a concentrated solution are separated by a semi-permeable membrane (which allows water molecules to pass through it, but not other solutes), water will flow from the dilute solution into the concentrated one. Under these conditions, a substantial difference in height between the two water columns may be established (Figure 7).

The difference in pressure exerted on the semi-permeable membrane by the two columns is called the osmosis pressure, π

$$\pi = cRT,$$

where c is concentration expressed in moles per liter, R is the universal gas constant (in $L \, atm \, mol^{-1} \, K^{-1}$), T is Kelvin, and the osmosis pressure has the units of atmospheres. "Concentration" in this equation is the total concentration of all solutes, regardless of their identities. Thus, for example, a $0.06 \, mol \, L^{-1}$ solution of magnesium chloride has osmolar concentration $0.18 \, mol \, L^{-1}$, because dissociation affords $0.06 \, mol \, L^{-1}$ of Mg^{2+} and $0.12 \, mol \, L^{-1}$ of Cl^-.

In conventional osmosis, the natural tendency of two solutions, initially of differing concentrations, to equalize their concentrations leads to a difference in pressure across

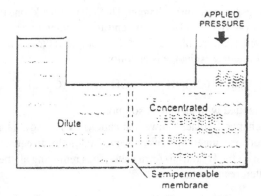

Figure 7. Osmotic pressure across a semi-permeable membrane (Bunce, 1994).

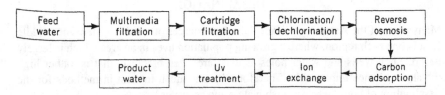

Figure 8. Flow diagram for the setup of a desalination plant employing reverse osmosis, including pre- and post-treatment steps (Bhattacharyya et al, 1999).

the semi-permeable membrane. In reverse osmosis, pressure is applied externally to one side of a semi-permeable membrane, which separates two solutions, initially of equal concentrations. Water flows through the membrane from the high-pressure side to the low-pressure side, establishing a difference in concentration of the solutions on either side of the semi-permeable membrane. Hence the technique of reverse osmosis consists of forcing a high-salt solution such as seawater through a semi-permeable membrane; the purified water passes through the membrane, leaving the more concentrated solution behind (Bunce, 1994).

Desalination of seawater and brackish water has been and, as of the end of the 1990s, is the primary use of reverse osmosis. Driven by a need for potable water in areas where there is a shortage, this industry has developed significantly. Desalination involves the reduction of the total dissolved solids (TDS) concentration, which is equal to more than 30,000 ppm, to less than 200 ppm.

The largest seawater desalination plant in the World operates in Jeddah, Saudi Arabia. It has a capacity of 56,800,000 liters per day, and the TDS content of the seawater is approximately 44,000 ppm. Pretreatment modules for this plant include chlorine treatment, a dual media filter, and a cartridge filter for the coagulation and filtration of dead cells of microorganisms (Figure 8).

FURTHER READING

1. Bunce N., 1994. *Environmental Chemistry.* second edition, Wuerz Publishing Ltd, Winnipeg, Chapter 5.

2. ESCAP, 1995. *State of the Environment in Asia and Pacific.* United Nations, NY, Chapter 8.

3. Radojevic M. and Bashkin V., 1999. *Practical Environmental Analysis.* Royal Society of Chemistry, UK, Chapter 4.

4. Baird C., 1999. *Environmental Chemistry.* second edition, W. H. Freeman and Company, NY, Chapter 9.

5. Richardson S. D., 1999. Drinking water disinfection by-products. In: Meyers R. A. (Ed.), *Encyclopedia of Environmental Pollution and Cleanup*, John Wiley, NY, 454–468.

6. Bhattacharyya D., Mangum W. C. and Williams W. E., 1999. Reverse osmosis. In: Meyers R. A. (Ed.), *Encyclopedia of Environmental Pollution and Cleanup*, John Wiley, NY, 1444–1450.

7. ESCAP, 2000. *State of the Environment in Asia and Pacific.* United Nations, NY, Chapters 18.

WEBSITES OF INTEREST

1. The World Wide Water home page at the University of California, Davis, has many links to other sources of information about water issues, http://pubweb.ucdavis.edu/documents/gws/envissues/george_fink/masterw.htm

QUESTIONS AND PROBLEMS

1. Discuss the level of raw water treatment for drinking in various Asian countries. Point out the differences between rural and urban populations.

2. What are the main steps in treating raw water for drinking water distribution systems in Asia? Do all of them need to be applied?

3. Explain the chemical reactions that occur at the stage of aeration during water treatment.

4. What chemical and physical-chemical reactions are of great importance at the stage of coagulation? Discuss the stoichiometry of these reactions.

5. Why must excessive phosphate be removed during water purification and what methods are used at this step?

6. Discuss the list of disinfectants that are applied for traditional and advanced technologies of raw water treatment.

7. Characterize the chemical reactions that occur during dissolution of chlorine in treated water.

8. Discuss the advantages and disadvantages of chlorine application for treatment of raw water in the Asian region.

9. Characterize the common substituents of chlorine for water disinfection and discuss the application of these disinfectants in your country.

10. Discuss the conditions that enhance the application of combinations of different disinfectants for raw water treatment.

11. Describe the economic parameters of using different disinfectants and compare international cost estimates with those for your city/country.

12. Characterize the chemical pollutants that are common for treated water in various Asian countries; draw attention to the sources of these pollutants in your country.

13. Discuss the sources of aluminum in the drinking water for different regions in Asia and its toxic effects.

14. Discuss the toxicity of heavy metals in drinking water; use additional materials from Chapter 12 for the answer.

15. What are the reasons for wide application and sales of bottled drinking water in the Asian region?

16. Discuss the common physical-chemical aspects of bottled drinking water production and highlight the methods which are of special use in your country.

17. Characterize the application of reverse osmosis for drinking water production and point out the physical-chemical details of this method.

PART III

TOXIC SUBSTANCES

CHAPTER 11

ECOTOXICOLOGICAL CHEMISTRY

1. INTRODUCTION

In the Asian region, most pollutants and hazardous chemical species must be of special concern because of their toxic effects. In this chapter we will consider the general aspects of these effects. Attention will be paid to a very short review of biochemical and physiological processes that are essential for understanding of toxic interactions within the organism. We will discuss the bioconcentration and biomagnification of pollutants in food webs. Toxicological chemistry of polluting species that are most widely distributed in the Asian region is also addressed.

2. BIOCHEMICAL ASPECTS OF TOXICANT BEHAVIOR IN LIVING ORGANISMS

In the Asian region, most contaminants or toxicants (see below) are of public concern due to their toxic effects. For better understanding these effects we should discuss briefly biochemical and physiological aspects of toxicant behavior in living organisms. Biochemistry is the science that studies the chemical reactions and materials in living organisms and physiology deals with the description of process inside the organisms. We will summarize very briefly here the main subjects and reactions in biochemistry and physiology.

2.1. Principal Structural Ingredients and Biochemical Molecules

The cell

Figure 1 shows the major structural units of the *eukaryotic cell*, which is the basic compartment of biochemical processes in animals.

These structural units include the following:

- *Cytoplasm*, which fills the cell and in which several important kinds of cell structures are contained;

- *Mitochondria*, which mediate energy conservation and utilization in the cell;

- *Ribosomes*, which particulate in protein synthesis;

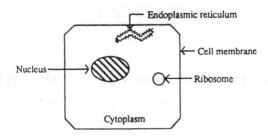

Figure 1. Some major features of the eukaryotic cell.

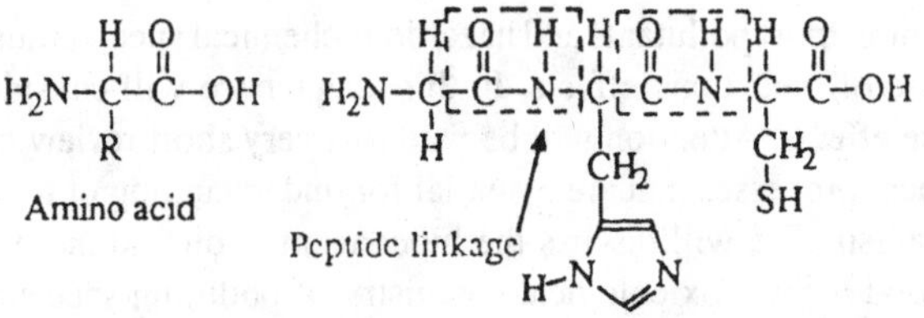

$$H_2N-\underset{\underset{R}{|}}{\overset{\overset{H}{|}}{C}}-\overset{\overset{O}{||}}{C}-OH$$

Amino acid

Peptide linkage

Figure 2. A tripeptide formed by the linking of three amino acids. The peptide linkages with which the amino acids are joined are outlined by dashed lines.

- *Endoplasmic reticulum*, which is involved in the metabolism of some toxicants by enzymatic processes;

- *Cell membrane*, which encloses the cell and regulates the passage of ions, nutrients, metabolic products, and toxicant metabolites into and out of the cell interior; when its membrane is damaged by toxic substances, a cell may not function properly and the organisms may be harmed;

- *Cell nucleus*, the control center of the cell, which contains *deoxyribonucleic acid*, through which the nucleus regulates cell division.

Proteins

Proteins are the basic building blocks of biological material that constitute enzymes and most of the cytoplasm inside the cell. Proteins are composed primarily of biopolymers of *amino acids*, whose general structure is shown in Figure 2.

Let us consider the protein structure shown in Figure 2. Here R represents a group ranging from an H atom (in the amino acid glycine) to moderately complex structures. The amino acids in proteins are joined at *peptide linkages* outlined by dashed lines.

The structures of protein molecules determine the behavior of proteins in crucial areas such as the processes by which the body's immune system recognizes substances that are foreign to the body. *Proteinaceous enzymes* depend upon their structures for

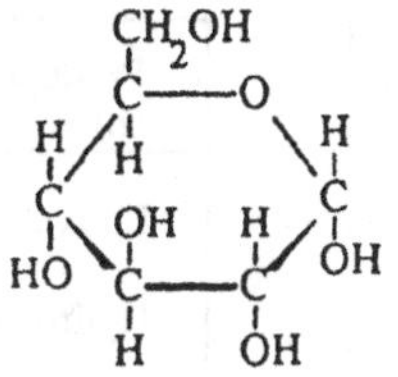

$$\begin{array}{c} CH_2OH \\ | \\ C\!-\!\!-\!\!-O \\ H/\!\!\!| H \qquad\qquad H \\ C \qquad\qquad\qquad | \\ | \qquad OH \quad H \quad C \\ HO\!\!\setminus C\!-\!\!-\!\!-C/\, OH \\ | \qquad\quad | \\ H \qquad OH \end{array}$$

Figure 3. Structural composition of glucose molecule.

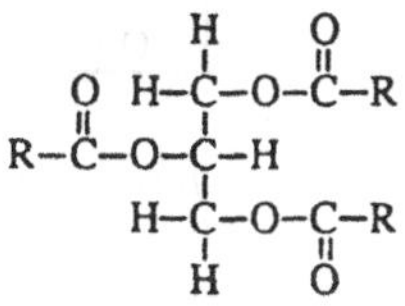

$$\begin{array}{c} \qquad\quad H \quad\;\; O \\ \qquad\quad | \quad\;\;\; \| \\ O \;\; H\!-\!C\!-\!O\!-\!C\!-\!R \\ \| \qquad\;\; | \\ R\!-\!C\!-\!O\!-\!C\!-\!H \\ \qquad\qquad | \\ \;\; H\!-\!C\!-\!O\!-\!C\!-\!R \\ \qquad | \quad\;\; \| \\ \qquad H \quad\; O \end{array}$$

Figure 4. General formula of triglycerides.

the very specific functions of the enzymes. The order of amino acids in the protein molecule determines its primary structure. Secondary and tertiary protein structures depend on the ways in which the polypeptide molecules are bent and folded. The loss of a protein secondary and tertiary structure is called *denaturation*. Denaturation may be caused by heat or the action of foreign chemicals, like corrosive poisons.

Carbohydrates

Carbohydrates have the approximate simple formula CH_2O and include a diverse range of substances composed of simple sugars like glucose (Figure 3).

High-molecular-mass *polysaccharides*, such as *starch and glicogen*, are biopolymers of simple sugars. The major functions of carbohydrates are to store and transfer energy, a process which may be destroyed by toxic chemicals.

Lipids

Lipids are substances that can be extracted from plant or animal matter by organic solvents, such as toluene, chloroform, or diethyl ether. The most common lipids are fats and oils composed of *triglycerides* (Figure 4).

Triglycerides are formed from the alcohol *glycerol*, $CH_2(OH)CH(OH)CH_2(OH)$ and a long-chain fatty acid, such as *stearic acid*, $CH_3(CH_2)_{16}C(O)OH$. Numerous other biological materials, including waxes, cholesterol, and some vitamins, are also classified as lipids.

Lipids are toxicologically important for several reasons. Some toxic substances interfere with lipid metabolism, leading to detrimental accumulation of lipids. Many toxic organic compounds are poorly soluble in water, but are lipid-soluble, and lipids can store toxicants.

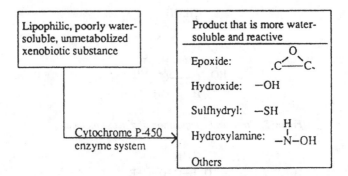

Figure 5. Scheme of phase I reactions (Manahan, 2000).

Enzymes

Enzymes are proteinaceous substances with highly specific structures that interact with particular substances or classes of substances called *substrates*. Enzymes act as catalysts enabling biochemical reactions to occur at body temperature, after which they are regenerated intact to take part in additional reactions (reaction (1))

$$\text{Enzyme} + \text{Substrate} \longrightarrow \text{Enzyme-substrate complex}$$
$$\longrightarrow \text{Products} + \text{regenerated enzyme.} \qquad (1)$$

Enzymes are named for what they do. As an example, *lipase* enzymes cause lipid triglycerides to dissolve and form glycerol and fatty acids.

Some toxic substances, such as cyanide, heavy metals, or herbicides, alter or destroy enzymes so that they function improperly or not at all. Toxicants can affect enzymes in several ways. Heavy metals, for example, tend to bind to sulfur atoms in enzymes, thereby altering the shape and function of the enzyme. On the other hand, insecticide parathion bonds covalently to the nerve enzyme *acetylcholinesterase*, which can then no longer serve to stop nerve impulses.

2.2. Enzyme-Catalyzed Reactions of Xenobiotic Substances

Xenobiotic chemical species, like many pesticides, are synthetic substances that are foreign to living systems and for which the natural biogeochemical cycles are absent. The processes by which living organisms metabolize the xenobiotics are enzyme-catalyzed phase I and phase II reactions. We will consider these reactions briefly.

Phase I reactions

Lipophilic xenobiotic species in the body tend to undergo *phase I reactions* that make them more water-soluble and reactive by the attachment of polar functional groups, such as $-OH$, $-SH$, $-N(H)OH$, etc. (Figure 5).

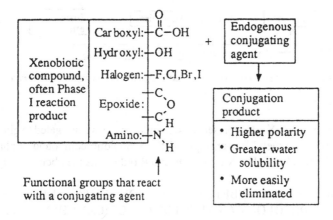

Figure 6. *Scheme of phase II reactions (Manahan, 2000).*

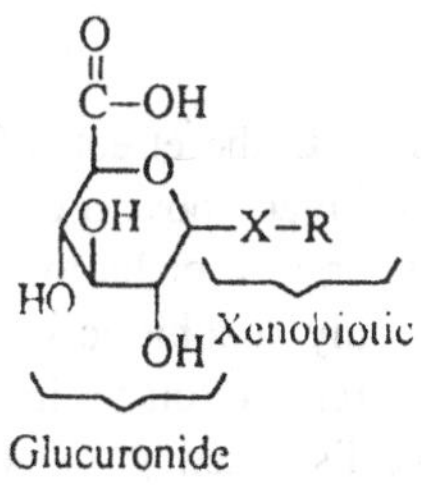

Glucuronide

Figure 7. *Glucuronide conjugate formed from a xenobiotic, HX — R.*

Most phase I processes are catalyzed by the *cytochrome P-450 enzyme* system associated with the endoplasmic reticulum of the cell and occurring most abundantly in the livers of vertebrates.

Phase II reactions

The polar functional groups attached to a xenobitic species in a phase I reaction provide reaction sites for *phase II reactions*. These are *conjugation reactions*, in which enzymes attach *conjugating agents* to xenobiotics, their phase I reaction products, and non-xenobiotic compounds (Figure 6).

The conjugation products of such a reaction are usually less toxic than the original xenobiotic species, less lipid-soluble, more water-soluble, and more readily eliminated from the body. The major conjugating agents and the enzymes that catalyze their phase II reactions are glucuronide, glutacione, sulfate and acetyl. The most abundant conjugation products are glucuronides. A glucuronide conjugate is illustrated in Figure 7.

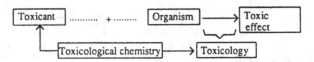

Figure 8. Interactions between toxicology, toxicological chemistry and environmental chemistry.

In this Figure 7, $-X-R$ represents a xenobiotic species conjugated to glucuronide and R is organic moiety. For example, if the xenobiotic compound conjugated is phenol, $HX-R$ is HOC_6H_5, X is O atom, and R represents the phenyl group, C_6H_5.

3. INTERACTIONS BETWEEN ENVIRONMENTAL CHEMISTRY, TOXICOLOGY AND ECOTOXICOLOGICAL CHEMISTRY

Here we will consider the interrelationship between toxicology, toxicological chemistry and environmental chemistry in the Asian region.

3.1. Toxicology

Toxicology is a science that deals with the effects of poisons on living organisms. A *poison* or *toxicant* is a substance that, above a certain level of exposure or dose, has detrimental effects on issues, organs, or biological processes. Many toxicants are xenobiotic materials, but not only xenobiotic species are toxicants. Among the toxicants, we should mention different pollutants in air, soil, water, plant and other environments in Asia that we have discussed in previous chapters.

The important aspects of toxicology are relationship between the demonstrated presence of a chemical or its metabolites in the body and observed symptoms of poisoning, mechanisms of biochemical transformation of toxicants in different species, the processes by which toxicants and their metabolites are eliminated from the organism, and treatment of poisoning with antidotes.

3.2. Toxicological Chemistry

Toxicological chemistry is the science that deals with the chemical nature and reactions of toxic chemicals, including their origins, uses, and chemical aspects of exposure, fates, and dispersal. Toxicological chemistry addresses the relationships between the chemical properties and molecular structures of molecules and their toxicological effects. Figure 8 shows the interactions between toxicology, toxicological chemistry and environmental chemistry.

Toxicities

The major variables in the way in which toxicants affect organisms are:

- toxicity of the same substance to different organisms;
- toxicity of different substances to the same organism;

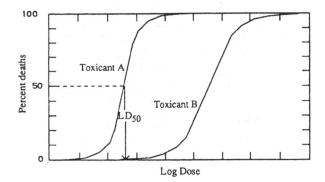

Figure 9. Dose-response curve, in which the response is the death of the organisms as a cumulative percentage at the Y axis.

- minimum levels for observable toxic effects;

- sensitivity to small increments of toxicant;

- levels at which most organisms experience the ultimate effects, particularly death;

- acute or chronic effects.

When a toxicant leaves no permanent effects, either through the action of the organism's natural defense mechanisms or the administration of substances to counteract the toxicant's action (antidote), it is a *reversible* effect. Effects that last after the toxicant is eliminated, such as the scar from a sulfuric acid burn on the skin, are termed *irreversible*. *Acute toxicity* refers to responses that are observed soon after exposure to a toxic substance. *Chronic toxicity* deals with effects that take a long time to be manifested. Chronic responses to toxicants may have latency periods as long as several decades in humans. Acute effects normally result from brief exposures to relatively high levels of toxicants and are comparatively easy to observe and relate to exposure to a poison. Chronic effects are often obscured by normal background maladies and tend to result from low exposure to a toxicant over relatively long periods of time. Chronic effects are much more difficult to study, but they are of great importance in dealing with hazardous wastes and pollutants in the Asian region.

Dose-Response relationship

Dose is defined as the degree of exposure of an organism to a toxicant, commonly in units of mass of toxicant per unit of body mass of the organism. The observed effect of the toxicant is determined as the *response*. A plot of the percentage of organisms that exhibit a particular response as a function of dose is known as a *dose-response curve*. The statistical estimate of the dose that would cause death in 50% of the subject is the inflection point of the S-shaped dose-response curve (Figure 9).

Table 1. Toxicity rating.

Rate of toxicity	Units, $mg\ kg^{-1}$ of body mass
Practically non-toxic	> 15,000
Slightly toxic	5,000–15,000
Moderately toxic	500–5,000
Very toxic	50–500
Extremely toxic	5–50
Super-toxic	< 5

The point of inflection is designated as LD_{50}, the concentration that leads to the death of 50% of the population. Figure 9 shows the LD_{50} values for two toxicants. We can see that the slopes of the dose-response curves may differ significantly.

A response to a very small level of toxicant is known as *hypersensitivity*, an allergic reaction, which is an exaggerated response of the body's immune system. On other hand, an observable response to only extremely high levels is called *hyposensitivity*.

Toxicity ratings

Substances may be assigned a toxicity rating, as shown in Table 1.

3.3. Physiological Aspects of Toxicants in the Body

The major routs and sites of absorption, metabolism, binding, and excretion of toxic substances in the body are shown in Figure 10

Toxicants in the body are metabolized, transported, and excreted; they have adverse biochemical effects and they cause manifestations of poisoning. We will divide these processes into two major phases, a kinetic phase and a dynamic phase.

Kinetic phase

In the *kinetic phase* a toxicant or the metabolic precursor of a toxic species (*pro-toxicant*) may undergo absorption, metabolism, temporary storage, distribution, and excretion (Figure 11).

A toxicant which is absorbed may be passed through the kinetic phase unchanged as an *active parent compound*, metabolized to a *detoxified metabolite* that is excreted, or converted to a *toxic active metabolite*.

Dynamic phase

In the dynamic phase, as shown in Figure 12, a toxicant or toxic metabolite interacts with cells, issues, or organs in the body to cause some toxic response.

The dynamic phase includes three major groups of effects, as follows:

- Primary reaction with a receptor or target organ;

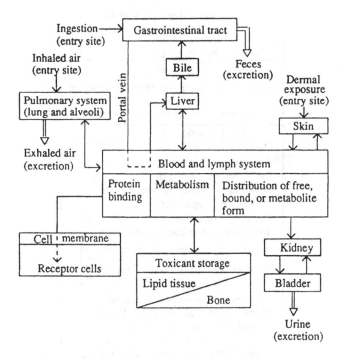

Figure 10. Physiological pathways of toxicants in the body (Manahan, 1994).

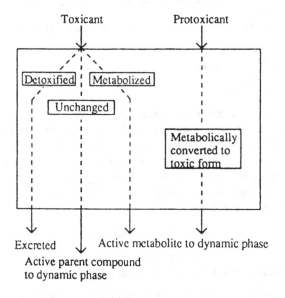

Figure 11. Physiological processes involving toxicants or protoxicants in the kinetic phase.

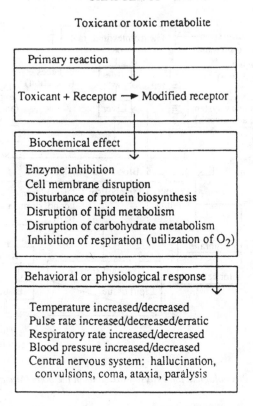

Toxicant or toxic metabolite

| Primary reaction |
| Toxicant + Receptor → Modified receptor |

Biochemical effect

Enzyme inhibition
Cell membrane disruption
Disturbance of protein biosynthesis
Disruption of lipid metabolism
Disruption of carbohydrate metabolism
Inhibition of respiration (utilization of O_2)

Behavioral or physiological response

Temperature increased/decreased
Pulse rate increased/decreased/erratic
Respiratory rate increased/decreased
Blood pressure increased/decreased
Central nervous system: hallucination,
   convulsions, coma, ataxia, paralysis

Figure 12. The dynamic phase of toxicant action (Manahan, 1994).

- A biochemical response;

- Observable effects.

A toxicant or an active metabolite reacts with a receptor to cause a toxic response. Such a reaction occurs, for example, when benzene epoxide forms an adduct with a nucleic acid unit in DNA resulting in alteration of the DNA. This is an example of an irreversible reaction between a toxicant and a receptor. A reversible reaction that can result in a toxic response is illustrated by the reaction (2), as the binding between carbon monoxide and oxygen-transferring hemoglobin, O_2Hb, in blood. Hemoglobin lost its ability to transfer the oxygen,

$$O_2Hb + CO \longleftrightarrow CoHb + O_2. \tag{2}$$

The binding of a toxicant to a receptor may result in some kind of biochemical effect. The major effects are the following (Manahan, 2000):

- Impairment of enzyme function by binding to the enzyme, coenzymes, metal activators of enzymes, or enzyme substrates;

- Interference with regulatory processes mediated by hormones or enzymes;

- Interference with lipid metabolism resulting in excess lipid accumulation with formation of so-called *fatty liver*;

- Alteration of cell membrane or carries in cell membranes;

- Interference with carbohydrate metabolism;

- Interference with respiration, the overall process by which electrons are transferred to molecular oxygen in the biological oxidation of energy-yielding substrates;

- Stopping or interfering with protein biosynthesis by the action of toxicants on DNA.

Physiological responses to toxicants

Prominent among the more chronic responses to toxicant exposure are mutations, cancer, and birth defects and effects on the immune system. Other observable effects, some of which may occur soon after exposure, include skin abnormalities (rash, dermatitis), cardiovascular disease, hepatic (liver) disease, gastrointestinal illness, renal (kidney) malfunction, and neurologic symptoms (central and peripheral nervous systems).

Among the more immediate and readily observed manifestations of poisoning are alterations in the vital signs of temperature, pulse rate, respiratory rate, and blood pressure. Symptoms of poisoning manifested in the eye include *miosis* (excessive or prolonged contraction of the eye pupil), *mydriasis* (excessive pupil dilation), *conjuctivitis* (inflammation of the mucus membrane that covers the front part of the eyeball and the inner lining of the eyelids) and *nystagmus* (involuntary movement of the eyeballs). Central nervous system poisoning may be manifested by convulsions, paralysis, hallucinations, and ataxia (lack of coordination of voluntary movements of the body), as well as abnormal behavior (agitation, disorientation, hyperactivity). Severe poisoning by some chemical species, such as carbamates and organophosphates, causes *coma*, the term used to describe a lowered level of consciousness (Manahan, 2000).

4. EFFECTS OF TOXICANTS ON CARCINOGENESIS, TERATOGENESIS, MUTAGENESIS, AND THE IMMUNE SYSTEM

We should point out here that often the effects of toxicant exposure are subclinical in nature. The most common of these are damage to the immune system, chromosomal abnormalities, and modification of functions of liver enzymes.

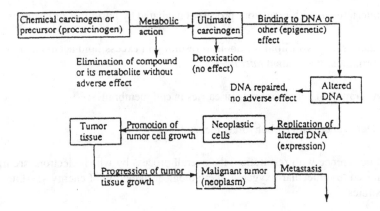

Figure 13. Outline of the process of carcinogenesis (Manahan, 1994).

4.1. Carcinogenesis

Chemical *carcinogenesis* is the term that applies to the role of species foreign to the body in causing the uncontrolled cell replication commonly known as cancer (Manahan, 2000). The biochemistry of cancer toxicants is considered in Box 1.

Box 1. Biochemistry of carcinogens (After Manahan, 2000)

Cancer is a condition characterized by the uncontrolled replication and growth of the body's own cells (somatic cells). *Carcinogenic agents* may be categorized as follows:

1. Chemical agents, such as many chemical organic and inorganic compounds;

2. Biological agents, such as hepadnaviruses or retroviruses;

3. Ionizing radiation, such as X-rays;

4. Genetic factors, such as selective breeding.

Clearly, in many cases, cancer is the result of the action of synthetic and naturally occurring chemical species. The role of chemicals in causing cancer is called *chemical carcinogenesis*. It is often regarded as the single most important facet of toxicology and clearly the one that receives the most publicity.

Large expenditures of time and money on the subject in recent years have yielded a much better understanding of the biochemical bases of chemical carcinogenesis. The overall process for the induction of cancer may be quite complex, involving numerous steps. However, it is generally recognized that there are two major steps in carcinogenesis: an *initiation stage*; followed by a *promotional state*. These steps are further subdivided as shown in Figure 13.

Initiation of carcinogenesis may occur as a reaction of a DNA-reactive species with DNA or by the action of an *epigenetic carcinogen* that does not react with DNA and is carcinogenic by some other mechanism. Most DNA-reactive species are genotoxic carcinogens because they are also mutagens. These substances react irreversibly with DNA. They are either electrophilic or, more commonly, metabolically activated to form electrophilic species, as is the case with electophilic $^+CH_3$ generated from dimethylnitrosoamine (see below Section 5 and Chapter 13 Biogeochemistry of nitrogen). Cancer-causing substances that require metabolic activation are called *procarcinogens*. The metabolic species actually responsible for carcinogenesis is termed an *ultimate carcinogen*. Some species that are intermediate metabolites between procarcinogens and ultimate carcinogens are called *proximate carcinogens*. Carcinogens that do not require biochemical activation are categorized as *primary* or *direct-acting carcinogens*. Most substances classified as epigenetic carcinogens are *promoters* that act after initiation. Manifestations of promotion include an increased number of tumor cells and decreased length of time for tumors to develop or in other words, shortened latency period.

A description of studies used in assessing carcinogenic risk is given below.

Epidemiological studies

Historically, epidemiology was the study of epidemics. One of the earliest studies was British physician John Snow's identification of contaminated water from the Thames River as the source of cholera during 1853–1854. This led to the eradication of cholera epidemics in London.

During the latter half of the 20th century, epidemiology moved beyond epidemics and infectious diseases. Its techniques have been adapted to identifying a variety of sources of diseases, including chemically induced cancer. Presently, *epidemiology* is defined as the study of the distribution and determination of health-related states and events in specified populations and the application of this study to the control of health problems.

The preferred source for assessing human health risks from environmental pollution is epidemiologic data. Using human data avoids the uncertainties resulting from species extrapolation (see below). Large numbers of exposed individuals are often available, allowing detection of effects at lower concentration than animal studies and thus practically overcoming uncertainties from low dose extrapolation. However, unlike animal studies, humans are generally exposed to a variety of environmental impacts, often making it difficult to assess the effects of the agent in question. Moreover, in many cases, lack of accurate exposure data precludes quantification of cancer risk.

Animal bioassays

For many chemicals thought to be toxic to humans, no direct quantitative data are available about the human dose-response curve. Thus, tests conducted on animals must be used instead. In most cases, even where there are data, the human will have been briefly exposed to high doses that result in acute effects (e.g., from accidents) and not exposure for long periods, which is often the concern in risk of cancer development due to pollution.

Whether the data are of human or animal origin, there is extreme biologic variation in the response at increasingly lower doses, such as may be actually encountered in the Asian environments. Thus, in order to detect reliably an increase of one cancer case per 10,000 animals (which would be considered a fairly high risk in a human population), many hundred thousands of test animals would have to receive the same low dose and be followed for their lifetimes. Because of the cost of such experiments at low doses, the frequency of health effects can only be estimated from studies conducted at high doses. Even so, relative to the number of potentially toxic chemicals in the Asian countries, few chemicals have been fully tested under rigorous conditions.

A number of methods (mathematical models) are available for extrapolating the results at high doses to estimate the frequency of effects at low doses. The data for a given chemical that has been tested on animals are used to fit a mathematical model, using regression techniques. That is, for each dose the total number of animals treated is divided into numbers of animals that showed specific effects (e.g., development of tumor). This fraction or probability of response for each dose comprises the raw data for the model. The results of the available animal studies are collected in various updated databases.

One of several kinds of mathematical relationships can be used to relate dose and effect in order to translate these data to risk estimates at low doses, e.g., linear, quadratic, logit, Weibull, one-hit, multi-hit. Each model is based on certain biochemical and physiological assumptions and has advantages and disadvantages. No one method has been shown to be better than the others. Often they all show a good fit to the experimental data available for different chemicals at higher doses.

Unfortunately, however, at low doses these different models often predict very different response frequencies. Differences can be as great as 4 or 5 orders of magnitude (Figure 14).

Since the linear model consistently predicts the highest response frequency per unit dose in very low dose range, it is usually the most conservative or least likely to underestimate human health risk. It is often recommended by regulating agencies to determine the risk of known or suspected human carcinogens. For non-carcinogens, the quadratic model that predicts a threshold dose at which there is no effect is often used.

By using these models, a prediction can be made as to how many additional negative health effects (e.g., excess cases of cancer) will occur, given various low doses of exposure by humans over a life time, for a specific chemical species (Box 2).

Box 2. Estimation of cancer risk of chemicals in drinking water (after ADB, 1992)

We can consider the following example. A known carcinogen is discovered in the drinking water supply, and a search is made of the database to find the potency of

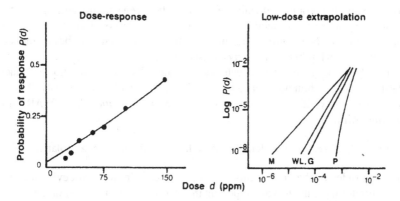

Figure 14. Comparative dose-response extrapolations for a carcinogen: M, multistage model; W, Weilbull model; L, logit model; G, gamma multi-hit model; P, probit model. Note how models that fit this data equally well at high doses can produce very different results when extrapolated to low doses (ADB, 1992).

this carcinogen (i.e., the frequency of cancers observed in animal tests—assuming no human data—at various high doses). These data are used in a linear regression model to fit a line and extrapolate to lower doses similar to the drinking water concentrations. Supposing that we have not measured actual exposures, only water concentration; we need to make exposure assumptions. Assume that people drink two liters of water per day for 70 years (lifetime). In terms of equivalent milligrams of chemical per kilogram of animal body mass per day for the lifetime exposure of the animal, two animals out of 10,000 were expected to develop cancer (also assuming zero additional cancer in the control animals). The estimate of risk for human beings is thus 2×10^{-5} excess cancers over a lifetime of exposure. In this example, it is assumed that the linear model accurately extrapolates to lower dose responses and that there are no differences between animal and human responses to this carcinogen. Weight is used to relate animal size to human size but body surface area could have been used instead with very different results. The effect of other exposures that could either reduce or increase the response to this carcinogen was not considered.

The reader must bear in mind that estimating dose-response relationships is a complex task. Only some basics have been addressed here.

Structure-activity relationships

A *structure-activity relationship* (SAR) is an empirical means of connecting the chemical reactivity or biological potency of a series of chemically related compounds to aspects of chemical structure. As a simple example, a structure-activity relationship exists between water solubility and the chain length of aliphatic alcohols and this can be formulated as follows: the longer the carbon chain, the lower the water solubility.

Another example is connected with a relationship between carcinogenicity and structure of certain nitrogen-containing organic species, like the conclusion that structures containing the $N-N=O$ group and known as nitrosoamines, tend to be carcinogenic. The SAR concept has been refined by the assignment of empirical numerical values to structural features of a molecule, such as size, shape, electron distribution, aromacity, polarity, hydrogen bonding capacity, hydrophilicity and hydrophobicity. Multiple linear regression yields a *quantitative structure-activity relationship* (QSAR) of the following form:

Property of interest $= a \times$ length $+ b \times$ hydrophobicity $+ c \times$ polarity $+ \cdots$.

The magnitudes of the coefficients, $a, b, c, \ldots$ and their standard deviations reveal which structural parameters are important in determining the intensity of the property of interest. We should relate these features as follows: a large coefficient means an important factor, a small coefficient or wide confidence interval is connected with a relatively unimportant factor.

SAR and QSAR are of great importance in predicting the properties of newly discovered or synthesized chemical species. Chemists can use these values in optimizing biologically active structures for pesticides and other agricultural chemicals, pharmaceuticals, and new substitutes for CFCs. Environmental scientists can apply SAR and QSAR in predicting the biological and toxicological properties of substances newly released or likely to be released into the environment.

Relationships among chemical structure, mutagenicity, and carcinogenicity were determined for 301 chemicals (Pepelko and Valcovic, 1999). The following parameters are useful in comparing a species with its structural analogues and congeners that produce tumors and affect related biological processes, such as receptor binding, activation, mutagenicity, and general toxicity:

- Nature and reactivity of the electophilic moiety(ies) present;

- Potential to form electrophilic reactive intermediate(s) through chemical, photochemical, or metabolic activation;

- Contribution of the carrier molecule to which the electophilic moiety(ies) is attached;

- Physical-chemical properties, like physical state, solubility, octanol-water partition coefficient, half-life in aqueous solution, etc;

- Structural and substructural parameters, such as electronic, steric, molecular, geometric, etc;

- Metabolic pattern (e.g., metabolic pathways and activation and detoxication ratio);

- Possible exposure route(s) of the agent.

Suitable analyses of non-DNA-reactive chemicals and of DNA-reactive chemicals that do not bind covalently to DNA require knowledge or postulation of the probable mode or modes of action of closely related carcinogenic structural analogs (e.g., receptor-mediated, cytotoxicity-related). Then examination of the physical-chemical and biochemical properties of the agent may provide the rest of the information needed to assess the likelihood of the agent's activity by the mode of action.

Biomarkers

A *biomarker* is a biological endpoint signaling a critical pretoxicological or precarcinogenic agent. The use of biological responses (biomarkers) in organisms present in a polluted environment is an approach that has caught the attention of specialists in ecotoxicology and toxicological chemistry. One of the primary reasons to conduct tests on organisms in situ is to provide a more realistic exposure regime. Chemical and physical agents are used to elicit measurable and characteristic biological responses in exposed organisms. Such evidence can provide a link between exposure and effect.

The specificity of biomarkers to chemicals varies greatly. The current status of measuring biomarkers, given in order of decreasing specificity, is summarized in Table 2.

At one end of the spectrum the enzyme aminolevulinic acid dehydratase (ALAD) is highly specific, being inhibited only by lead. At the other end, mixed function oxydase enzymes and stress proteins are induced by many different classes of chemicals. Both specific and nonspecific biomarkers have their place in environmental assessments. A nonspecific biomarker can tell one that a pollutant is present in a meaningful concentration but does not tell one which chemical is present. Based on this information, a more detailed chemical investigation can be justified. In contrast, a specific biomarker tells one which chemical is present, but gives no information on the presence of other pollutants.

The most commonly investigated biomarkers for carcinogenesis are protein or DNA adducts. These *adducts* are chemicals bound to the protein or DNA molecule, usually covalently. Unless repaired before cell division occurs, DNA adducts cause errors in transcription. As a result, the normal activity of the transcribed protein is often changed or inactivated.

Biomagnification and bioconcentration

Some chemicals, such as organochlorines (see Chapter 14 also), are relatively inert to both oxidation and hydrolysis reactions and they are considered as persistent organic pollutants, POPs. In addition, organochlorines are rather non-polar, lipophilic substances. It means that they are soluble both in non-polar organic solvents and in the lipids of organisms (Bunce, 1994).

The *octanol-water partition coefficient*, K_{ow}, is the measure of the lipophilicity of POPs. Octanol is an organic solvent that serves as a model for typical lipids. Highly lipophilic substances, such as DDT and the PCBs (polychlorinated biphenyls)

Table 2. Some biomarkers and their status (Peakall and Shugart, 1999).

Biomarker	Measure of toxic effects	Pollutant	Status
Inhibition of ALAD	Yes	Lead	Sufficiently reliable to replace chemical analysis; can be related to mortality
Induction of metallothionein	No	Cadmium	No advantage over chemical analysis; not related to mechanism of action
Egg shell thinning	Yes	DDT, DDE, Dicofol	Wide variation in sensitivity; related to reproductive success
Anticoagulant; clotting proteins	Yes	Rodenticides	Has been related to mortality, risk assessed from blood protein level
Porphyrin profile	No	Organochlorine species	Levels of porphyrin found in environmental samples are well below those causing adverse effects
Depression of plasma, retinol, and thyroxide	Yes	3,4,3',4"Tetra-chlorbiphenyl	Dermal and epithelia lesions; binding to specific protein has been shown
Inhibition of AchE	Yes	OPs and carbamates	More reliable than chemical analysis
Induction of vitellogenin	No	Alkylphenols	Directly related to mortality; mechanism related to environmental estrogens
Induction of mixed function oxidases	Yes	Organochlori-nes, PAHs	Analysis of TCDD-EQ has been related to reproductive success; Inhibition of P450s related to specific chemicals
DNA and hemoglobin adducts	No	PAHs	Good monitor of exposure, especially for PAHs
DNA integrity	Yes	Metals, OCs, PAHs, and genotoxicants	DNA damage is serious indication of harm; relationship to effects tenuous
Other serum enzymes	No	Metals, OCs, OPs	A considerable number of enzymes are altered by pollutants, but relationship to effects are not clear
Immune response	Yes	Metals, OCs, PAHs	Proper functioning of immune system is critical to health, but system has considerable reserve
Stress proteins	No	Metals and OCs	Difficult to separate effects of chemicals from other stressors

AchE—acetylcholinesterase; ALAD—aminolevulinic acid dehydrotase; DDE—2,2-bis(p-phenol)-1,1,-dichloroethylene; DDT — 1,1,1-trichloro-2,2-bis (chlorophenyl) ethane; Dicofol—2,2,2,-trichloro-1,1,-di(4-chlorophenyl) ethanol; OCs—organochlorines; OPs—organophosphates; PAHs—polycyclic aromatic hydrocarbons; TCDD-EQ—dioxin equivalents.

have $K_{ow} > 10^6$ and have a strong tendency to partition into lipids out of water. These compounds are generally found at higher concentration in the issues of aquatic organisms than in the water, in which these organisms live.

This phenomenon is known as *bioconcentration (bioaccumulation)* and is common for substances having $K_{ow} > 1000$. The relative phenomenon, *biomagnification* through the food chain, occurs when predator species ingest the xenobiotic burdens of their prey, resulting in higher pollutant concentration in the predator than the prey.

Thus, we can determine the octanol-water partition coefficient as

$$K_{ow} = \frac{\text{concentration of solute in octanol}}{\text{concentration of solute in water}}.$$

Compounds with large K_{ow} such as DDT, PCBs, dioxins, and some other pesticides are found at elevated levels in aquatic species taken from the surface water reservoirs, lakes and rivers in the Asian region.

The bioconcentration factor, BCF, is the ratio of the average concentration of the solute in the whole organism to the concentration of the solute in water

$$BCF = \frac{\text{concentration of solute in organism}}{\text{concentration of solute in water}}.$$

By taking octanol as a model solvent for fat, and assuming that the fatty tissues of an aquatic organism have reached equilibrium with the surrounding water, the BCF of a lipophilic solute can be related to K_{ow} accounting for the percent by weight of fat in the organism

$$BCF = K_{ow} \times \% \text{ by weight of fat.}$$

If we assume that the average concentration of fat in aquatic organisms is about 5%, then

$$BCF = 0.05 \, K_{ow}.$$

Although bioconcentration causes the level of a potentially toxic substance to be much higher in an aquatic organism than in the water in which it lives, equilibrium is not always reached. For example, the weight-adjusted body burden of PCBs in lake trout captured from the Great Lakes (North America) increases with size/age (Bunce, 1994). See more details in Box 3.

Box 3. Application of kinetic constants for BCF calculations (after Bunce, 1994)

Bioconcentration factor, BCF, can be related to the rate constants of uptake and depuration, including metabolism and excretion of a solute

$$BCF = \frac{k(\text{uptake})}{k(\text{depuration})}.$$

All substances prone to bioconcentration are taken from the environment much faster than they are depurated. Indeed, the true criterion for bioconcentration is a faster rate constant for uptake than for depuration, rather than lipophilicity. This criterion explains the bioconcentrating properties of both lipophilic organic compounds and cumulative metallic toxicants.

We can apply kinetic principles to the phenomenon of bioconcentration. It is possible to describe the uptake (rate constant k_1), clearance (also called depuration, rate constant k_2), and metabolism (rate constant k_3) of non-polar pollutants into organisms such as fish by means of kinetic equations. Analysis of the dependence of the concentration of the toxic substance with time is called *toxicokinetics*

$$C \text{ (aq)} \longrightarrow c \text{ (fish)} \quad k_1,$$
$$C \text{ (fish)} \longrightarrow c \text{ (aq)} \quad k_2,$$
$$C \text{ (fish)} \longrightarrow c \text{ (metabolic products)} \quad k_3.$$

Further we should use the language of calculus to describe the rates of these processes mathematically. First, let us look at the rate of each process separately

$$\text{rate}(1) = k_1 \cdot c \text{ (aq)},$$
$$\text{rate}(2) = k_2 \cdot c \text{ (fish)},$$
$$\text{rate}(3) = k_3 \cdot c \text{ (fish)}.$$

Equation (3) describes the dependence of c (fish) with time; the uptake process tends to increase the concentration of xenobiotic in the fish, whereas depuration and metabolism both tend to decrease this concentration

$$\frac{dc\text{(fish)}}{dt} = k_1 \cdot c \text{ (aq)} - (k_2 + k_3) \cdot c \text{ (fish)}. \tag{3}$$

For aquatic life living in a large lake, we can assume that c (aq) is constant, i.e., uptake of the pollutant by aquatic life does not appreciably change the concentration in the water. This will be true for the fish in a large lake, but might not be true for a fish in an experimental aquarium.

Equation (3) simplifies when equilibrium has been reached, because then the rates of uptake and elimination are equal. Under these circumstances dc (fish)/dt $= 0$, and equation (3) simplifies to equation (4),

$$k_1 \cdot c \text{ (aq)} = (k_2 + k_3) \cdot c \text{ (fish, equilibrium)}. \tag{4}$$

By rearrangement of equation (4), we can relate the bioconcentration factor to the rate constants for uptake and clearance of the toxicant. This may be compared with the earlier definition of BCF based on equilibria; the advantage of the kinetic definition is that the BCF may be deduced in cases where equilibration has not been achieved, provided that the rate constants can be estimated

$$\text{BCF} = \frac{c \text{ (fish, equilibrium)}}{c \text{ (aq)}} = k_1(k_2 + k_3).$$

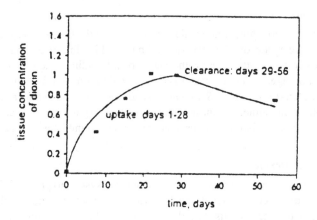

Figure 15. Uptake and clearance of 1,3,6,8-tetrachloro-p-dioxin by juvenile trout in water (Bunce, 1994).

Figure 15 shows an example of the uptake of a chlorinated dioxin by young trout from water in a laboratory experiment.

Uptake continued for 28 days with the dioxin concentration in the water kept constant during that time. After 28 days, the trout were transferred to clean water, and the loss of the toxic substances from the trout was followed for a further 28 days. For this toxicant, k_3 was essentially zero (very slow metabolism), and from these data, it is possible to determine the rate constants k_1 and k_2 and the BCF. The values, to one significant figure, were $k_1 = 200 \, \text{day}^{-1}$ and $k_2 = 0.1 \, \text{day}^{-1}$, leading to BCF $= k_1 / k_2 = 2000$.

4.2. Teratogenesis and Mutagenesis

Teratogenesis

Teratogens are chemical species that cause birth defects. These usually arise from damage to embryonic or fetal cells. However, mutations in germ cells (egg or sperm cells) may cause birth defects, such as Down's syndrome. The biochemical mechanisms of teratogenesis are varied. These include enzyme inhibition by xenobiotics; deprivation of the fetus of essential substrates, such as vitamins; interference with energy supply; or alteration of the permeability of the placental membrane.

Mutagenesis

Mutagens alter DNA to produce inheritable trails. Although mutation is a natural process that occurs even in the absence of xenobiotic substances, most mutations are harmful (Manahan, 2000). The mechanisms of mutagenicity are similar to those of carcinogenicity and mutagens often cause birth defects as well. Therefore, mutagenic hazardous substances are of major toxicological concern.

Bruce Ames test

Mutagenicity used to infer carcinogenicity is the basis of the *Bruce Ames* test, in which observations are made on the reversion of mutant histidine-requiring *Salmonella* bacteria back to a form that can synthesize its own histidine. The test makes use of enzymes in homogenized liver issue to convert potential procarcinogens to ultimate carcinogens. Histidin-requiring *Salmonella* bacteria are inoculated onto a medium that does not contain histidine, and those mutate back to a virgin form that can synthesize histidin and establish visible colonies that are assayed to indicate mutagenicity.

4.3. Immune System Response

The *immune system* acts as the body's natural defense system to protect it from xeno-biotic chemicals; infectious agents, such as viruses or bacteria; and neoplastic cells, which give rise to cancerous tissue. Adverse effects on the body's immune system are being increasingly recognized as important consequences of exposure to dangerous species. Pollutants can cause *immunosuppression*, which is the impairment of the body's natural defense mechanisms. Xenobiotics can also cause the immune system to lose its ability to control cell proliferation, resulting in leukemia or lymphoma.

Some toxicants like beryllium, chromium, nickel, formaldehyde, different pesti-cides, resins and plasticizers are known as the chemical species, which cause *allergy* or *hypersensitivity* of the immune system. This kind of response results when the immune system overreacts to presence of a foreign agent or its metabolites in a self-destructive manner.

Allergy is very common now in the polluted regions and cities of the Asian countries.

5. ECOTOXICOLOGICAL PROPERTIES OF PRINCIPAL POLLUTANTS

This section discusses toxicological aspects of different chemical species, which are of great concern in the various environmental compartments in the Asian region. The description of toxicological features of other chemical compounds can be found in S. Manahan (2000) and Chapters 12–14 of this book.

5.1. Toxic Inorganic Compounds

Ozone

Ozone, O_3 (see Chapter 3) has several toxic effects. Air containing 1 ppmv of ozone has a distinct odor. Inhalation of ozone at this level causes severe irritation and headache. Ozone irritates the eyes, upper respiratory system, and lungs. Inhalation of ozone can cause fatal pulmonary edema. Chromosomal damage has been observed in subjects exposed to ozone. Ozone generates free radicals in tissue. These reactive species can cause lipid peroxidation, oxidation of sulfhydryl ($-SH$) groups, and other oxidation processes. Compounds that protect organisms from the effects of ozone include radical scavengers, antioxidants, and compounds containing sulfhydryl groups.

Heavy metals

Heavy metals are the metals with atomic mass > 56 units. They (see also Chapter 12) are particularly toxic in their chemically combined forms and some, notably mercury, are toxic in the elemental and organic forms. The toxic properties of some of the most hazardous heavy metals and metalloids are discussed in this sub-section.

Beryllium (atomic mass 9.01) is not truly a heavy metal, but this is one of the more hazardous toxic elements. Its most serious toxic effect is berylliosis, a condition manifested by lung fibrosis and pneumonitis, which may develop after a latency period of 5–20 years. Beryllium exposure also causes skin granulomas and ulcerated skin and is a hypersensitizing agent.

Elemental *mercury* vapor can enter the body through inhalation and be carried by the bloodstream to the brain, where it penetrates the blood-brain barrier. It disrupts metabolic processes in the brain causing tremor and psychopathological symptoms such as insomnia, shyness, depression, and irritability. Divalent ionic mercury, Hg^{2+}, damages the kidney. Organometallic mercury compounds such as dimethylmercury, $Hg(CH_3)_2$, are also very toxic.

Cadmium affects adversely several important enzymes. It can cause also painful bone disease, osteomalacia, and kidney damage. Inhalation of cadmium oxide dusts and fumes results in cadmium pneumonitis, which are characterized by edema and pulmonary epithelium necrosis.

Lead, widely distributed as metallic lead, inorganic and organometallic species, has a number of toxic effects, including inhibition of the synthesis of hemoglobin and adversely effects on central and peripheral nervous systems and kidneys.

Arsenic is a metalloid, which forms a number of toxic compounds, and the most toxic is arsenite oxide, As_2O_3. This As^{3+} is absorbed through the lungs and intestines. In biochemical processes, As acts to coagulate proteins, forms complexes with coenzymes, and inhibits the production of adenosite triphosphate (ATP) in essential metabolic reactions.

Cyanide

Both *hydrogen cyanide*, HCN, and *cyanide salts*, are rapidly acting poisons. A dose of only 60–90 mg is sufficient to kill a human. Cyanide bonds to Fe^{3+} in an iron-containing ferrocytochrome oxidase enzyme, preventing its reduction to Fe^{2+} in the oxidative phosphorylation process, by which the body utilizes oxygen. The crucial enzyme is inhibited because ferrocytochrome oxidase, which is required to react with O_2, is not formed and utilization of oxygen in cells is prevented so that metabolic processes cease.

Carbon monoxide

Carbon monoxide, CO, is a common cause of accidental poisoning (Table 3).

After entering the blood through the lung, carbon monoxide reacts with hemoglobin and carboxyhemoglobin prevents hemoglobin from carrying oxygen to body issues.

Table 3. CO *concentration in air and toxic effects.*

CO concentration in air, ppm	Toxic effects
10	Visual perception
100	Dizziness, headache, weariness
250	Loss of consciousness
1000	Rapid death

Asbestos

Asbestos is the name of a group of fibrous silicate minerals, typically those of the serpentine group, for which the approximate formula is $Mg_3P(Si_2O_5)(OH)_4$. For many decades, asbestos has been widely used in structural materials, brake linings, insulation, and pipe manufacture. Inhalation of asbestos may cause asbestosis (a pneumonia condition), mesothelioma (tumor of the mesothelial tissue lining the chest cavity adjacent to the lung), and bronchogenic carcioma (cancer originating with the air passages in the lungs). At present, many applications of asbestos have been phased out and widespread programs undertaken to remove the material from the buildings.

Sulfur dioxide

Sulfur dioxide, SO_2, dissolves in water and accordingly, largely removed in the upper respiratory tract. It is an irritant to the eyes, skin, mucous membranes and respiratory tract. Some individuals are hypersensitive to sodium sulfite, Na_2SO_3, which is used as a chemical food preservative. Sulfur dioxide is especially hazardous for ecosystems due to acid rains (see Chapter 4).

5.2. Organometallic Compounds

The toxicological properties of some organometallic species—pharmaceutical organoarsenicals, organomercury fungicides, and tetraethyllead antinock gasoline additives —that have been used for many years are well known. However, similar toxicological experience is lacking for many relatively new organometallic compounds that are now in semiconductors, as catalysis, and for chemical synthesis, so they should be treated with great caution until proven safe.

Many organometallic compounds often behave in the body in ways totally unlike inorganic forms of the same metals. Their increasing toxicity is mainly due to larger lipid solubility.

Organolead compounds

Tetraethylled, $Pb(C_2H_5)_4$, is the most toxic among organolead compounds. This is colorless, oily liquid that was used as an octane-boosting gasoline additive. TEL has a strong affinity for lipids and can enter the body by inhalation, ingestion, and absorption through the skin. It affects the central nervous system with symptoms such as fatigue, weakness, restlessness, ataxia, psychosis, and convulsions. Recovery from severe TEL poisoning tends to be slow and in case of fatal poisoning, death has occurred as soon as one or two days after exposure.

At present, TEL has been phased out in many Asian countries and other additives are in use.

Methylcyclopentadiynyl manganese tricarbonyl (MMT)

MMT is an organic derivative of manganese used to increase the octane level of gasoline, replacing lead compounds. MMT can be absorbed into the body by inhalation, through the skin and by ingestion. It irritates the eyes, the skin and the respiratory tract. The substance may cause effects on the central nervous system, resulting in tissue lesion. Acute symptoms of MMT hazard via inhalation and ingestion are dizziness, headache, nausea, labored breathing, and abdominal pain. In concentrated form it is highly toxic by all routes of exposures, approximately 5–15 ml within 3–5 minutes. A harmful contamination of the air can occur rather quickly on evaporation of this substance at 20 °C.

Methyl tertiary butyl ether (MTBE)

MTBE is a volatile organic chemical. It is the most widely used oxygenate in Asian countries for national reformulation gasoline programs, for instance in Thailand, as extension of the oxyfuels program. MTBE can reduce carbon monoxide and ozone levels in the urban atmosphere because it promotes more complete burning of gasoline. The anesthetic concentration (AC_{50}) and lethal concentration (LC_{50}) in mice are 1.0 and 1.6 mMol/L, respectively. The lethal dose (LC_{50}) of oral MTBE in rats is 4 mg/kg. Acute animal tests have demonstrated MTBE to have low acute toxicity via inhalation and moderate acute toxicity via ingestion. Acute inhalation exposure can resulted in ataxia and abnormal gait in rats.

Chlorinated cyclopentadienes

Cyclopentadiene is an abundant byproduct of petroleum refining. As its name implies, there are two double bonds in each molecule. When fully chlorinated (Figure 16), it can be combined with one of several other organic molecules to produce a whole series of insecticidal compounds with properties such as environmental persistence that made them superficially attractive. Most of these compounds are toxic for humans and animals.

Organotin compounds

Among various organotin compounds, tributyltin chloride and related tributyltin, TBT, are of the most importance. These compounds have bactericidal, fungicidal, and insecticidal properties and have particular environmental significance because of

cyclopentadiene   perchlorocyclopentadiene

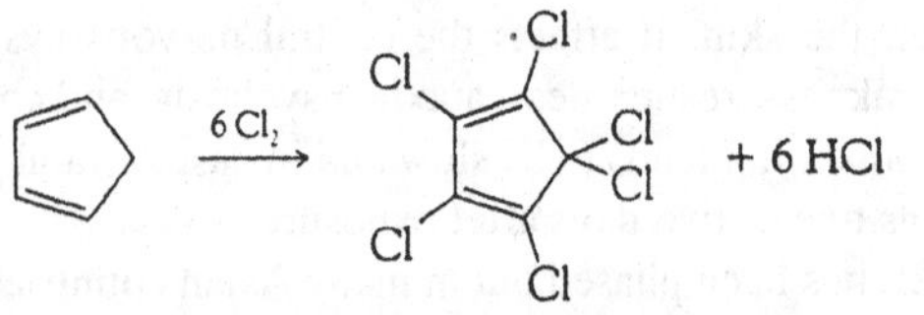

Figure 16. Conversion of cyclopentadiene to perchlorocyclopentadiene (Manahan, 2000).

their increasing applications as industrial biocides. Organotin compounds are readily absorbed through the skin, sometimes causing a skin rash. They bind with sulfur groups on proteins and interfere with mitochondrial functions.

Carbonyls

Metal carbonyls, especially nickel carbonyl, $Ni(CO)_4$, and iron pentacarbonyl, $Fe(CO)_5$, are extremely hazardous species. They are volatile and readily taken into the body through the respiratory tract or through the skin. The carbonyls affect tissue directly and they break down to toxic carbon monoxide and products of metal, which have additional toxic effects.

5.3. Toxic Organic Compounds

Benzene and aromatic hydrocarbons

Inhaled benzene is readily absorbed by blood, from which it is strongly taken up by fatty tissues. For the non-metabolized compound, the process is reversible and benzene is excreted though the lungs. Benzene is converted to phenol by a Phase I oxidation reaction (see Section 2) in the liver. The benzene epoxide intermediate in this reaction is responsible for the unique toxicity of benzene, which involves damage to bone marrow. Benzene is a skin irritant, and progressively higher local exposures can cause skin redness (erythema), burning sensations, fluid accumulation (edema) and blistering. Inhalation of air containing 7 g m^{-3} of benzene causes acute poisoning within an hour, because of a narcotic effect upon the central nervous system manifested progressively by excitation, depression, respiratory system failure, and death. Inhalation of air containing $> 60 \text{ g m}^{-3}$ of Benzene can be fatal within a few minutes.

5.4. Polycyclic Aromatic Hydrocarbons, PAHs

Benza(a)pyrene (see Chapter 3) is the most studied of the PAHs. Some metabolites of PAH compounds, particularly the 7,8-diol-9,10 epoxide of benz(a)pyrene shown in Figure 17, are known to cause cancer. There are stereoisomers of this metabolite, both of which are known to be potent mutagens and presumably can cause cancer.

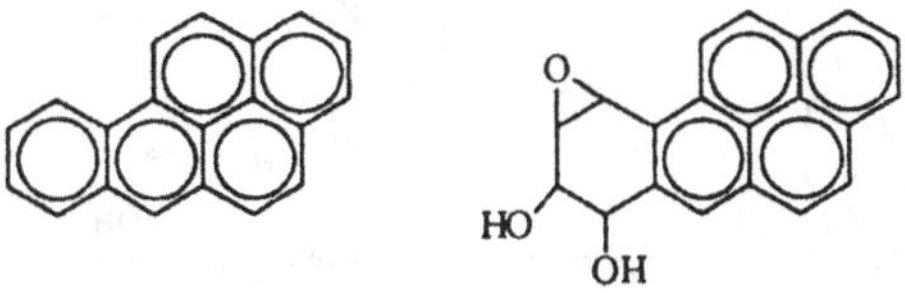

Benzo(a)pyrene   7,8-Diol-9,10-epoxide of benzo(a)pyrene

Figure 17. Benza(a)pyrene and its metabolic product.

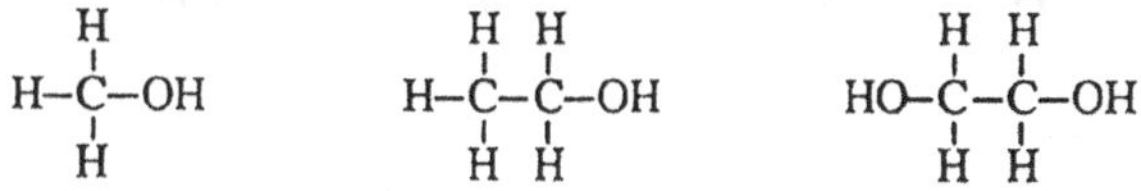

Methanol   Ethanol   Ethylene glycol

Figure 18. Light alcohols.

Alcohols

Human exposure to the three light alcohols shown in Figure 18 is common because they are widely used industrially and in consumer products.

Alcohols are oxygenated compounds in which the hydroxyl functional group is attached to an aliphatic or olefinic hydrocarbon skeleton.

Methanol, which has caused many fatalities when ingested accidentally or consumed as a substitute for beverage ethanol, is metabolically oxidized to formaldehyde and formic acid. In addition to causing acidosis, these products affect the central nervous system and the optic nerve. Acute exposure to a lethal dose causes an initially mild inebriation, followed in about 10–20 hours by unconsciousness, cardiac depression, and death. Subletal exposures can cause blindness from deterioration of the optic nerve and retinal ganglion cells. Inhalation of methanol fumes may result in chronic, low level exposure.

Ethanol is usually ingested through the gastrointestinal tract, but can be absorbed as vapor by the alveoli of the lungs. Ethanol is oxidized metabolically more rapidly than methanol, first to acetaldehyde, then to CO_2. Ethanol has numerous acute effects resulting from central nervous system depression. These range from decreased inhibitions and slowed reaction times at 0.05% blood content of ethanol, through intoxication, stupor and—at more than 0.5% blood ethanol—death.

Despite its widespread use in automobile cooling systems, exposure to *ethylene glycol* is limited by its low vapor pressure. However, inhalation of droplets of ethylene glycol can be very hazardous. In the body, ethylene glycol initially stimulates the central nervous system, then depresses it.

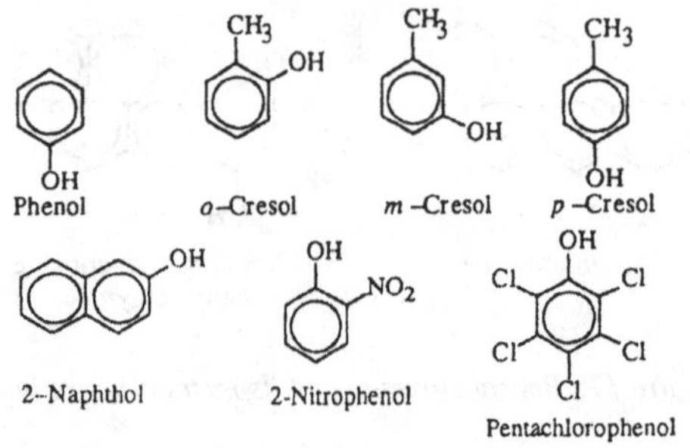

Figure 19. Some phenols and phenolic compounds.

Phenols

Figure 19 shows some of the more important phenolic compounds, aryl analogs of alcohols, which have properties much different from those of the aliphatic and olefinic alcohols.

Nitro- groups ($-NO_2$) and halogen atoms (particularly Cl) bonded to the aromatic rings strongly affect the chemical and toxicological behavior of phenolic compounds.

Although the first antiseptic used on wounds and in surgery, phenol is a protoplastic poison which damages all kinds of cells and is alleged to have caused an astonishing number of poisonings since it came into general use (Manahan, 2000). The acute toxicological effects of phenol are predominantly upon the central nervous system and death can occur as soon as one half hour after exposure. Acute poisoning by phenol can cause severe gastrointestinal disturbances, kidney malfunction, circulatory system failure, lung edema, and convulsions. Fatal doses of phenol may be absorbed though the skin. Key organs damaged by chronic exposure to phenol include the spleen, pancreas, and kidneys. The toxic effects of other phenols shown in Figure 19, are similar to those of phenol.

Nitrosoamines

N-nitroso compounds (*nitrosoamines*) are characterized by the $-N-N=O$ functional group. They are monitored in many materials to which human beings may be exposed, like whiskey, beer and cutting oil used in machining. Dimethylnitrosoamine, $(CH_3)_2N-N=O$, once widely used as an industrial solvent, was shown to be carcinogenic. These nitrosoamines are formed also in the body when uptake of excessive nitrates is accompanied with some drugs containing secondary amines (see Chapter 13 also).

Isocyanates and methyl isocyanate

Compounds with the general formula $R-N=C=O$, *isocyanates*, are widely used industrial chemicals noted for the high chemical and metabolic reactivity of their

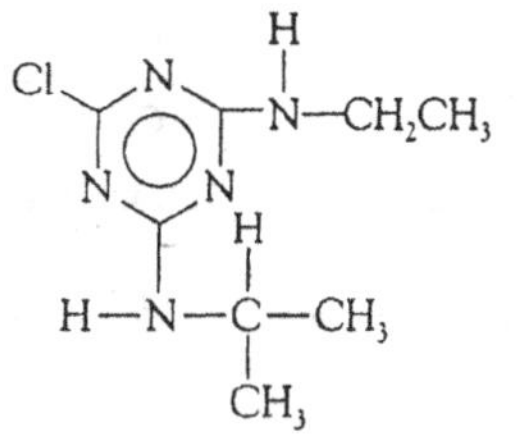

Figure 20. Examples of organonitrogen pesticides.

characteristic functional group. *Methyl isocyanate*, $H_3C-N=C=O$, was the toxic species responsible for the famous catastrophic industrial poisoning in Bhopal, India in 1984. In this accident, several tons of methyl isocyanate were released, killing more than two thousand and affecting about one hundred thousand people. This seems to be the worst industrial incident in history. The lungs of injured people were attacked; survivors suffered long-term shortness of breath and weakness from lung damage and nausea and bodily pains.

Organonitrogen pesticides

The structural skeleton of carbamic acid outlined by the dashed box in the structural formula of atrazine (Figure 20) is a characteristic feature of pesticidal carbamates.

Widely used on lawns and gardens, insecticidal carboryl has a low toxicity to mammals. Highly water-soluble carbofuran was shown to be taken up by the roots and leaves of plants and to poison insects that feed on the leaves. The toxic effect to animals is connected with an inhibition of acetylcholinesterase. Herbicidal paraquat has the toxicity rating 5 ($5–50 \text{ mg kg}^{-1}$). Dangerous or even fatal exposures can occur by all pathways, including ingestion, inhalation of spray, and skin contact. Paraquat is a systemic poison that affects enzyme activity and is damaging to various organs. Acute exposure may cause variations in the levels of catecholamine, glucose, and insulin. The most prominent initial symptom of poisoning is vomiting, followed within a few days by dyspnea, cyanosis, and evidence of impairment of the kidneys, liver, and heart. In a fatal case, pulmonary fibrosis, often accompanied by pulmonary edema and hemorrhaging, is manifested.

Alkenyl halides

The most significant alkenyl or alefinic organohalides are the lighter chlorinated compounds, such as *vinyl chloride* and *tetrachlorethylene* (Figure 21).

The numerous acute and chronic toxic effects of organohalides were recorded since these compounds are of wide use in industry and wasted in landfills. The toxic effects of vinyl chloride exposure are connected with the damage to central nervous, blood, lymph, and respiratory systems. This compound is a proved carcinogen, causing a rare

Vinyl chloride   Tetrachloroethylene

Figure 21. Structural formulas of vinyl chloride and tetrachlorethylene.

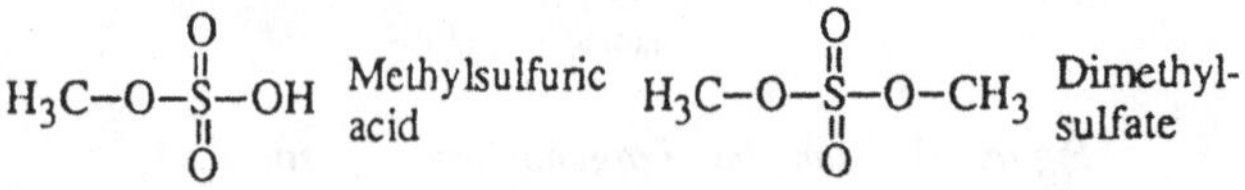

$H_3C-O-\underset{O}{\overset{O}{S}}-OH$ Methylsulfuric acid $H_3C-O-\underset{O}{\overset{O}{S}}-O-CH_3$ Dimethyl-sulfate

Figure 22. Structural formulas of methylsulfuric acid and dimethylsulfate.

angiosarcoma of the liver. This deadly form of cancer has been observed in workers chronically exposed to vinyl chloride while cleaning autoclaves in the polyvinylchloride, PVC, manufacturing. Tetrachlorethylene damages the liver, kidneys, and central nervous system. It is a suspected carcinogen.

Polychlorinated biphenyls

Because of their once widespread use in electrical equipment, as hydraulic fluids, and in many other applications, polychlorinated biphenyls, PBCs (see Chapter 14) became widely distributed, extremely persistent environmental pollutants in the Asian region. PCBs have a strong tendency to undergo bioaccumulation in lipid tissue. More details on the toxicological features of PBC will be discussed in Chapter 14.

Organochlorinated pesticides

This is a very wide class of pesticides, from famous DDT and HCH to 2,4,5-T or Agent Orange, which contains the dioxins as impurities. All these compounds are very toxic and we will spend more time for discussion in Chapter 14.

Organosulfur compounds

Despite the high toxicity of H_2S, not all organosulfur compounds are practically toxic. Their hazards are often reduced by their strong, offensive odors that warn of their presence. Inhalation of even very low concentrations of the alkyl thiols, such as *methanethiol*, H_3CSH, can cause nausea and headache; higher levels can cause increased pulse rate, cold hands and feet, and cyanosis. Like H_2S, the alkyl thiols are precursors of cytochrome oxidase poisons. An oily water-soluble liquid, *methylsulfuric acid* is a strong irritant to skin, eyes, and mucous tissue. Colorless, odorless *dimethylsulfate* is highly toxic and is a primary carcinogen. The structural formulas of these species are shown in Figure 22.

Skin or mucous membranes exposed to dimethylsulfate develop conjunctivitis and inflammation of nasal tissue and respiratory tract mucous membranes following an

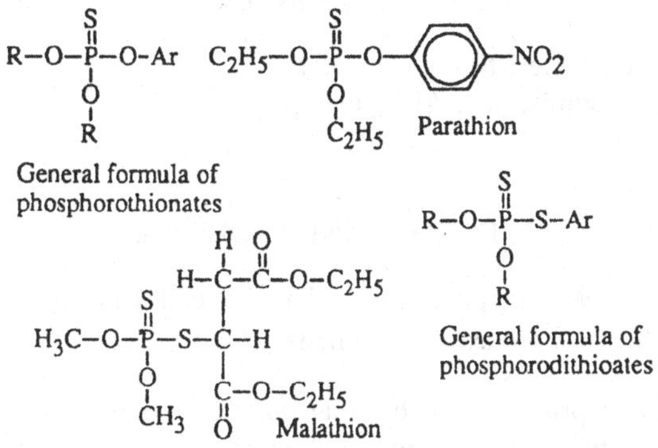

Figure 23. Phosphorothionate and phosphorodithioate ester insecticides. Malathion contains hydrolyzable carboxyester linkages (Monahan,2000).

initial latent period during which few symptoms are observed. Damage to the liver and kidney, pulmonary edema, cloudiness of the cornea, and death within 3–4 days can result from heavier exposures.

Phosphorothionate and phosphorodithioate ester insecticides

The first commercially successful phosphorothionate and phosphorodithioate ester insecticide was *parathion*, O,O-diethyl-O-p-nitrophenylphosphorothionate, licensed for use in 1944. This insecticide has a toxicity rating of 6 ($< 5\,\mathrm{mg\,kg^{-1}}$), supertoxic. Since its use began, several hundred people have been killed by parathion, including 17 of 79 people exposed to contaminated flour in Jamaica in 1976. As little as 120 mg of parathion is enough to kill an adult human. Most accidental poisonings have occurred by absorption through the skin. *Malathion* is the best known of the phosphorodithioate insecticides. It has a relatively high insect : mammal toxicity rate because its two carboxyester linkages (Figure 23), which are hydrolyzed by carboxylase enzymes (possessed by mammals, but not insects) to relatively non-toxic products. For example, although malathion is a very effective insecticide, its LD_{50} for adult male rats is about 100 times that of other similar compounds.

FURTHER READING

1. Manahan S. E., 2000. *Environmental Chemistry*. seventh edition, Lewis Publishers, NY.

2. Baird C., 1999. *Environmental Chemistry*. second edition, W. H. Freeman and Company, NY.

3. Bunce N., 1994. *Environmental Chemistry*. second edition, Wuerz Publishing Ltd, Winnipeg, Chapter 9.

WEBSITES OF INTEREST

1. Toxicological characteristics of chemicals,
   http://www.rff.org/misc_docs/risk_book.pdf

QUESTIONS AND PROBLEMS

1. Describe major structural units of the eukaryotic cell, which is the basic compartment of biochemical processes in animals.

2. Characterize the principal biochemical units of living organisms. Present the functions of proteins, hydrocarbons and lipids.

3. Present the definition of enzyme and discuss the characteristic features of these compounds in living organisms.

4. Discuss enzyme-catalyzed reactions of xenobiotic substances; point out phase I and phase II reactions.

5. Describe the principles of toxicology and toxicological chemistry and note the relationship with environmental chemistry.

6. Characterize major variables in the way in which toxicants affect organisms. Focus your attention on acute and chronic effects of contaminants.

7. Discuss the dose-response curve and the main parameters of these curves, including LD_{50} values.

8. Characterize toxicity rating and present a few examples of chemicals with different toxicity ratings.

9. Discuss physiological aspects of toxicants in the body and characterize kinetic and dynamic phases.

10. What are the major effects of binding a toxicant to a receptor, which may result in some biochemical effects?

11. Discuss biochemistry of carcinogenesis and present relevant definitions for primary and proved carcinogens, suspected carcinogens and procarcinogens.

12. Discuss the epidemiological approaches to assessing the risk of various carcinogenic compounds.

13. Characterize the approaches to assessing the carcinogenic effects of chronic and subletal doses of chemicals. Present the application of various models for this assessment.

14. Describe the application of biomarkers in carcinogenic studies and give a characteristic examples.

15. Discuss structure-activity and quantitative structure-activity relationships and application of SAR and QSAR methods for toxicological chemistry.

16. Assess the applicability of the octanol-water partition coefficient, K_{ow}, as the measure of the lipophilicity of persistent organic pollutants.

17. Characterize the methods for estimating bioconcentration and biomagnification and present relevant examples of calculations for assessing BCF.

18. Describe the toxicological properties of various heavy metals and stress the role of speciation in toxicology of these compounds.

19. Characterize different organometallic compounds and their toxicological properties with special attention to carbonyl species.

20. Give the examples of chronic and acute toxicity of phenol and its derivatives and present examples.

21. Discuss the toxicity of organochlorinated biphenyls; use the relevant data from various chapters of this textbook.

22. Characterize the carcinogenicity of PVC and present the toxicological parameters for these compounds.

23. Give examples of organosulfur compounds and present the main toxicological features of these chemicals.

24. Discuss the ecotoxicological chemistry of parathion and malathion insecticides.

CHAPTER 12

ENVIRONMENTAL CHEMISTRY
OF HEAVY METALS

1. INTRODUCTION

In many previous chapters, the discussion of questions regarding various environmental aspects of environmental chemistry in air, water, and soil compartments touched upon the problems of heavy metals. However, we should pay more attention to these pollutants, which are of crucial environmental concern in the Asian region. In this chapter, the emphasis will be given to heavy metal emissions from coal (including lignite) burning power plants, and to the specific aspects of environmental behavior of the most dangerous contaminants, like arsenic, mercury and lead. The problems of heavy metal site remediation will be considered in Chapter 16.

2. EMISSION OF HEAVY METALS FROM COAL-BURNING POWER PLANTS

Combustion processes are among the significant contributors of heavy metals as airborne emissions. The global emission of heavy metals is shown in Table 1. On the global scale, combustion processes are responsible for the emission of about 30% from total anthropogenic and natural sources, from 4.4% for lead and 75% for vanadium. Among the anthropogenic processes only, the combustion inputs are the most significant contributors, especially for V, Hg, and Ni.

Heavy metals of consequence in combustion systems include (for example) antimony, arsenic, barium, beryllium, cadmium, chromium, copper, lead, manganese, mercury, nickel, selenium, silver, strontium, and zinc. These metals may exist in very low concentrations in such fuels as wood and in higher concentrations in various ranks of coal. During the combustion process, these metals behave in a variety of ways, depending upon the presence or absence of specific reactants (e.g., oxidants such as chlorine, other metals), the general conditions of the combuster, and the temperatures at which combustion occurs. These metals inevitably concentrate in one or more of the various solid products of combustion (e.g., kiln or grate ash, fly ash), in scrubber water and scrubber sludge if a wet scrubber is employed, or may pass through the entire system and become airborne emissions (Figure 1).

Table 1. Annual airborne emissions of heavy metals from combustion sources, all anthropogenic activities and all natural sources, $\times 10^3$ tons per year (After Tillman, 1994).

Heavy metal	Combustion sources only	All anthropogenic sources	All natural sources	Total	Combustion from total emission, %
As	2.5	18.9	12.2	31.1	8.1
Cd	1.6	7.6	1.1	9.0	17.8
Cr	13.5	30.5	43.3	73.8	18.3
Cu	9.8	35.3	28.1	63.4	15.5
Hg	3.5	3.9	2.5	6.4	54.7
Mn	20.4	38.2	316.9	355.1	5.7
Ni	42.4	51.6	29.3	80.9	52.4
Pb	15.1	332.3	12.2	344.5	4.4
Sb	2.0	3.5	2.6	6.1	32.7
Se	4.0	6.3	10.3	16.6	24.1
V	85.2	86.0	27.7	113.7	75.0
Zn	18.6	131.8	44.7	176.5	10.5

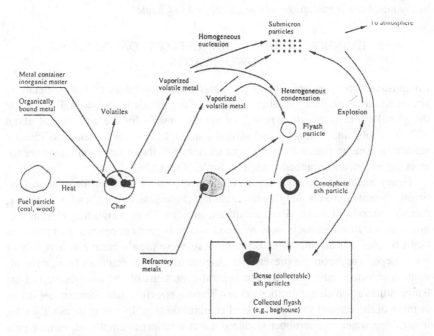

Figure 1. Mechanical depiction of the fate of heavy metals from conventional fuels in combustion systems (Tillman, 1994).

In many Asian countries where coals (including lignite) are widely used for power generation, the power sector is among significant contributors of suspended particulate matter (SPM) pollution to the environment. The SPM is considered to be a minor part of fly ash, which remains in flue (waste) gas passing the ash-collecting devices of a power generation unit, and thereafter is carried by flue gas to the atmosphere. At the same time, heavy metals (HM) and other trace elements, such as As, Ba, Be, Cd, Co, Cr, Cu, Hg, Mn, Mo, Ni, Pb, Sb, Se, Sn, Sr, V and Zn (Tillman, 1994), originally presented in coals at very small concentrations, are eventually transferred to the environment with SPM and dispersed over the area around the particular power plant, as the source of pollution.

We will consider the role of coal-burning power plants in emission of heavy metal using the example of Mae Moh power plant in the northern Thailand (Bashkin et al, 2000).

2.1. Case Study of Thailand

At present, Thai lignite is one of the major energy sources in Thailand. Like any other coal, lignite structurally consists of aromatic rings connected by bridges of carbon, sulfur, nitrogen, and other heteroatoms, and also includes the "grains" of mineral matter, basically consisting of alumosilicates (or clays), sulfides, carbonates, oxides (usually quartz) and chlorides. The heavy metals are incorporated into the coal structure as either organometallic compounds or inorganic materials, closely associated with the fuel matter. During combustion, fuel water is vaporized, and the process simultaneously occurs with fuel devolatilization followed by thermal destruction of mineral matter and its conversion to oxides. In conventional boiler furnaces, the resultant ash particles are basically composed of seven oxides (SiO_2, Al_2O_3, Fe_2O_3, CaO, MgO, Na_2O and K_2O) as well as traces of HM presented in various species. When fluidized-bed combustion is used, some other compounds (for example, $MgCO_3$ and $CaSO_4$) are contained in SPM leaving the stacks of the power plant.

Almost all the chemical elements have been found in different types of solid fuel with some exception for a few extremely rare elements, such as polonium, astatine, francium, and actinium (Finkelman, 1993). In fly ash of coals, the content of different elements may vary in the range from the parts per trillion (ppt) to more than 50 wt.%. The content of HM in fly ash for different coals varies from less than 0.5 ppm (for Cd, Hg) up to several thousand ppm (for Ba, Zn) as indicated by Tillman (1994). Respectively, the rates of pollution for various HM are different, whereas the particular heavy metal is characterized by its own toxicity (with respect to the ecosystem). Thus, the environmental impact of each element is to be considered from both quantitative and qualitative aspects. Presently, around 70% of Thai lignite is being mined in Northern Thailand and directly utilized by the Mae Moh Power Plant, in effect, the only lignite-fired power plant in the country. This fuel-power complex is located in the Mae Moh Valley, 25 km east of the city Lampang in the North of Thailand. The existing project comprises a large open pit lignite mine and 13 operating lignite-fired units with a total output capacity of 2,625 MW_e (near 15% of the total installed capacity in the country).

The operation of lignite-fired boilers of the power plant is followed by a significant emission of SPM into the atmosphere. By the early 1990s, the annual pollution rate of SPM from the 100%-loaded power plant had approached the maximum, estimated as much as about 1 Mton/year (see more details in Chapter 1).

It might be specifically emphasized that Thai lignite is a highly sulfurous fuel containing 1.7–3.0% (in some samples up to 7.0%) of sulfur in fuel mass as-received. The share of SO_2 emission from the power generation sector accounts for 60–65%, the predominant part of which belongs to the Mae Moh Power Plant. The contribution shares of other pollutants (NO_x, CO_2, CO) formed in utility and industrial boilers of power and other sectors are much lower (Chungpaibulpatana et al, 1997).

The elevated SO_2 pollution into the atmosphere and accumulation of HM in the ecosystem provide synergetic environmental impacts (in particular, on soils around the power plant). These impacts result in acceleration of the biogeochemical migration of most HM under acid conditions caused by the SO_2 emission from different industrial sectors, including power production. As a negative result, the acidification of tropical soils due to sulfur acidity undoubtedly facilitates the accumulation of a large number of HM in food chains of both terrestrial and aquatic ecosystems (Bashkin & Park, 1998). Modern studies have already documented the occurrence of acid rains over Northern Thailand due to the increase in use of lignite as an energy source. Based on the critical load concept, Kozlov & Towprayoon (1998) have shown that the critical load values of sulfur for terrestrial Tropical Rain Forest ecosystems of the region are rather low (less than 500 eq/ha/yr). As a result of both the high SO_2 emission rate and high sensitivity of ecosystems, the excessive input of acidity was calculated for more than 75% of the area considered (see also Chapter 15).

Brief geographical description of northern Thailand and the Mae Moh valley

The considered area with the Mae Moh Power Plant is located in Northern Thailand. The area is bound between 15 °N to 20.5 °N and 97.5 °E to 101.5 °E (Figure 2.)

The vegetation types from Tropical Wet Forest to Tropical Dry Forest are very typical in the region. Northern Thailand is characterized by a humid tropical maritime climate with extremely wet summers and dry winters. The average annual temperature for more than 30 years registered in the Lampang province of Northern Thailand ranges from 24 °C (Chiang Mai city) to 27 °C (Tak city). The monthly temperature variation between the warmest month (April) and the coldest month (December or January) is only about 7–10 °C.

Annual precipitation generally varies from 1050 mm to 1400 mm, although it is higher in some particular areas. The dry season (from November to April) is associated with southeast and north winds. The dominant wind directions depend on the season: southwest during monsoon season and prevalent northwest during the dry season. The rainy season (from May to October) driven by the southwest monsoons provides about 90% of the annual precipitation. For the most part of the year, the dominant high altitude winds are directed from the south and southwest. Meso-scale transport is characterized by low wind speed, increasing the pollutant residence time to about 24 hours (Schultz Institute, 1991).

Figure 2. Geographical location of Mae Mon valley.

Soil development reflects the prevailing climate. The strong evapotranspirative demand coupled with generally sparse precipitation (over much of the year) results in a rather specific moisture regime in the region. Local soils fall into two broad groups. Upland soils developed on dissected erosion surfaces, and old alluvial fans and terraces. Alluvial soils have developed on recent alluvium, associated with active sedimentation of rivers or streams. Upland soils are variable, with characteristics reflecting the local geology and moisture regime. Soils developed on old, well-drained surfaces are Ultisols (e.g., Paleustults). These are highly leached, with relatively little silica and relatively abundant hydrous oxides of aluminum and iron. These soils are moderately acidic (pH_{H_2O} ranges from 5.0 to 6.2), with low base saturation (less than 35%), low cation exchange capacity, but high anion exchange capacity. As a result, they have a high sulfur absorption capacity. Upland soils developed on more surfaces or soils developed in poorly drained sites, in which leaching has not been so pronounced, are mostly Alfisols. These, by definition, are less leached, and have a higher base saturation (> 35%), than the Ultisols. However, soil acidity is more variable than that for Ultisols (pH_{H_2O} ranges from 4.5 to 7.5). Some Alfisols have developed from parent materials rich in carbonates, such as limestone. The accumulation of HM in the upper humus horizons is the characteristic feature for many of these soils (Kaewruenrom, 1990; Vearasilp and Songsawad, 1991; Dobrovolsky, 1994).

The climate characteristics and also the vegetation and soil properties of Northern Thailand were used for calculation of pollutants deposition rates and comparison with natural HM content in different soil/ecosystem combinations.

The Moe Moh valley is about 15–20 km wide, 50 km long and is aligned from northeast to southwest. The valley terrain is flat with an average elevation between 320 and 360 m above sea level (a.s.l.). Two ridges parallel the valley on either side. To the northwest, the hills average 700 m a.s.l. and to the southeast 900 m a.s.l. To the northeast the valley is also enclosed by hill terrain. To the southwest the valley is open to Lampang province. Like other areas of Northern Thailand, this valley is located in a monsoon zone, which is influenced mainly by two monsoons, the southwest and northeast. The annual temperature and precipitation patterns are similar to those for the whole region of interest. Southerly and southwesterly winds are predominant in the annual wind pattern. Wind speeds are very low, averaging about 2.0 m/s at Lampang (Climatological Division, 1994).

Fuel characteristics and ash analysis

The intensive consumption of Thai lignite as a source for power generation dates from the early 1980s, when the installation of the first three 75 MW_e units of the Mae Moh Power Plant was completed. Since then, the power generation capacity of the plant has been gradually expanded by erection of four 150 MW_e and six 300 MW_e lignite-fired units. Approaching the 2,625 MW_e total capacity by the early 1990s, the daily rated fuel consumption by all the plant's boilers (at the 100% load, for averaged fuel) accounts for about 70,000 tons of lignite (Khummongkol, 1999).

As mentioned above, Thai lignite is classified as high sulfur, low-rank coal. Moreover, year by year the fuel quality is deteriorating. Lignite supplied to the power plant from the nearby different mines, is currently characterized (on mass as-received) by low content of carbon (20–35%), medium moisture content (26–35%), variable ash content (17–41%), and high sulfur content (1.7–3%). Nitrogen content varies in the range from 0.8 to 1.25%, whereas the oxygen content from 7.0 to 10.7% and hydrogen content from 1.8 to 3.2%. The lower heating value for Thai lignite is estimated (on average) to be 10.4 MJ/kg (Chungpaibulpatana et al, 1997).

An example of fly ash of Thai lignite sampled from two different units of the power plant is shown in Table 2 (Wongsiri et al, 1999).

Table 2. Representative composition of fly ash from units of the Mae Moh Power Plant.

Unit	Oxides, wt.%									
	SiO_2	Al_2O_3	Fe_2O_3	MgO	CaO	Na_2O	K_2O	TiO_2	P_2O_5	SO_3
No. 1	47.53	20.52	9.37	2.72	11.35	0.61	1.77	0.52	0.08	0.59
No. 2	53.70	24.54	7.55	1.72	4.72	0.33	2.85	0.57	0.10	0.32

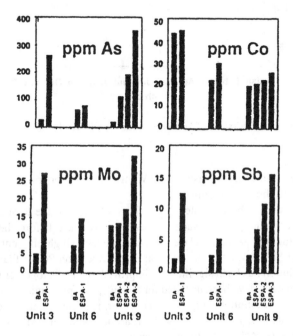

Figure 3. Variability of heavy metals measured in fly-ash samples collected from three power-generating units of Mae Moh plant (Hart et al, 1995).

The ash composition is completed with data on three more oxides (namely, TiO_2, P_2O_5 and SO_3) in addition to those listed above.

We can see also that the content of heavy metals in fly-ash is different for various units of the Mae Moh plant. This is connected with different trapping of SPM from passing gases. The content of different heavy metals in fly-ash is represented in Figure 3.

Emission of SPM from the Mae Moh power plant

The SPM emission from the power plant is of particular interest in this study, since this input data is used for calculation of the HM emission. The rate of the SPM emission depends on many factors, such as boiler efficiency and load, fuel quality and (particularly) ash content, efficiency of ash-collecting devices and some minor

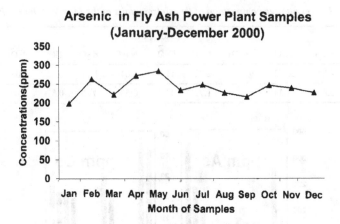

Figure 4. Emissions of As from the Mae Moh Power Plant in 2000.

factors. The relevant data (by Bashkin and Wongyai, 2002) for arsenic emission with SPM in 2000 are shown in Figure 4.

SPM emission can be estimated as much as 1/5–1/3 of the total ash emission from the power plant. In fact, such a situation took place before the year 1995, when units of the "old generation" were not equipped with high-efficient electrostatic precipitators (ESP). Recently, the situation in the Mae Moh Power Plant associated with SPM pollution has been significantly improved; the ESPs together with flue gas desulfurization units have been installed for all units, except of units 1–3 of relatively small capacities (Khummongkol, 1999). However, previous extensive emissions of various pollutants were accompanied with the accumulation of HM in soils and other components of biogeochemical tropical chains.

Heavy metals emission from power generation in 1981–1995

The calculated values of HM emissions for selected years of the period of 1981–1995 are represented in Table 3.

In this study we consider the toxic metals (according to US EPA classification), namely As, Ba, Cr, Ni, Sb, and some others, the emitted values of which are expected to be high. One can see that the emissions of heavy metals depend proportionally on both the SPM emitted from the power plant and the HM content in fly ash.

The prevailing values of HM emission (more than 650 tons per year) were found to be for Barium and Strontium in 1990, i.e., at the peak of lignite use for power generation in the country. In this year, the emissions of Vanadium, Rubidium, and Arsenic were found to be in the range between 120 and 202 tons. Relatively small emissions were calculated to be for Cesium, Chromium, Nickel, Cerium and Lanthanum (25–63 tons per year). The smallest emissions were found to be for Uranium, Thorium, and Antimony, the value of which lay within the range from 6 to 8 tons per year.

Table 3. Emission of various HM from power generation in Northern Thailand, tons/year.

Element	1981	1986	1990	1995
As	29.394	84.561	201.285	196.599
Ba	95.22	273.93	652.05	636.87
Cr	9.177	26.4005	62.8425	61.3795
Ni	7.3554	21.1601	50.3685	49.1959
Sb	1.1868	3.4142	8.127	7.9378
Ce	7.4658	21.4777	51.1245	49.9343
Cs	3.588	10.322	24.57	23.998
La	4.5264	13.0216	30.996	30.2744
U	0.85698	2.46537	5.86845	5.73183
Rb	19.872	57.168	136.08	132.912
Sr	96.6	277.9	661.5	646.1
Th	2.3598	6.7887	16.1595	15.7833
V	18.078	52.007	123.795	120.913

Such an impact was substantially sustained during the first half of the 1990s included in the period of analysis.

The environmental impact of HM, however, depends not only on their emission, but also on their deposition rates, the relative increase in HM content in soil and some other factors considered below.

Assessment of heavy metal deposition rates

The general concept of the atmospheric transport and deposition computational method is that the concentration of any substance determined on the basis of its emissions, is subsequently transported by (averaged) wind flow and dispersed over the impacted area due to atmospheric turbulence. Basically, the rate of substance removal from the atmosphere by wet and dry deposition and photochemical degradation is described by general model algorithms. The transport and dispersion of HM in the atmosphere are assumed to be similar to those for other air pollutants, for instance, such as SO_2 and smog compounds (Pacina et al, 1993; de Leeuw, 1994, EMEP/MSC-E, 1996; Dutchak et al, 1998). Based on such an approach, the computational results of sulfur deposition over the area of interest obtained by other authors might be particularly used for the estimation of HM depositions.

The RAINS-ASIA computational model has been applied for the area in the North of Thailand scaled with $1° \times 1°$ grid for different time/fuel scenarios based on sulfur deposition predicted data (World Bank, 1994). Some new results have been reconsidered recently using updated input data (Kozlov and Towprayoon, 1998). According to them, the sulfur deposition rate for different areas in the region of interest varies from 89 to 4749 mg m^{-2} per year.

Much more detailed studies were carried out by Doolgindachbaporn (1995) and Ross et al (1998) for the Moe Moh valley, the relatively small area in Northern Thailand were the Mae Moh Power Plant is located. The resolutions were selected to be in the range of 5 to 10 km cells. Differing in some details, these models indicate the area of the most polluted zone to be about 100 km^2. Similar pollution areas from individual sources with stack heights of 75–125 m, were shown for many other regions and reviewed in workshop reports on HM deposition (Pacina et al, 1993; de Leeuw, 1994, EMEP/MSC-E, 1996).

The total amount of emitted HM (see Table 3) was deposited proportionally to SO_2 over the area of 12321 km^2 in the most polluted 1° × 1° Lola grid cell, 100 °E–18°N. The deposition rates of different environmentally dangerous heavy metals in 1995 are shown in Table 4.

As follows from the computational results, the deposition rates of HM would have varied in 1995 from 174 g/km^2/yr for U up to 19665 g/km^2/yr for Sr for the impacted area of 12321 km^2. The intermediate values were shown for As (5984 g/km^2/yr), Cr (1868 g/km^2/yr) and Ni (1497 g/km^2/yr).

Assessment of heavy metals accumulation in ecosystems

In spite of rather approximate estimation of HM depositions in this work, it would be very interesting and useful to compare the cumulative HM concentrations with their natural contents for the soils of Northern Thailand. Most heavy metals considered in this work (Ba, Cr, Ni, Cs, La and V) are known to be active biogeochemical migrants. During the period of power plant operation, they have accumulated in the upper humus soil layer in amounts significantly greater than natural contents in local soils and geological rocks, or clarks (Dobrovolsky, 1994). The natural HM contents are shown (for reference) in Table 5, together with the calculated values of the increment of HM contents in the humus layer and related to clarks for both selected areas of the environmental impact in Northern Thailand. Data are for the year 1995 and obtained for the area of LoLa grid cell of 18 °N–100 °E and the most polluted impact area surrounding the Mae Moh plant. In calculations, the average thickness of the layer of HM accumulation was assumed to be 20 cm. As follows from Table 5, the annual increment of the content of some heavy metals (As, Ni, Cs, La and V) in comparison with clarks, lies in the range from 0.2 to 0.5%.

The maximum value of the increment is found to be 13.5% for Barium. Meanwhile, most of the metals are biogeochemically active elements, and their accumulation is much more profound in the upper humus layer. The relative increment in this layer is significantly less, and ranges between 0.005–0.05% of the respective values. Thus, the danger might be associated with the impact of As, which is considered to be one of the most important pollutants ecologically (see below). According to our calculation, As is accumulating in soils at an annual rate of 0.5% (with respect to the natural content, referred to as clark).

Table 4. Deposition rates for selected heavy metals emitted from Mae Moh Power Plant in 1995, g/km^2/year.

Lo,°E	La,°N	As	Ba	Cr	Ni	Sb	Ce	Cs	La	U	Rb	Sr	Th	V
98	16	16	112	364	35	5	28	13	17	3	76	368	9	69
	17	144	467	45	36	6	36	18	22	4	98	472	12	88
	18	225	831	80	64	10	65	31	39	7	173	839	20	157
	19	112	364	35	28	4	28	14	17	3	76	367	9	69
99	16	303	987	95	76	12	77	37	47	9	205	996	24	186
	17	686	2233	214	172	28	174	84	106	20	464	2255	55	422
	18	2409	7841	753	603	97	612	294	371	70	1629	7918	193	1481
	19	335	1091	105	84	13	85	41	52	10	227	1101	.27	206
100	16	543	1766	169	136	22	138	66	84	16	337	1783	44	334
	17	941	3064	294	236	38	239	115	145	27	636	3094	75	579
	18	5984	19475	1868	1497	242	1520	730	921	174	4045	19665	479	3680
	19	734	2389	229	184	30	186	90	113	21	496	2412	59	451
101	16	351	1142	110	88	14	89	43	54	10	237	1154	28	216
	17	447	1454	140	112	18	114	54	69	13	302	1468	36	275
	18	941	3064	240	236	38	239	115	145	27	636	3094	75	579
	19	447	1454	140	112	18	114	55	69	13	302	1468	36	275
102	16	319	1038	100	80	13	81	39	43	9	216	1049	26	196
	17	287	935	90	72	12	73	35	44	8	194	945	23	177
	18	319	1038	100	80	13	81	39	49	9	216	1049	26	196
	19	225	830	80	70	10	65	31	39	7	173	839	20	157

The situation is enhanced by the synergetic influence of acidification loading from sulfur compounds, which is also dramatic in this area. Increasing acidity of soils and surface waters is known to be accompanied by the increasing mobility of most HM. This, in turn, leads to possible accumulation of HM in the food chains of both terrestrial and aquatic ecosystems (Bashkin and Wongyai, 2002).

Table 6 represents the amounts of HM accumulated by the year 1995 in the soil-biogeochemical fluxes occurring in Tropical Wet Forest ecosystems surrounding the Mae Moh Power Plant.

The assumption was made that the entire quantity of the deposited HM might have accumulated in the upper soil layer and could migrate with soil-biogeochemical fluxes of these trace elements. This is reasonable, taking into account the high content of Ca and Mg in fly ash from the Mae Moh Power Plant (see Table 2). One can see that for practically all HM (except of Sr), the 15-year period of the power plant's operation (1981–1995) has led to a significant accumulation of HM in the upper soil layer.

Table 5. Natural content (ppm) and its annual increment (%, with respect to the natural content) for selected HM in soils of the impacted area around the Mae Moh Power Plant (calculated for the year 1995).

| Element | Natural content, ppm | | Annual increment (%) related to natural contents | | | |
| | | | Impacted area 12321 km^2, cell of 18°N–100°E | | Impacted area 100 km^2, evenly polluted | |
	Humus layer	Clark	Humus layer	Clark	Humus layer	Clark
As	9.0	5.0	0.4	0.50	50.0	103.00
Ba	432	0.6	0.05	13.50	6.43	2781.94
Cr	197	5.8	0.003	0.12	0.82	27.70
Ni	84	3.2	0.008	0.22	0.16	40.23
Sb	—	1.0	—	0.09	—	20.83
Cs	20	1.8	0.018	0.20	3.12	34.72
La	80	2.0	0.005	0.20	0.99	39.58
U	—	1	—	0.07	—	16.88
Rb	—	100	—	0.02	—	3.48
Sr	—	300	—	0.02	—	5.64
Th	—	6.0	—	0.03	—	6.88
V	212	3.2	0.006	0.48	1.49	98.96

— no data available

The values of accumulation might lie in the range from 1.8 ppm (for U) to 488.0 ppm (for Ba), and these values have to be added to the natural content of heavy metals in soils (see Tables 5 and 6). The resulting (cumulative) values for rapidly accumulated HM (As, Ba, Cr, Ni and V) exceed the existing environmental quality criteria values for HM content in soils (Radojevich and Bashkin, 1999) shown in Table 7.

Furthermore, under certain conditions, the contents of HM are greater than the requirements (limits) established for soil remediation, even in cases of commercial and industrial land use.

3. ENVIRONMENTAL CHEMISTRY OF ARSENIC

3.1. Arsenic Speciation

In the Asian region, in the case of environmental contamination of various media by heavy metals, it is the element itself which is toxic (see Chapter 11); however the element speciation is very important for some metals, including mercury, lead, and especially arsenic. Organometallic compounds are often of greatly different toxicity than simple inorganic ions, and mercury and lead are the typical examples (see Sections 4 and 5). The opposite situation exists for arsenic, which enters the environment

Table 6. Accumulation of HM in different types of soils in 1981–1995 (calculated for the impacted area of 100 km^2).

Element	Natural content, ppm		Accumulation in humus layer due to HM emission, ppm
	Upper humus layer	Clark	
As	15.0	5.0	50.0
Ba	432	0.6	488.0
Cr	197	5.8	19.4
Ni	84	3.2	15.5
Sb	—	1.0	2.51
Ce	—	—	15.9
Cs	20	1.8	7.6
La	80	2.0	9.5
U	—	1	1.8
Rb	—	100	41.9
Sr	—	300	203.5
Th	—	6.0	5.0
V	212	3.2	38.1

Table 7. Selected environmental quality criteria for HM content in soils, ppm (Radojevic and Bashkin, 1999).

Element	Assessment criteria	Remediation criteria for land use		
		Agriculture	Residential	Commercial
Arsenic	5	20	30	50
Barium	200	750	500	2000
Chromium (total)	20	750	250	800
Nickel	20	150	100	500
Tin	5	5	50	300
Vanadium	25	200	200	200

through burning coal, lignite and oil, in which it is a trace element (see above). The order of toxicity is as follows: organoarsenics < arsenic (V), which predominates under aerobic conditions < arsenic (III), which is formed by reduction of arsenic (V) in sediments. For example, fish and shellfish store arsenic as arsenolipids, which are almost non-toxic, to the extent that moderate quantities of these fishes can safely be eaten.

3.2. *Arsenic Biogeochemistry in South and Southeast Asia*

High arsenic-containing geological rocks occupy the vast area between the Industan and Indo-China peninsulas, and the *arsenic biogeochemical sub-region of the bio-sphere*, in accordance with biogeochemical mapping (Bashkin and Howarth, 2002), is shown in India, Bangladesh, Thailand and Malaysia. For understanding biogeo-chemistry of arsenic in food webs, the biogeochemical mapping of these vast areas should be carried out. This mapping is a tool to synthesize biogeochemical parameters and provide the information necessary for strategic risk assessment and ecological management of the local and regional landscapes in South and Southeast Asia. The understanding of biogeochemical peculiarities of cycling in different ecosystems will help in recognizing the sofest places for industrial projects and rehabilitation of polluted sites, like those in Mae Moh Valley after lignite mining. Many regional biogeochemical standards will be based on this mapping, such as critical loads of pollutants (heavy metals) on terrestrial and aquatic ecosystems. Biogeochemical mapping of As should include also the relevant results from human biogeochemistry.

Human biogeochemistry is a rapidly developing branch of modern biogeochem-istry dealing with the quantitative assessment of relationships between migration of chemical species in food webs in natural and technogenic biogeochemical provinces and human health. In spite of existing knowledge about endemic diseases in natural and anthropogenic biogeochemical provinces, these data are still very limited. Human biogeochemistry must be enlarged in various Asian countries for understanding hu-man illnesses, first of all cancer and cardiovascular diseases, human diet and human adaptation to As polluted environment. Some examples can be shown.

Millions of people in rural areas of Bangladesh are being slowly poisoned as they drink water contaminated with small but potentially fatal quantities of arsenic. Estimates by the World Bank claim that from 18 to 50 million people out of a total population of about 120 million in the country are at risk. Thousands are already showing symptoms of poisoning. Nineteen rural districts covering an area of $500\,km^2$ near the border of Bangladesh and India have arsenic-contaminated wells. Of the 20,000 tube wells tested so far, 25% have dangerous levels of arsenic, 40% have unsafe levels and only 35% were safe or below $0.01\,mg\,L^{-1}$ of arsenic. The World Health Organization recommends a level of $0.01\,mg\,L^{-1}$ of arsenic (10 ppb) but the governments of Bangladesh and India regard $0.05\,mg\,L^{-1}$, a level five times higher than the WHO standard, as acceptable.

Similar results can be shown for the whole belt between India and Thailand. Biogeochemical mapping in this vast area is a great challenge for present and future research.

These calculated data are in agreement with HM monitoring results that have been shown for the beginning of the 1990s in soils surrounding the Mae Moh Power Plant (Huyen, 1996).

Arsenic is the most dangerous element emitted from various industrial activities in the Mae Moh complex. The biogeochemical cycle of As includes its migration in the air-water-sediment-soil-plant system. The averaged value of the As concentration

Table 8. Content of arsenic in surface and ground runoff waters in the industrial area of Mae Moh electric power plant, Northern Thailand.

Water	Total As, ppb
Surface	25–450
Ground	48–1200

Table 9. Biogeochemical migration of As in aquatic ecosystem of the Mae Moh Power Plant settling pond ($\mu g/kg$, on dry basis).

Samples	As Content
Water, $\mu g/L$	25
Sediment	107,773
Eichhornia crassipes	52,430
Ipomoea aquatica	78,080
Typha angustifolia	54,020

in surface waters of the Mae Moh reservoir is about 1.0 $\mu g/L$, whereas in the power plant settling ponds they are 4.0–11.0 $\mu g/L$, in the northeast wetland 0.8–5.0 $\mu g/L$ and in the southwest wetland 2.0–5.0 $\mu g/L$. The dynamic concentrations do not exceed 50 $\mu g/L$.

The As content in the bottom sediments is rather high, specifically in the power plant settling ponds, for which the As content was determined to be 107.7 ppm (referred to dry basis). This data greatly exceeds the maximum permissible concentration (5.9 ppm in Canada, as reported in Radojevich and Bashkin (1999).

The highest results were shown for the industrial areas of Thailand (Table 8).

An excessive accumulation of As in the water column and bottom sediments of the Mae Moh reservoir, as well as in the settling ponds, leads to increasing the uptake of this element by aquatic plants and, consequently, to enhancing migration in biogeochemical food chains (Table 9).

4. ENVIRONMENTAL GEOCHEMISTRY OF MERCURY

Mercury occupies a unique and infamous place in environmental biogeochemistry. It was the first chemical species for which a direct connection was proven between relatively low concentrations in a natural water system, bioaccumulation up the biogeochemical food webs, and a serious health impact on a human population at the top of the food chain.

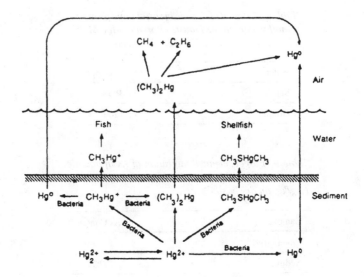

Figure 5. The mercury cycle in air-water system (Butcher et al, 1992).

Speciation of mercury

The biogeochemistry and bioavailability of mercury is strongly influenced by speciation. Mercury can exhibit a variety of aqueous and particulate species (Figure 5).

The usual form of mercury in aqueous solution is the Hg^{2+} ion. Mercury has two oxidation states, Hg(I) and Hg(II), but the first of these, which contains the unusual ion $^+Hg - Hg^+$, is stable only as insoluble salts such as Hg_2Cl_2. It disproportionates in solution as follows from reaction (1)

$$Hg_2^{2+} \text{ (aq)} \longrightarrow Hg^{2+} \text{ (aq)} + Hg \text{ (l)}. \tag{1}$$

We can see from the reaction that reduction of Hg^{2+} under anaerobic conditions, for example in bottom sediments, gives the metal in liquid form.

Mercury(II) is a very soft Lewis acid, which forms stable complexes preferentially with soft Lewis bases such as sulfur ligands. You should remember here that the major natural form of mercury is sulfides. Increasing the pH of the aqueous solution due to pollution or river water discharge to marine water leads to precipitation of HgO. We know that HgO has finite solubility in water, and the solution may be described in terms of mercury(II) hydroxide as the following reactions (2)–(4)

$$HgO(s) + H_2O \longrightarrow Hg(OH)^\circ, \tag{2}$$

$$Hg(OH)^\circ + H^+ \longrightarrow HgOH^+ + H_2O, \tag{3}$$

$$HgOH^+ + H^+ \longrightarrow Hg^{2+} + H_2O. \tag{4}$$

The major complicating factor in environmental biogeochemistry of mercury and its speciation is the biological methylation of Hg^{2+} to CH_3Hg^+ and $(CH_3)_2Hg$. This process converts inorganic mercury to organo-mercury, which is both more lipophilic and toxic (see below).

Mercury is methylated in nature by the attack of methylcobaltamin (vitamin B_{12}) upon Hg^{2+}. Methylcobaltamin contains a methyl group bonded to a central cobalt atom, making a methyl group somewhat carbanion-like. Representing methylcobaltamin as $L_5Co - CH_3$, the simplified equation is shown below from reaction (5)

$$L_5Co-CH_3 + Hg^{2+} \longrightarrow L_5Co^+ + CH_3Hg^+. \qquad (5)$$

CH_3Hg^+ occurs mostly as CH_3HgCl. In shellfish, CH_3HgSCH_3 is also found, since mercury has a strong affinity for sulfur (Bunce, 1994). While CH_3Hg^+ derivatives predominate when mercury is methylated in sediments below pH 7, further methylation to $(CH_3)_2Hg$ becomes important as the pH rises.

The carbon-mercury bond is intrinsically weak, about 200 KJ/mol. However, it is almost completely nonpolar. Neither nucleophiles nor electrophiles react readily with such a center, and so organomercurials tend to be kinetically unreactive.

Reduction of Hg^{2+} to $Hg°$ and alkylation to form methyl- or dimethylmercury can both be viewed as detoxication process, because all of the products are volatile and can be lost from the aqueous phase (see Figure 4). Organisms can also convert the methylated forms to $Hg°$, which is more volatile and less toxic. However, both the methylated and reduced species are more toxic to humans and other mammals than is Hg^{2+}.

Most biogeochemical processes of mercury speciation are driven by microbial activities. Bacteria can facilitate the mobilization of mercury from mineral deposits by oxidizing sulfide and thereby allowing mercury, which has been sequestered in the extremely insoluble solid cinnibar (HgS), to dissolve.

Anthropogenic mercury loading

Mercury is a relatively rare chemical element. In the lithosphere it occurs mainly as sulfides, HgS. Mercury sulfide comes in two forms: cinnibar, which is black, and vermillion. In some places mercury exists in small proportions as a free chemical species.

Mercury refining involves heating the metal sulfide in air in accordance with the following reaction (6)

$$HgS + O_2 \longrightarrow Hg + SO_2. \qquad (6)$$

Gaseous mercury is condensed in a water-cooled condenser and redistilled for sale. At present industrial mercury uses are connected with electric batteries, electric tungsten bulb, pulp bleaching and agrochemical production.

Mercury batteries are used widely in everyday life, in applications such as cameras and hearing aids. About 30% of U.S. production of mercury is used in this way, the reason being the constancy of the voltage in the mercury battery, almost to the point of complete discharge.

The electrical uses of mercury include its application as a seal to exclude air when tungsten light bulb filaments are manufactured. Fluorescent light tubes and mercury arc lamps used for street lighting and as germicidal lamps, also contain mercury.

Mercury is consumed in the manufacture of organomercurials, which are used in agriculture as fungicides, e.g., for seed dressing.

Ecotoxicological effects of mercury

There is no known biochemical reaction in organisms that applies Hg as an essential element. Mercury is the only metal which is a liquid at ordinary temperatures. The boiling point of this metal is 357 °C. This temperature is relatively low for metals and its vapor pressure is significant even at room temperature. The threshold limit value (TLV) of elemental mercury is 0.05 mg/m^3, a value that is less than the equilibrium vapor pressure at ambient temperatures. However, in the mercury mines in Sicily, where mercury occurs in shales, the mines are exposed to elemental mercury vapor, which content in the air may reach toxic levels of about 5 mg/m^3. Another source of exposure in mines is mercury-containing dust.

We can see accordingly, that in Sicilian mines the TLV was exceeded to 10 fold. The TVL of organic mercury is set at 0.01 mg/m^3, in recognition of their greater toxicity.

Mercury-containing mineral, vermillion, has for centuries been used as a pigment for oil based paint. Mercury poisoning among artists has occurred as a result of licking the brush to get a fine point.

Organic derivatives of mercury are more hazardous than the simple inorganic salts because they are lipid soluble and hence bioconcentrate. These species are able to cross the blood-brain barrier, thereby causing the neurological symptoms associated with mercury intoxication.

The greater toxicity of the lipophilic and bioconcentratable forms of mercury is shown in the following values of LD$_{50}$, a lethal dose for 50% of the exposed population of birds: HgCl$_2$, 5000 mg/kg: C$_6$H$_5$HgOAc (seed dressing), 1000 mg/kg; C$_2$H$_5$HgCl, 20 mg/kg. Alkylmercury compounds, especially short chain alkylmercury derivatives, are able to cross the blood-brain barrier, and this explain why mercury poisoning is accompanied with mental disturbance.

Elemental mercury is mainly hazardous as the vapor. There is less danger of absorbing the metal for the digestive tract. Like the alkylmercurials, elemental Hg affects the central nervous system, accompanied with such symptoms as tremors, irritability, and sleeplessness. Kidney damage is also reported as a result of the influence of inorganic mercury salts due to a complexation of mercury by the protein metallothionein, which accumulates in the renal tubules.

Consumption of mercury in the diet leads to the accumulation of this metal in the body (Figure 6).

Mercury acts as a cumulative poison because the rate of clearance of mercury from the body is low. The single intake at a specified level may cause no ill effects, however the same concentration of metal in the meal of a steady diet can lead to sickness or even death.

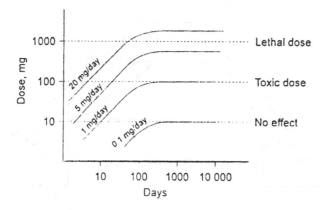

Figure 6. Accumulation curves for different levels of mercury in daily intake (Bunce, 1994).

Environmental biogeochemical cycling of mercury

Unlike most heavy metals, the natural and anthropogenic cycles of mercury are dominated by atmospheric transport. We can show this a few characteristic with examples. First, metallic mercury has the highest vapor pressure of any heavy metal, and it is released in biogeochemically significant amounts by volcanic eruptions, volatilization of methyl- and alkyl-mercurials from land and ocean surfaces. Volatilization processes are also of great importance for industrial emissions such as smelting of minerals and burning of fossil fuels.

Mercury has been used by man for about 2000 years. The comparison of pre-industrial (natural) and anthropogenically modified biogeochemical cycles of mercury allows us to make some general remarks. First, the fluxes between atmosphere and both terrestrial and oceanic ecosystems are much greater than between terrestrial and oceanic ecosystems. Second, human activity changed significantly the pre-industrial balance fluxes between land and atmosphere. Modern fluxes from terrestrial ecosystems to the atmosphere are about 40% higher than pre-industrial ones. In the most extreme case, the riverine transport has been changed, up to 4-fold. The results of Figures 7 and 8 show that the average residence times for mercury in the atmosphere, terrestrial soils, oceans, and oceanic sediments are approximately 11 days, 1000 years, 3200 years and 2.5×10^8 years, respectively.

These up-to-date results can be compared with those of the late 1970s. A review of natural sources of mercury emission has been carried out in some researches. Table 10 presents values of natural emissions obtained by J. Nriagu in 1989.

Data of other authors differ appreciably from the estimates shown in Table 10. For example, Thornton et al (1995) believed that the integral value of natural emissions to the atmosphere reaches 180×10^2 ton/year, whereas those of Rasmussen (1994) were 350×10^2 ton/year. According to data of the Geological Survey of Canada (GSC, 1995) the natural inflow of mercury to the atmosphere can exceed even 1000×10^2 ton/year.

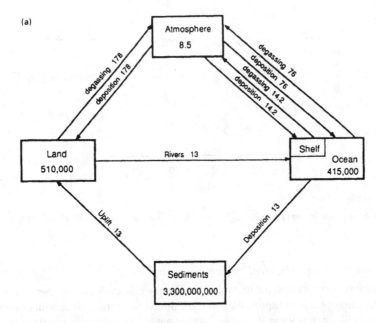

Figure 7. The pre-industrial global biogeochemical cycle of mercury. Units are 10^2 tons (pools) and 10^2 ton/year (fluxes). (Adapted from National Academy of Sciences, 1978).

Table 10. Mean median values and a possible scattering range (in brackets) of mercury emissions to the atmosphere from different natural sources (After Nriagu, 1989).

Natural source	Mercury emissions, 10^2 ton/year
Wind-borne soil particles	5(0–10)
Seasalt spray	2(0–4)
Volcanoes	1(0.3–20)
Wild forest fires	2(0–5)
Biogenic particulates	3(0–4)
Continental volatiles	61(2–120)
Marine volatiles	77(4–150)
Total natural emission	25(0.1–49)

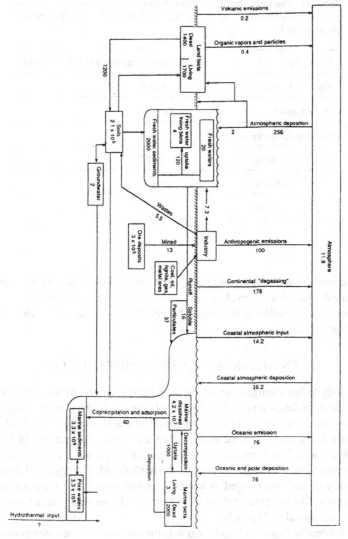

Figure 8. The global biogeochemical cycle of mercury at the present environment. Units are 10^2 tons (pools) and 10^2 ton/year (fluxes). (Adapted from National Academy of Sciences, 1978).

A similar comparison can be conducted with results of the environmental biogeochemical cycle of mercury. The most important anthropogenic emission sources of mercury are listed in Table 11.

We can see the significant range of scattering in Table 11. Some authors presented significantly narrower ranges of the global emission estimates for mercury, from 36 to 45 × 10^2 ton/year.

Table 11. The most important anthropogenic emission sources of mercury to the atmosphere on the global level in 1983 (After Nriagu and Pacina, 1988).

Source type	Mercury emission flux, 10^2 ton/year
Fossil fuel combustion	7–35.0
Wood burning	0.6–3.0
Metallurgical processes	0.5–2.0
Waste incineration	2.0–22.0
Total — median value	36.0
— scattering range	9.0–62.0

These modern data allow us to make some corrections in the environmental biogeochemical cycle of mercury; however, these can not change the general conclusion on the dramatic alteration of the natural biogeochemical cycle of this element under anthropogenic activity.

5. ENVIRONMENTAL GEOCHEMISTRY OF LEAD

Human activity has changed the intensity of natural biogeochemical fluxes of lead during industrial development. However, the history of lead use is the longest of any metal. The period of relatively intensive production and application of lead is about 5000 years. Lead has been used as a metal at least since the times of the Egyptians and Babylonians. The Romans employed lead extensively for conveying water in the elaborate water distribution systems. Through the Middle ages and beyond, the malleability of lead encouraged its use as a roofing material for the most important constructions, like the great cathedrals in Europe. The modern production of lead is $n \times 10^6$ tons annually (Figure 9).

The noticeable changes of lead content in the environment are dated as long ago as about 2700–3000 years. These results come from glacier and peat monitoring using deep drilling.

Speciation of lead
The common compounds of lead derive from the +2 oxidation state. As a member of the periodic group IV-A, lead also forms tetravalent compounds which are covalent. The most important are the tetraalkylleads, which are used as gasoline additives. The Pb − C bond is very non-polar, and the organolead species tend to be kinetically inert, like organomercurials.

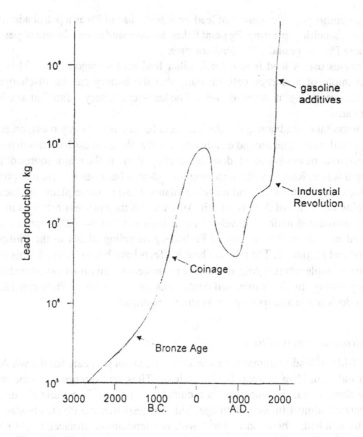

Figure 9. Historical production and consumption of lead (Bunce, 1994).

The speciation of lead(II) in aqueous solution involves several polymeric hydroxo-complexes. Below pH 5.5, Pb^{2+} (aq) predominates. However, with increasing pH, $Pb_4(OH)_8^{4+}$, and $Pb_3(OH)_4^{2+}$ form consequently with the following deposition of $Pb(OH)_2(s)$.

Anthropogenic lead loading

Lead occurs in nature as the sulfide, galena, PbS. Lead is more electropositive than mercury, and roasting the sulfide in air forms lead oxide, reaction (7)

$$2\,PbS + 3\,O_2 \longrightarrow 2\,PbO + 2\,SO_2. \tag{7}$$

The oxide is then reduced to a metal with coke. The impure metal is refined by electrolysis.

Major anthropogenic sources of lead include the use of Pb as a petrol additive, Pb mining and smelting, printing, Pg paint flakes, sewage sludge and the use of pesticides containing Pb compounds, like lead arsenate.

A famous use of lead is also the familiar lead-acid storage battery. This device is an example of a storage cell, meaning that the battery can be discharged and recharged over a large number of cycles. The lead-acid battery is familiar as a battery in your car.

An important disadvantage of the lead-acid battery is its heavy mass, on account the high Pb density. The second disadvantage is that the used car batteries distribute a lot of lead into the environment; despite recycling, they are the major source of lead in municipal waste. Recently, the recycling of lead-acid batteries has created problems in the local environment around recycling plants. Most of these plants are located in developing countries of Asia and Latin America and they process batteries imported from industrialized nations. Levels of Pb as high as 60,000–70,000 ppm have been measured in soils in the vicinity of Pb-battery recycling plants in the Philippines, Thailand and Indonesia. The relevant health effects have been observed. This appears to be one example where trying to conserve resources and minimize pollution has gone seriously wrong. In California, soil contaminating 1000 ppm of Pb is considered to be hazardous waste and its disposal is strictly regulated.

Ecotoxicolocal effects of lead

The half-life of lead in humans is estimated to be about six years for the whole body burden and from 15 to 20 years for the skeleton. Thus, an excretion from the skeleton is very slow. Lead, like mercury, is a cumulative poison. The skeletal burdens of lead increase almost linearly with age. This suggests that the Pb steady state is not normally reached. Chelation of Pb^{2+} with ethylendiaminetetraacetic acid (ADTA) has been found beneficial in reducing Pb body burden for clinically affected patients.

Lead, like mercury, causes neurological diseases. The organolead compounds are more toxic than mineral lead salts, since they are non-polar, lipid-soluble, and more readily cross the blood-brain barrier. This disease is related to child mental retardation, lower performance on I.Q. tests, and hyperactivity. Severe exposure in adults causes irritability, sleeplessness, and irrational behavior. Some have gone as far as to blame anti-social behavior and criminality on sub-clinical Pb poisoning. A correlation between Pb in blood and Pb in air, dust and soils has been observed in many studies. The U.S. Center for Disease Control has proclaimed a goal of reducing blood lead contents in children below $10\,\mu g/100$mL.

Lead is a well-known poison, but the effects of exposure to lower levels have been contentious. There is growing evidence of sub-clinical Pb poisoning, especially among young children who play in polluted parks, gardens and streets. Contaminated soil or dust particles may be transferred to children's hands and ingested accidentally. Humans are exposed to Pd from various sources and road dust and soils can contribute to the total lead exposure. Approximately one half of lead ingested in food is absorbed.

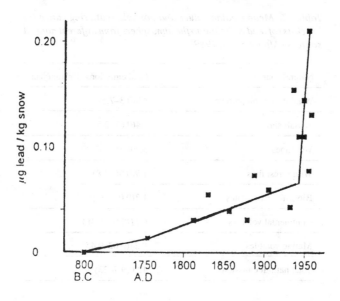

Figure 10. Increase of lead in Greenland snow, 800BC to present. (Bunce, 1994).

Environmental biogeochemistry of lead

The long-term uses of lead explain why this element should be so widely dispersed in the environment. In this relation we should answer the question as to what is the natural background level of lead. At present this is a controversy. Lead levels in modern people are frequently 10% of the toxic level. Some analyses of ancient bones and ancient ice cores seem to suggest that this relatively high level is not new and has existed in the environment for thousands of years. Accordingly, the assumption was made that life has evolved in the presence of this toxic element.

However, recent researches have challenged this viewpoint, claiming that these lead analyses in ancient samples are the results of inadvertent contamination of the samples during their collection and analysis. Dr. C. C. Patterson of the California Institute of Technology argues, for example, that the ice cores are contaminated by lead from drilling equipment. His data of chemically careful Pb analysis on Greenland ice cores show the increasing trend of lead pollution (Figure 10).

Similar data reported on the content of lead in meticulously preserved old skeletons contain 0.01 to 0.001 times as much lead as contemporary skeletons. A different perspective is provided in the analysis of pre-industrial and contemporary Alaskan Sea otter skeletons. The total concentrations of lead in the two groups of skeletons were similar, but their isotopic compositions were different. The pre-industrial skeletons contained lead with an isotopic ratio corresponding to natural deposits in the region, while the ratio in the contemporary ones was characteristic of industrial lead from elsewhere (Smith et al, 1990).

Table 12. Mean median values and possible scattering range (in brackets) of lead emissions to the atmosphere from different natural sources (After Nriagu, 1989).

Natural source	Lead emissions, 10^3 ton/year
Wind-borne soil particles	3.9(0.3–7.5)
Seasalt spray	1.4(0.02–2.8)
Volcanoes	3.3(0.54–6.0)
Wild forest fires	1.9(0.06–3.8)
Biogenic particulates	1.3(0.02–2.5)
Continental volatiles	0.21(0.01–0.38)
Marine volatiles	0.24(0.02–0.45)
Total natural emission	12.0(9.7–23)

Global mass balance of lead

We can compare the natural and anthropogenic emissions of lead in the global cycle. Table 12 shows the natural sources and Table 13 the major anthropogenic ones.

Data of other researchers differ appreciably from the data of Table 12. For example, Thornton et al (1995) considered the integral natural emission of lead as much as 35×10^3 ton/year. According to data of the Geological Survey of Canada (GSC, 1995), the natural inflow of lead to the atmosphere can reach 330×10^3 ton/year.

As evident from Table 13, the uncertainty level of the estimates is significant. The mid-1990s estimates are given by Thornton et al (1995) and the global anthropogenic emission was calculated as much as 450×10^3 tons per year. The latest estimates were made for various continents by Pacina et al, 1993 (Table 14).

We can see that anthropogenic emissions of lead are most important in developed industrial countries because they are mainly associated with fossil fuel combustion for power generation and with transport. The Asian region is ahead of other regions on a basis of total emission data. The highest emission per individual is in Australia owing to non-ferrous industry emissions predominating on this continent. As to the pollution density per area unit, however, Europe is ahead of Australia, Africa and South America by an order of magnitude. The total global emission was estimated as much as 210×10^3 tons per year with averaged emission of about 40 g/capita per year.

Thus, the contemporary anthropogenic emission of lead exceeds the natural emission in global scale almost 30 times.

Table 13. The most important anthropogenic emission sources of lead to the atmosphere on the global level in 1983 (After Nriagu and Pacina, 1989).

Source type	Lead emission flux, 10^3 ton/year
Fossil fuel combustion	2.7–18.4
Wood burning	1.2–3.0
Metallurgical processes	31.1–83.8
Waste incineration	1.6–3.1
Mobile sources	248.0
Other human activities	4.0–19.6
Total — median value	332
— scattering range	288.7–376.0

Table 14. Anthropogenic lead emissions to the atmosphere as of 1989 on different continents (After Pacina et al, 1993).

Continent	Emission, 10^3 tons per year	Relative emission per:	
		Capita, g/yr/capita	Area unit, g/km^2/yr
Africa	17.5	30	600
Asia	74.3	20	1700
Australia	5.4	200	700
Europe	69.6	70	7000
North America	36.8	90	1500
South America	15.1	50	800
World	208.6	38	1390

FURTHER READING

1. Bunce N., 1994. *Environmental Chemistry.* second edition, Wuerz Publishing Ltd, Winnipeg, Chapter 9.

2. Butcher S. S., Charlson R. J., Orians G. H. and Wolfe G. V. (Eds.) 1992. *Global Biogeochemical Cycles.* Academic Press, London et al, Chapter 15.

3. Moldan B. and Cherny J. (Eds.) 1994. *Biogeochemistry of Small Catchments*, John Wiley and Sons, Chapter 13.

4. Dobrovolsky V. V., 1994. *Biogeochemistry of the World's Land*. Mir Publishers, Moscow and CRC Press, Boca Raton; Ann Arbor; Tokyo; London, 206–211.

5. Radojevic M. and Bashkin V., 1999. *Practical Environmental Analysis*. Royal Society of Chemistry, UK, Chapter 5.13.

6. Bashkin V. and Howarth R., 2002. *Biogeochemistry*. Chapter 8, Kluwer Academic Publishers.

WEBSITES OF INTEREST

1. Toxicological characteristics of chemicals,
   http://www.rff.org/misc_docs/risk_book.pdf

QUESTIONS AND PROBLEMS

1. Present the role of combustion in total heavy metal emission. Describe this problem regarding your country.

2. Discuss the scheme of heavy metal transformation in the combustor of a coal-burning power plant. Point out the stages of heavy metals emissions.

3. Describe the case study of heavy metals emissions in northern Thailand. Indicate the characteristic features of the Thai lignite.

4. How can you calculate the emission of heavy metals with fly ash? Use the described approach for similar power plants in your country.

5. What types of assumption can you apply for calculating heavy metals depositions when the relevant models are absent?

6. Characterize the ecological consequences of heavy metals accumulation due to coal burning power plant operations. Compare the results in Mae Moh Valley with similar ones for your country.

7. Discuss the role of speciation in biogeochemical and environmental behavior of arsenic.

8. Characterize the specific features of biogeochemical cycling of arsenic in the southern and east-southern sub-regions of Asia.

9. Discuss the behavior of mercury in the modern environment. Why is this metal of special attention for environmental chemistry?

10. Describe the role of speciation in the biogeochemical cycle of mercury. What forms of mercury are the most dangerous and why?

11. Describe the ecotoxicology of mercury and focus your attention on the various LD_{50} for different mercury species.

12. Compare the pre-industrial and modern cycles of mercury on a global scale. What are the consequences of disturbance of the natural biogeochemical cycle of mercury?

13. Discuss geochemical characteristics of lead and the distribution of this element in ancient and modern environments. Present evidence of accelerated migration of lead during the last centuries.

14. Compare the ecotoxicology of mercury and lead and discuss their migration in biogeochemical food webs.

15. Consider the global fluxes and pools of lead. What processes are of main importance in natural and man-altered biogeochemical cycles of lead?

16. Discuss the emission fluxes of lead in various continents of the Earth. Compare total data, and the relevant comparison based on per area unit and per capita data.

CHAPTER 13

ENVIRONMENTAL CHEMISTRY OF NITROGEN

1. INTRODUCTION

The major input of nitrogen to the terrestrial and aquatic (fresh and marine water) ecosystems in the Asian region is application of nitrogen fertilizers. At the end of the 1990s, the total use of mineral fertilizers in the Asian region was about 60 million tons, as a sum of nitrogen, phosphorus and potassium fertilizers, and nitrogen fertilizers accounted for 65–70%. The demand for food and other agricultural products is expected to rise by 30 to 100% between now and 2020 in parallel with a growing population. Yet, the net cropland may decrease in many countries over the same period, partly owing to a competing demand for arable land from on-going urban growth. To some expert, this implies a greater reliance on technology and chemical inputs, first of all, nitrogen fertilizers, to realize higher target yields, merely to maintain the current per capita food availability. Thus, nitrogen is an essential nutrient for the Asian agriculture, however, extensive application of nitrogen is accompanied by losses of this element due to denitrification as nitrous oxide, leaching as nitrate and ammonium to surface and ground water and accumulation in foodstuffs. These processes are of great environmental concern, as we have discussed earlier. Here we will consider the biogeochemical cycle of nitrogen in the Asian region and its anthropogenic disturbance.

2. BIOGEOCHEMICAL CYCLE OF NITROGEN

Nitrogen is an essential element for all forms of life and its biogeochemical cycle is one of the most important in the modern biosphere. It is a structural component of amino acids from which proteins are synthesized. Animal and human tissue (muscle, skin, hair, etc.), enzymes and many hormones are composed mainly of proteins.

2.1. Nitrogen Cycling Processes

The biogeochemical cycling of nitrogen has been extensively studied in different ecosystems and the main processes are listed below

- *Fixation* is the conversion of atmospheric N_2 to organic N;

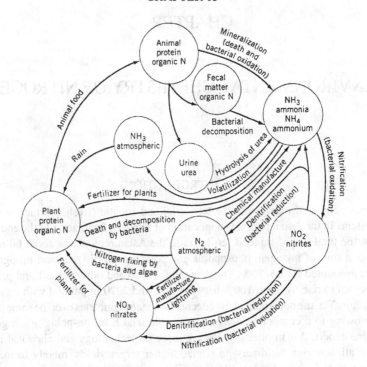

Figure 1. The nitrogen cycle (Bosch and Wolfe, 1999).

- *Mineralization* is the conversion of organic N to inorganic N;

- *Nitrification* is the oxidation of NH_4^+ to nitrite (NO_2^-) and nitrate (NO_3^-);

- *Denitrification* is the conversion of inorganic N to atmospheric N_2O and N_2;

- *Assimilation* is the conversion of inorganic N to organic N.

Nitrogen chemistry and cycling in the environment are quite complex due to the great number of oxidation stages. A large number of chemical, biochemical, geochemical, and biogeochemical transformations of nitrogen are possible, since N is found at valence stages ranging from -3 (in ammonia, NH_3) to $+5$ (in NO_3^-). Microbial species possess the transformations between these stages, and use the energy released by the changes in redox potential to maintain their life processes. Generally these microbial reactions manage the global nitrogen biogeochemical cycling (Figure 1).

The most abundant form of nitrogen at the Earth's surface is molecule nitrogen, N_2, and this form is the least reactive species. Various above mentioned processes convert atmospheric nitrogen to one of the forms of fixed nitrogen and can be used by living organisms. The reverse process of denitrification returns nitrogen to the atmosphere as N_2. Generally we consider the 0.781 atm of nitrogen in the atmosphere almost as inert filler, because elemental nitrogen is quite non-reactive. The known aspects of N biogeochemistry lead us to the conclusion that the atmospheric content of nitrogen, like carbon, is regulated principally by biological processes.

The atmosphere contains about 3.9×10^{15} tons of elemental nitrogen. Biological nitrogen fixation and production of NO in combustion and thunderstorms are the major natural sinks (reaction (1)). Finally, NO deposits as HNO_3 in rainwater. In the Haber process N_2 fixes industrially (reaction (2))

$$N_2(g) + O_2(g) \xrightarrow{\text{high temperature}} 2\,NO(g), \tag{1}$$

$$N_2(g) + 3H_2(g) \xrightarrow{\text{catalyst, 450\,°C}} 2\,NH_3(g). \tag{2}$$

Terrestrial and aquatic nitrogen in the form of NH_4^+ and NO_3^- are cycled through the biosphere to make proteins and nucleic acids. The processes of decay return the nitrogen to the atmosphere as N_2 and as N_2O by the action of denitrifying bacteria. Almost all the nitrogen fixed by the Haber process is used as fertilizer so that increased fertilizer uses increase the rate of return nitrogen to the atmosphere through biological denitrification. Indeed, as much as a quarter of all nitrogenous fertilizers applied to crops is denitrified during crop vegetation. This effect is reflected in the gradual increase in the global atmospheric levels of N_2O, which have risen from 0.29 to 0.31 ppmv over the past 25 years. Nitrous oxide, N_2O, is rather non-reactive and has a residence time of $\sim 20\,\text{yr}$. This species contributes to ozone depletion potential (see Chapter 6).

Main nitrogen species

Table 1 lists the most common nitrogen species that exist in nature. This table presents the oxidation stage, the boiling points (b.p.) for each species and its heat of formation [$\Delta H°(f)$] and free energy of formation [$\Delta G°(f)$]. The same figures are shown for water for comparison.

General characterization of biogeochemical cycling processes of N

Let us consider the various ways that N is processed by the biosphere. These ways are important for both terrestrial and oceanic nitrogen cycles. They are shown schematically in Figure 2.

All the processes shown in Figure 2 are driven by different microbes. Some of these processes are energy-producing, and some of these occur symbiotically with other organisms. We will start the consideration of biological N transformations from fixation, since this is the only natural process that can bring nitrogen from the atmosphere pool to ecosystems.

1. *Biological nitrogen fixation* is any process in which N_2 in the atmosphere reacts to form any N compounds. This process is enzyme-catalyzing reduction of molecule nitrogen ammonia (NH_3), ammonium (NH_4^+), and various organic nitrogen forms. In natural processes, biological nitrogen fixation is the ultimate source

Table 1. The list of natural nitrogen species.

Species	Valence	Boiling point, °C	$\Delta H°$(f), kJ/mol, 298K	$\Delta G°$(f), kJ/mol, 298K
HN_3(g)	−3	−33	−46	−16.5
NH_4^+(aq)	−3		−72	−79
NH_4Cl(s)	−3		−201	−203
CH_3NH_2(g)	−3		−28	28
N_2(g)	0	−196	0	0
N_2O(g)	+1	−89	82	104
NO(g)	+2	−152	90	87
HNO_2(g)	+3		−80	−46
HNO_2(aq)	+3		−120	−55
NO_2(g)	+4	21	33	51
N_2O_4(g)	+4		9	98
N_2O_5(g)	+5	11	115	
HNO_3(g)	+5	83	−135	−75
$Ca(NO_3)_2$(s)	+5		−900	−720
HNO_3(aq)	+5		−200	−108
H_2O		100	−242	−229

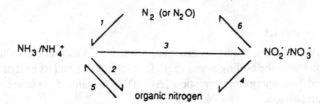

Figure 2. Biological transformation of N species. 1—nitrogen fixation, 2—ammonia assimilation, 3—nitrification, 4—assimilatory nitrate reduction, 5—ammoniafication, 6—denitrification.

of nitrogen for all biota. A number of symbiotic and non-symbiotic microbes, both bacteria and algae, have the ability of N fixation. The symbiotic species are quantitatively more significant. There are two major limitations to biological N fixation. The first is related to the required sources of energy to splitting the nitrogen triple bond in N_2, $N \equiv N$. The free energy for formation of NH_3 from N_2 and H_2 ($\Delta G°$) is negative at 25°C (Table 1); however, only some organisms with highly developed catalytic systems are able to fix nitrogen. The second limitation is connected with obligatory anaerobic conditions for N fixation since this is a reductive process. Accordingly, only those organisms that live in an anaerobic environment or can create this environment will fix nitrogen.

In terrestrial ecosystems the symbiotic bacteria, particularly strains from the genus *Rhizobium*, play the most important role in N fixation. These bacteria are found on the roots of many leguminous plants (soybeans, clever, chickpeas, etc.). There are other symbiotic diazotrophs (nitrogen-fixing microbes), but till now the *Rhizobium* has been the most extensively studied.

The cyanobacteria are considered the main N-fixing organisms in both freshwater and marine ecosystems. Some estimates have shown that these species are responsible for up to $\sim 80\%$ of total fixed nitrogen in freshwaters. Cyanobacteria are widespread so that significant nitrogen fixation is possible on local, regional and global scales.

2. *Ammonia assimilation* is the process by which fixed ammonium transforms to the organic form. This process is of significant importance for those organisms that can directly utilize nitrogen from NH_3/NH_4^+. Direct ammonia assimilation yields significant energy saving for those organisms which have an ability to utilize this form of nitrogen. A free ammonium ion does not exist for long in any aerobic environment where it is easily nitrified in a nitrification process.

3. *Nitrification* is the oxidation of NH_3/NH_4^+ to nitrite and further to nitrate to gain energy for living microorganisms. Nitrates are the prevalent form of nitrogen in aerobic waters and soils. Nitrification consists of two conjugated stages, which both give an energy yield. The first step is oxidation of ammonium to nitrite, reaction (3), and the second one is the subsequent oxidation of formed nitrites to nitrates, reaction (4). We can write the following typical equations

$$NH_4^+ + 3/2\,O_2 \longrightarrow NO_2^- + H_2O + 2\,H^+, \quad \Delta G^\circ = -290\,\text{kJ/mol}, \quad (3)$$

$$NO_2^- + 1/2\,O_2 \longrightarrow NO_3^-, \quad \Delta G^\circ = -82\,\text{kJ/mol}. \quad (4)$$

The first step in this reaction consequence is processing generally by bacteria genus *Nitrosomonas*, and the second step by *Nitrobacter*. Both these microbes are autotrophic organisms. As carbon sources they utilize CO_2 and gain energy from ammonium oxidation. The role of heterotrophic bacteria, which utilize organic compounds rather than carbon dioxide, is less significant quantitatively, however, these bacteria can also perform the nitrification process. Some intermediates like hydroxylamine (NH_2OH), NO and N_2O were determined in these reactions. The production of N_2O may serve some environmental problems both with ozone depletion and global warming effects (see Chapter 6). Two different pathways can be monitored for nitrates in terrestrial and aquatic ecosystems. The first is related to assimilatory nitrate reduction and the second to denitrification.

4. *Assimilatory nitrate reduction* is simultaneous reduction of nitrate and uptake of N into the biomass of any organism. This process might be dominant when nitrogen is in low supply, which is typical in aerobic soil and water column conditions. We can consider this as a primary N input to many microorganisms and plants.

Most plants can assimilate both reduced and oxidized forms of nitrogen, even though there is an energy cost in first nitrate reduction.

5. *Ammoniafication* is the other major source of reduced nitrogen for living organisms. Ammoniafication can be defined as the breaking down of organic nitrogen compounds with realizing ammonia or ammonium. Decomposition of soil or aquatic organic matter is the typical example of this process and heterotrophic bacteria are principally responsible for it. During ammoniafication, microbes get the carbon source from dead plant or animal biomass and yield the NH_3/NH_4^+ system as the additional products. Most of this reduced nitrogen will be conserved in a biological cycle but a small fraction of it may be volatilized. The more significant source of ammonia is the volatilization during the breakdown of animal excreta and in some regions these values are comparable with losses of nitrogen in the denitrification process.

6. *Denitrification* is the reduction of nitrates to any gaseous N species, generally N_2 and partly N_2O. This is the only process in the biogeochemical cycle of nitrogen which provides the removal of fixed N compounds to the atmosphere. Under most conditions, the N_2 is the prevalent reaction end-product. However, under natural acid reaction of soil or under acidification of terrestrial and especially aquatic ecosystems resulting from acid deposition, the formation of nitrous oxide can be preferential. Microbes use nitrate as a terminal electron sink (oxidant) in the absence of oxygen (O_2), i.e., in anaerobic soil or water conditions. The overall process is both oxidizing an organic compound and reducing the nitrate. This process leads to energy yielding for approximately 17 genera of facultative anaerobic bacteria. These bacteria can utilize nitrate as an oxidizing agent. The intermediate products of denitrification reactions are nitrite, nitrogen oxide (NO) and nitrous oxide (N_2O). The ratio of $N_2 : N_2O$ in the denitrification product is of environmental concern. We can note only that N_2 accounts for 80–100% of the nitrogen realized or the $N_2 : N_2O$ ratio may be equal to 16 : 1 for global N budget calculations.

It is generally agreed that in terrestrial ecosystems, denitrification is a significant source of both N_2 and N_2O under natural and agricultural conditions (Figure 3).

Denitrification is still insufficiently quantitatively understood in aquatic ecosystems, especially by comparison with terrestrial ecosystems. There are different views on whether the oceans are a source or sink of nitrous oxide. Various data indicate that the ocean is, on average, supersaturated with respect to N_2O, and that N_2O supersaturation is positively correlated with NO_3^- and negatively correlated with O_2. A number of studies have suggested that denitrification is a major sink for fixed nitrogen in the oceans. Recent studies suggest that denitrification losses in the oceans are on the order of 0.13×10^9 tons/yr (9.2×10^{12} mol N per year) which exceed known oceanic N inputs. More than half of this denitrification 0.067×10^9 tons/yr (4.8×10^{12} mol N per year) takes place in sediments, with the remainder in pelagic oxygen minimum zones (Bashkin and Howarth, 2002).

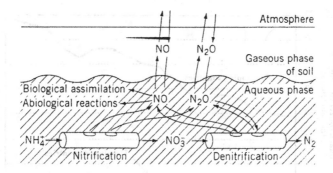

Figure 3. The "hole-in-the-hole" conceptual model of N_2O formation trough nitrification and denitrification (Oonk and Kroeze, 1999).

The considered processes relate to N transformation in various chains of the biological cycle. The biological N cycle itself is only part of a total global biogeochemical turnover of this element. The overall cycle is the interaction of biotic and abiotic processes.

Global nitrogen cycle

Prior to estimating the global N cycle, let us consider the nitrogen reservoirs in the biosphere. The effect of human activity on the global cycling of nitrogen is great, and furthermore, the rate of change in the pattern of use is much greater even for many other elements (Galloway et al, 1995). The single largest global change in the nitrogen cycle comes from increased reliance on synthetic inorganic fertilizers, which accounts for more than half of the human alteration of the nitrogen cycle (Vitousek et al, 1997). The process for making inorganic nitrogen fertilizer was invented during World War I, but was not widely used until the 1950s. The rate of use increased steadily until the late 1980s, when the collapse of the former Soviet Union led to great disruptions in agriculture and fertilizer use in Russia and much of eastern Europe. These disruptions resulted in a slight decline in global nitrogen fertilizer use for a few years (Matson et al, 1999). Use as of 1996 was approximately 83×10^6 ton N per year. Approximately half of the inorganic nitrogen fertilizer that was ever used on Earth has been applied during the last 15 years.

By 1997/98, in Asia the regional use of inorganic nitrogen fertilizer was again growing rapidly, with much of the growth driven by use in China, where production of nitrogen fertilizers at the end of the 1990s was about 27 million tons annually (Figure 4).

However, other human-controlled processes, such as combustion of fossil fuels and production of nitrogen-fixing crops in agriculture, convert atmospheric nitrogen into biologically available forms of nitrogen. Overall, human fixation of nitrogen (including production of fertilizer, combustion of fossil fuel, and production of nitrogen-fixing agricultural crops) increased globally some two- to three-fold between 1960

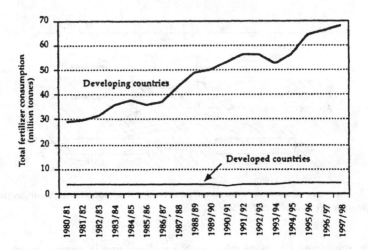

Figure 4. Fertilizer consumption trend in Asia (FAO, 1999).

to 1990 and continues to grow (Galloway et al, 1995). By the mid 1990s, human activities made new nitrogen available at a rate of some 140×10^6 ton N per year (Vitousek et al, 1997), matching the natural rate of biological nitrogen fixation on all the land surfaces of the world (Vitousek et al, 1997; Cleveland et al, 1999). Thus, the rate at which humans have altered nitrogen availability globally far exceeds the rate at which humans have altered the global carbon cycle. A similar conclusion can be derived from the Asian data (see below).

Countries such as China have been largely self sufficient in food production for the past two decades, in part because of increased use of nitrogen fertilizer. The use of fertilizer in China is now very high—almost 10-fold greater than in the United States—and further increases in fertilizer use are less likely to lead to huge increases in food production, as they have in the past. Therefore, if China's population continues to grow it may once again be forced to import food from the United States and other developed countries, leading to more use of nitrogen fertilizer here.

3. ENVIRONMENTAL BIOGEOCHEMISTRY OF NITROGEN IN THE ASIAN REGION

3.1. Alteration of Nitrogen Biogeochemical Cycling

Human activity has greatly altered the nitrogen cycle on land, in aquatic systems, and in the atmosphere. Currently, global fixation of atmospheric N_2 for fertilizer, in combustion of fossil fuels, and by leguminous plants, exceeds that by all natural sources, and changes in land use cause large additional amounts of nitrogen to be released from long-term reservoirs to both vegetation and soil. These disturbances have been linked to a number of environmental concerns, including excessive accumulation of nitrates in groundwaters, inland and coastal water eutrophication, acidification of freshwater lakes and rivers, climate changes, saturation of forest ecosystems and decline of forest biodiversity and productivity. However, some of these environmental problems can be considered as positive. For instance, accelerated nitrogen cycling has also been suggested to increase the sequestration of carbon in forest ecosystems, slowing the rise of atmospheric carbon dioxide (see Chapter 2).

The foundations of environmental biogeochemistry were based in part on studies of small-scale watersheds (less than 500 ha) in Europe and North America. Recently, much more attention has been given to evaluating material fluxes through the landscape at larger scale. However, we are only beginning to understand the rules that govern regional exchanges of nitrogen between atmospheric, terrestrial and aquatic ecosystems. These exchanges are especially critical to the functioning of freshwater and coastal ecosystems, as even relatively small changes in the processing and retention of nitrogen applied to terrestrial ecosystems could have a large impact on the downstream aquatic ecosystems. In fact, several studies have noted much greater fluxes of nitrate in rivers over the past few decades. Nitrate concentrations in the Mississippi River, USA, have more than doubled since 1965 and that in many European rivers (Danube, Rein, Volga, Seine, et al) has probably increased 5- to 10-fold during the 20th century. Similar trends are now shown for main East Asian rivers (Yellow, Yangtze, Mekong, Chao-Phraya, Han, et al).

The pollution of atmosphere and transboundary transport of nitrogen (and sulfur) species is a great concern for the whole North Hemisphere. This leads to an increased atmospheric deposition of nitrogen.

Increased use of nitrogen fertilizers, cultivation of nitrogen fixing crops and depositions are accompanyed by enforced denitrification and production of N_2O. The latter is known as a species that accelerates ozone destruction in the stratosphere (see Chapter 6).

These are the milestones of problems related to the environmental biogeochemistry of nitrogen. We will consider the environmental biogeochemistry of nitrogen in various scales, from local and regional up to the global. Examples will be given for the most altered Asian regions, such as the East Asian region.

3.2. Alteration of Nitrogen Cycling in the East Asian Region
Regional biogeochemical budget approach

At present the estimation of regional fluxes of pollutants gives a powerful tool for understanding of man-made changes in different scales, from landscape to global. The characteristic examples are the calculation of nitrogen cycling in the North Atlantic Ocean and its watershed (Howarth, 1996), in the Baltic (Bashkin, 1997a) and Mediterranean (Bashkin, 1997b) drainage basins.

Nitrogen is a key element of many biogeochemical processes and can be both a nutrient limiting the productivity of terrestrial and aquatic ecosystems and a pollutant, for which excessive accumulation in biogeochemical food chains leads to many environmental problems. The main anthropogenic sources of nitrogen pollution are related to fertilizer application, waste production and emission of gaseous species.

Modern projections indicate that potentially large increases in emissions may occur during the next 25–50 years namely in East Asia in accordance with planned development patterns. According to the present estimates, at the current growth rate of energy consumption, by the year 2020 NH_x, NO_y and SO_2 emissions will surpass the emissions of North America and Europe combined. The primary man-made source of acidifying and greenhouse compounds in East Asia is fossil fuel of low quality, with high content of sulfur (up to 7% in Thai lignite, Chinese brown coal etc.) and heavy oil. The multiple effect of acidification and increased N deposition in East Asia may cause decreases in base cations, leading to nutrient imbalances in forest vegetation. This increases their vulnerability to diseases, attacks of insects and parasites. Nitrogen deposition changes natural vegetation, bacterial and mycorrhizal composition into more nitrophilic communities, with less diversity of species, especially endemic ones. The NO_x compounds can transport over great distances in East Asia. The spatial scale of atmospheric NH_x pollution and its effects depend on its form: gaseous NH_3 is important near sources up to 100 km; aerosol NH_4 is transported over longer distances leading to effects up to 1000 km (Carmichael et al, 1997). At present, N_2O warrants also a priority for policy and research institutes. The increasing N_2O concentration in the atmosphere mainly results from nitrification and denitrification processes due to increasing application of mineral fertilizers in various regions of the Earth and especially in East Asia. This N species contributes to the greenhouse effect (about 6%) and to the destruction of stratospheric ozone (Erisman et al, 1999). There is agreement both nationally and internationally that long-range transboundary air pollution is not limited to the geographical limits of individual East Asian countries. It is known that during the winter the major weather patterns in East Asia facilitate the transboundary transport of air pollutants from west to east, from land to sea and the reverse in summer. Pollutants can thus be transported from country to country in the whole region of East Asia. For instance, it has been calculated that during the 1990s about 50% of oxidized and about 60% of reduced nitrogen deposition over the Republic of Korea was imported from abroad. It is therefore impossible for individual countries to solve the problem of air pollution and acid rain alone. There is need for regional intergovernmental cooperation. Currently, regional/sub-regional agreements on the issue of pollutant emission abatement strategy do not exist at all or are in the initial stages (Bashkin and Park, 1998).

The abatement strategy for reduction of N emission and deposition, as well as decreasing of undesirable losses from agroecosystems owing to excessive application of fertilizers, has to be based on calculations of its regional biogeochemical budgets with quantitative parameterization of different fluxes in terrestrial and aquatic ecosystems.

General description of East Asian domain

The waters of four oceans wash the Asia continent with respect to the oceanic slopes. The East Asian domain is bounded by a Pacific Ocean slope of $11,905,000 \, km^2$ or 27.4% of total area of Asia (see Figure 1, Chapter 8). The four main river basins are placed in this area: Amur, Huanghe, Yangtze and Mekong rivers. There are also

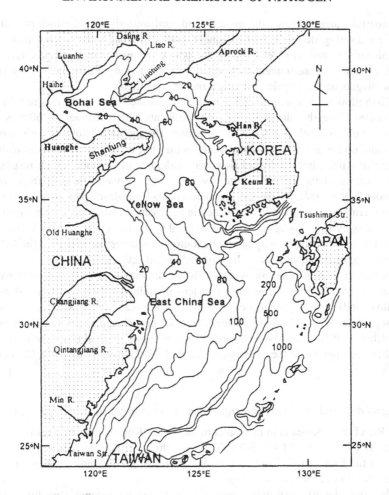

Figure 5. Bathymetry and geographic system including the east China Sea, the Yellow Sea and the Bohai Sea system.

plenty of middle-size and small rivers in East Asia. These rivers are drained into East (Japan), Yellow, East China and South East China Seas of the Pacific ocean.

The Yellow Sea is a typical epicontinental sea surrounded by the continent of China and the Korean peninsula and connected with the East China Sea to the south and with the Bohai Sea in the north (Figure 5).

The mean water depth is about 44 m and the surface area is 420,000 km^2. The rivers (the Huanghe, the Aprock, the Han, the Keum, the Haihe, the Luanhe) drain freshwater of about > 160 km^3 and suspended materials of 1.1×10^9 ton annually into the Bohai and Yellow Sea system. Many industrial complexes and large cities are along the coastlines, to which great quantities of pollutants are discharged into rivers or directly into coastal waters. Further, the westerly and northwesterly winds, which prevail in winter and spring, deliver a great amount of mineral dust and anthropogenic material. As this coastal system receives large amounts of terrestrial material produced by natural weathering and human activities, it is a suitable study site for the investigation of the biogeochemical cycle of nitrogen on a regional scale.

The Yellow Sea is a semi-enclosed basin. Wide coastal areas (< 40 m water depth) are located along shorelines nearby both continents with a channel (> 60 m water depth), which is developed in NW-SE direction. The southward flows in both coastal areas and northward flows of warm water in the channel are the general circulation pattern in winter, but northward flow may disappear in summer, which results in the formation of the Yellow Sea cold water in the channel. The tidal fronts near the boundary of shallow coastal area and channel are developed during summer when thermal stratification is established in the channel and vertically homogenous water mass by strong tidal currents is sustained in shallow coastal areas (Seung and Park, 1990). These fronts may affect the transport of terrestrial materials (including nitrogen) to offshore, and hence the biogeochemical activity.

A billion tons of terrigenous materials are discharged annually through rivers (Huaghe, Aprock, Han, Keum etc.) and tens of millions of tons of mineral dusts (Yellow sand) during a year are deposited into surface seawaters from the atmosphere. Most particles derived from the Huanghe river are deposited in the Bohai Sea (Cheng and Chao, 1985), and it was suggested that the basin-wide atmosphere dust flux might be comparable to the river input (Gao et al, 1992; Zhang et al, 1992; Choi et al, 1998).

General description of natural terrestrial ecosystems in the Republic of Korea

The Republic of Korea is in the southern part of the Korean mountain peninsula between 126°E–130°E and 34° N–38° N. Geographically this is the northern temperate zone of the Eastern Hemisphere. The overall area of the Republic of Korea (ROK) comprises 99,022 km^2.

In Korea,the principal native vegetation is Forest ecosystems. Very little of the original, virgin composition remains, mainly in the plateau of high mountain areas. The dominant vegetation consists of forest trees with varying undergrowth of shrubs and small plants.

The forest vegetation in the country is divided according to climate into two kinds, the temperate and subtropical zone forests. The secondary vegetation, which occurs in extensive areas in the western and southern regions, consists mostly of shrubs, grasses and conifers, associated with deciduous trees. Grassland with shrubs also appear commonly as a kind of climax vegetation in the higher elevation and plateau remnants.

In accordance with land use type, the following natural ecosystems are presented in the South Korean peninsula: Coniferous Forest, Deciduous Forest, Mixed Forest, Evergreen Forest, Plain Shrubs and Mountain Shrubs with Grasslands. The agricultural land use includes about 20% of total South Korean area.

Mountainous topography occupies 2/3 of the total South Korean area. The mountain range of Taebaeg on the Gangweon-Do runs southward along the east coast with lateral branches and spurs expanding in a southwesterly direction. The slopes to the east are steep while those to the west are gentle. The mountain range slopes towards the south, thus making the southern part of the country fairly level. In contrast, the northern part is mountainous and hilly land.

About 80% of the ROK are in regions where the altitude of the summit ranges from 300 to more than 1,000 meters. The land bears a strongly dissected relief, reworked by numerous erosion cycles. The low lands include both coastal plains clayey materials and the continental alluvial plains and valley flood plains of the interior.

According to USDA Soil Taxonomy, the soils in the Republic of Korea are classified as 6 orders and 14 suborders (Um, 1985). The dominant soils are Inceptisol order with 4 suborders: Andepts, Aquepts, Ochrepts and Umbrepts (5,840,441 ha). The Entisol order consists of 4 suborders (Psamments, Aquents, Fluvents and Orthents) and occupies 2,849,102 ha. The Altisol order is presented by Aqualfs and Udalts suborders on an area of 309,677 ha and Histosol order is divided between Saprist and Hemist suborders with total area of 384 ha.

Data sources

The statistical characteristics of different parts of the East Asian domain were extracted from both national and international sources (Environmental Statistics Yearbook, 1998; UNESCO, 1978; ESCAP, 1998). Data on content of nitrogen species in river waters were selected both from our own and literature studies.

3.3. Application of Biogeochemical Mass Balance Approaches to Regional Studies

Any type of budget for biogeochemical turnover of pollutants depends on the availability of data. As usual the more precise calculations can be made for the small watershed with homogenous (Moldan and Cherny, 1994) or variable (Bashkin, 1984; Gunderson and Bashkin, 1994) land use. With currently available data, it is not possible to fully account for the fate of both natural and anthropogenic nitrogen added to the East Asian domain as a whole and in its individual parts like Northeast Asia or the Yellow Sea basin. Relatively more accurate estimates might be carried out for the South Korean agroecosystems and for the total area of the Republic of Korea.

The SCOPE project constructed a mass balance for reactive nitrogen under anthropogenic influence on the regional scale (Howarth, 1996). For inputs, for the terrestrial ecosystems the SCOPE N analysis considered application of N fertilizers, N fixation by agricultural crops, if any, NO_y depositions, and import or export of N in food and animal feedstocks. Output items considered crop uptake, denitrification and volatilization and river discharge. For marine ecosystems, this analysis includes also seawater exchange.

Table 2. Links of nitrogen biogeochemical cycle accounting for N mass balance calculations in East Asia.

Links	ROK, Agroecosystem	ROK, country scale	Yellow Sea basin	East Asian domain
Deposition	+	+	+	+
Fertilizers	+	+	−	+
N fixation	+	+	+	+
Import	−	+	−	−
Riverine N	−	−	+	−
Crops	+	−	−	−
Denitrification	+	+	+	+
Volatilization	+	+	+	+
Runoff	+	+	−	+
Sedimentation	−	−	+	−
Sea exchange	−	−	+	−

Depending upon the region, the following items were taken into account for various types of regional biogeochemical mass balance of nitrogen (Table 2).

The input links of regional N biogeochemical mass balance in South Korean agroecosystems are assumed to include deposition, mineral and organic fertilizers, and biological fixation (Table 11, Chapter 8). Output items considered crop uptake, river discharge, denitrification and volatilization. All calculations were conducted for 1994–1997 (Bashkin et al, 2002).

N *deposition*

The nitrogen depositions were calculated by two models, HEMISPHERE (Sofiev, 1998) and MOGUNTIA (Dentener & Crutzen, 1994), and showed similar values (10.7–11.0 kg/ha/yr). Application of mineral fertilizers averaged 226 kg/ha/yr in South Korean agriculture, being the maximum input source in mass balance of nitrogen.

Biological N *fixation*

Assessment of biological fixation (nonsymbiotic only, since the area under symbiotically fixed crops was very small in the ROK) was carried out using the following data (Environmental Statistics Yearbook, 1998; Cleveland et al, 1999; Zhu et al, 1997)

Agricultural land use

Area of rice plantation—1009560 ha
Rate of fixation—45 kg/ha/yr
Annual flux—45430 tons

Area of other crops—966280 ha
Rate of fixation—15 kg/ha/yr
Annual flux—14494 tons

Forest land use

Forested area—5072600 ha
Rate of fixation—1 kg/ha/yr
Annual flux—5072 tons

TOTAL: 64996

Denitrification
Denitrification values were calculated on the basis of the following data (Environmental Statistics Yearbook, 1998; Freney, 1996; Lin et al, 1996; Mosier et al, 1998; Zhu, 1997):

Agricultural land use

Losses from fertilizers

Area of rice plantation—1009560 ha
Rate of denitrification—32% from fertilizer rate
Annual flux—73011 tons
Area of upland crops—966280 ha
Rate of denitrification—15% from fertilizer rate
Annual flux—31887 tons

Losses from manure

Rate of denitrification—13% from manure N rate
Annual flux—20528 tons

Losses from soils

Rate of denitrification—3 kg/ha/yr
Annual flux—5916 tons

Table 3. Annual accumulation of nitrogen in human and animal excreta in the Republic of Korea, mean 1994–1997.

Items	Rate, kg/capita/yr	Population, thousand	Tons/year
HUMAN			
Population*			
Urban	0.44	33745	14848
Rural	0.69	5494	3791
Subtotal			18639
LIVESTOCK			
Cattle	11.35	3267	37085
Horse	9.79	6.7	6552
Pig	3.22	6691	21525
Sheep	0.7	1.6	1
Goat	0.7	653	458
Poultry	1.0	85623	88483
Subtotal			154113
TOTAL			172752

* Only adults.

Agricultural recycled N

Agricultural recycled N was considered for the regional biogeochemical budget in South Korean agroecosystems as organic fertilizer nitrogen. The values of organic fertilizer N were assessed using statistical data on human and animal/poultry population and rates of N in excreta (Table 3).

Anthropogenic NH_3 emission

These values have been estimated by S.-U. Park (1998). The modified European calculation factors (IPCC, 1997) were applied. The average total value was 142123 tons and NH_3 emission from fertilizers was predominant (35% of the total value).

N *surface water discharge*

In addition to the input/output items for agroecosystems, we estimated the N fluxes with river runoff for calculating the N budget for the whole South Korean area. The mean annual water discharge was 61.6×10^{12} L. In accordance with statistical data, about half of wastewater was untreated in the Republic of Korea in 1994–1997. As a consequence, the content of reduced nitrogen in surface waters was almost the same as the content of oxidized N. Nitrite-N was also monitored in South Korean rivers and its mean content was 0.045 mg/L. The dissolved organic nitrogen (DON) content in the most monitored rivers and water reservoirs was negligible (< 1 mg/L) due to both intensive mineralization and algae uptake as well as low content of organic matter in

Table 4. Annual riverine fluxes of nitrogen from the area of the Republic of Korea, mean 1994–1997.

Species	Content, mg/L N	Fluxes, tons/year
$N - NO_3^-$	1.470	90552
$N - NO_2^-$	0.045	2784
$N - NH_4^+$	1.540	94864
$N - SPM$	0.085	5220
TOTAL		193142

South Korean soils. The fluxes of suspended matter were significant, totaling 1.1×10^9 ton/yr, especially during the summer monsoon period, with discharge-weighted mean N content of 0.085%. The total fluxes of DIN and N-SPM were 193142 tons per year in 1994–1997 (Table 4).

3.4. Regional mass balance of nitrogen in the Republic of Korea

The quantitative parameterization of different input and output links of the N biogeochemical cycle in the South Korean peninsula allows us to make up the calculation of two mass balance estimates, for agroecosystems ($\sim 2.0 \times 10^6$ ha) and for the whole area of the Republic of Korea. The input/output items for different types of mass balance were estimated in accordance with characteristic fluxes shown in Table 2.

Biogeochemical budget of nitrogen in South Korean agroecosystems
On a basis of the above-mentioned data, the N budget in South Korean agroecosystems is as follows (Table 5).

Since South Korean agriculture combines the features of developed countries (great amount of synthetic fertilizers applied) and developing ones (N recycling), it was of interest to compare the corresponding values of N mass balance in some European countries and China with local figures (Table 6).

The general comparison of mass balance values between South Korean agroecosystems and those for developed European countries confirms similar processes such as a great surplus of nitrogen (typically more than 100 kg/ha/yr). A similar situation exists in modern Chinese agriculture. However, N crop uptake efficiency in Asian countries is less than in European ones and the Republic of Korea is last in line with only 38% of efficiency. This really supports the idea that an increasing non-sustainability within agriculture, human nutrition and waste management complexes has occurred both in European and Asian countries. It has been leading to a disturbance of the N biogeochemical cycle.

In order to estimate the fate of nitrogen accumulated in agroecosystems, we assessed the annual N accumulation in municipal waste for the urban area of the country (Table 7).

Table 5. Nitrogen budget in agroecosystems of the Republic of Korea, 1994–1997.

Items	Ton/year	% from input
INPUT		
Fertilizers	446081	65.0
Manure	157904	23.0
Biological fixation	59924	8.7
Deposition	21692	3.3
Subtotal	685601	100
OUTPUT		
Crop production	259779	37.8
Denitrification	132211	19.3
NH_3 volatilization	142123	20.7
Subtotal	524113	77.8
BUDGET*	+151488	22.2

Note: Accumulated nitrogen is distributed among the following items: surface runoff; groundwater leakage, and increasing N content in crops, mainly vegetables.

Table 6. Nitrogen mass balance (kg/ha/yr) and N crop uptake efficiency (% from total input) in different developed and developing countries.*

Country	Agricultural area, 10^6 ha	Input	Output	Surplus	Crop uptake efficiency
Denmark	2.9	217	30	187	59
Germany	12	215	51	164	73
UK	18.1	127	17	110	—
Netherlands	2.3	463	96	365	63
Norway	1.0	147	80	67	71
Sweden	3.7	121	21	100	63
S.Korea	2.0	347	51	296	38
China	94.9	294	95	199	51

*Data for European countries from K. Isermann (1991) and for China from G. X. Xing and Z. L. Zhu (2002).

To calculate these values, the following data were used (Environmental Statistics Yearbook, 1998):

Population, thousand—46164;
Annual N accumulation in sewage & wastes—6 kg/capita;
Annual accumulation flux—276987 tons.

One can see that the main amount of N was deposit in landfill with subsequent transformation and leaching to surface and ground water as well as denitrification.

Table 7. Distribution accumulated N between various waste treatment types in the ROK.

Items	Tons/year	% from total
Landfill	207740	75
Incineration	13849	5
Agricultural use	5540	2
Recycling	27699	10
Damping at sea	22159	8

Table 8. Biogeochemical budget of nitrogen in South Korea.

Items	Ton/year	% from input
INPUT		
Deposition	108160	13.3
Fertilizers	446081	54.8
N fixation	64996	8.0
Import		
Foods	184110	22.6
Goods	10377	1.3
Subtotal	813724	100
OUTPUT		
River discharge	189124	23.2
Denitrification	132211	16.2
NH_3 volatilization	142123	17.5
Sea waste damping	22159	2.7
Subtotal	485617	59.7
BUDGET	+328107	+40.3

In its turn, this leads to further pollution of drinking water, eutrophication of surface waters, mainly, water reservoirs, and increasing input of N_2O to the atmosphere. For instance, the DIN contents in many South Korean water reservoirs are 4–10 mgN/L in summer season and most of these reservoirs are eutrophied (Environmental Statistics Yearbook, 1998).

3.5. Regional N Mass Balance in the Republic of Korea

Taking into account the values of various input/output items of biogeochemical N cycle as well as literature data (Environmental Statistics Yearbook, 1998; Park, 1998; Freney, 1996; Mosier et al, 1998; Zhu, 1997), the regional mass budget was calculated for the whole South Korean territory (Table 8).

Table 9. Distribution of accumulated nitrogen in the Republic of Korea.

ACCUMULATION	Ton/year	% from total
Landfill	207740	63.2
Forest uptake	16455	5.0
Groundwater leakage*	103912	31.8

* as a difference.

The dominant input items were related to the application of mineral fertilizers and import of food and goods (about 80% from total input). Deposition (about 55% from abroad through transboundary air pollution) and non-symbiotic N fixation were responsible for the other 20% of input. The output was connected with N volatilization via direct NH_3 volatilization and biological denitrification (33.7% from total input) and river discharge (23.2%). The sum of output was about 60% from input.

Thus, this budget was positive (+40.3% from total input) and the estimated distribution of accumulated N is shown in Table 9.

As it has been shown already during the analysis of N balance in agroecosystems, the main part of excessive nitrogen is stored at landfills with corresponding prolonged problems of environmental pollution. The values of N forest uptake were calculated on a basis of data on net primary productivity and N content in tree stems and branches. Groundwater leakage was assessed as a difference between total N accumulation and sum of annual landfill storage and plant uptake in forest ecosystems. We note that generally the decomposition of waste residues in landfills lasts longer than one year and this approach to estimating groundwater leakage can be used.

3.6. Nitrogen Mass Balance in the Yellow-Bohai Sea System

In accordance with approaches shown in Table 10, the total riverine N fluxes were assessed for the Yellow-Bohai sea system (UNESCO, 1978; Zhu, 1997; ESCAP, 1998; Choi, 1998; Cha et al, 1998). These data are shown in Table 10.

The analysis of the International SCOPE N Project suggested that without the influence of humans on the landscape, the flux of N from land to coastal waters would likely be of the order of 130 $kgN/km^2/yr$ when expressed per area of watershed (Lewis et al). In comparison, actual fluxes from the Yellow-Bohai Seas drainage basin (areas of China and the Koreas) were some 8-fold larger for the China part of the drainage basin and 13-fold larger for drainages from the South Korean area in 1994–1997. Due to a huge amount of SPM transport in the Yellow river, the N discharge from the China area was mainly composed of solid matter (87%) whereas the opposite picture was for the South Korean area: only 6% of total N flux was discharged with SPM and 94% was discharged as DIN.

Joining these data with those existing in the literature (Cha et al, 1998; Choi, 1998; Park, 1998; Nixon et al, 1996), we calculated the N fluxes in the Yellow Sea (Table 11).

Table 10. Assessment of total nitrogen riverine fluxes to Yellow-Bohai seas system.

River	Discharge, km^3/yr	N species	N content, mg/L	N fluxes, tons/yr
Han	25.0	NO$_3^-$	1.33	33188
		NH$_4^+$	1.31	32800
		NO$_2^-$	0.037	930
		N-SPM	0.064	1600
		Total		68518
Keum	6.4	NO$_3^-$	1.61	10317
		NH$_4^+$	1.54	9859
		NO$_2^-$	0.045	298
		N-SPM	0.056	1040
		Total		21514
Aprock	33.6	NO$_3^-$	0.20	6920
		NH$_4^+$	0.41	13840
		N-SPM	0.098	3460
		Total		24220
Liaohe	14.8	NO$_3^-$	0.20	2960
		NH$_4^+$	0.39	5920
		N-SPM	0.16	2664
		Total		11544
Haihe	22.8	NO$_3^-$	0.50	11400
		NH$_4^+$	0.65	14820
		N-SPM	0.18	4104
		Total		30324
Yellow	59.2	NO$_3^-$	0.87	51495
		NH$_4^+$	1.39	76947
		N-SPM	18.3	1083177
		Total		1211619
Yellow & Bohai seas	161.8	DIN		271614
		N-SPM		1096045
		Total		1367959

It has been shown that the main part of both soluble (53%) and particulate (> 99%) nitrogen has been flowing into the Yellow sea from Northern China in spite of only 30% excess of river water discharge from the Chinese area. Total riverine input from the watershed drainage basin was equal to 52% from input values (2581 kt/yr). Both wet and dry depositions were responsible for 20 and 22% from input values with relatively small values of N fixation in marine waters. Denitrification is the main output item (37%) with similar values of sedimentation and water exchange with the East China Sea (7 and 8% from total input) and with negligible values of losses as N$_2$O (< 1%).

Table 11. Sub-regional fluxes of nitrogen in Yellow-Bohai seas system.

Items	Tons/year	% from input
INPUT		
Wet deposition	504000	20
Dry deposition	557000	22
N fixation	152880	6
Riverine soluble N	271614	10
Riverine SPM- N	1096045	42
Subtotal	2581539	100
OUTPUT		
Denitrification	946680	37
Losses as N_2O	23520	1
Sedimentation	170000	7
Water exchange	212000	8
Subtotal	1352200	53
ACCUMULATION	+1229339	+47
N pool in marine water	1818000	
Residence time, yr	1.47	

So, annual accumulation of nitrogen in the Yellow sea was 1229 kt/yr (+47% from input) and the residence time of nitrogen was 1.5 year. This means that the N content in marine water was doubling every three years during 1994–1997. It led to excessive eutrophication and pollution of the Yellow Sea.

3.7. Preliminary Assessment of N Mass Balance in East Asian Domain

In addition to the main task of this section, preliminary estimates of N fluxes in the whole East Asian domain were also conducted. This regional mass balance was presented as a comparison of input (N deposition, agricultural N fixation and fertilizers only) and output values (river discharge, volatilization and denitrification in agroecosystems only). It was based upon three assumptions:

i) on a regional scale, anthropogenic N fixation during combustion and the Haber reaction process might be equal to volatile losses (denitrification and NH_3 volatilization) of N to atmosphere and crop uptake;

ii) biological N fixation is equal to denitrification of soil N; and

iii) N soil immobilization is equal to N soil mobilization.

These assumptions are workable under steady state mass balance calculations on a regional scale like the East Asian domain where information sources are scarce

Table 12. Assessment of nitrogen deposition in East Asian domain, 1990s.

Country/Region	Area, km^2	HEMISPHERE		MOGUNTIA	
		Rate, kg/km^2/yr	Flux, tons/yr	Rate, kg/km^2/yr	Flux, tons/yr
Russia*	1244000	140	174000	200	248800
China	9600000	830	8008000	1010	9696000
N.Korea	120410	1490	170000	800	96330
S.Korea	99390	1070	107000	1100	109320
Japan	377800	1140	429000	800	302240
Taiwan	36000	—	—	600	21600
Vietnam	330000	—	—	500	165000
Thailand	511000	—	—	400	204400

* Amur river basin only.

and uncertainty of available data is high. Statistical data for N fertilizer application was produced from national and international yearbooks, N fixation was assessed on the basis of Cleveland et al, (1999) data. Denitrification and N volatilization were estimated using national data for China (Xing G. X. and Zhu Z. L., 2002), Republic of Korea and Russia.

Since the main uncertainty is related to N deposition input and N riverine output, these data are considered separately.

N deposition input

The assessment of N deposition was also done on the basis of two existing model results (HEMISPHERA model—M. Sofiev, 1998; MOGUNTIA model—P. Zimmermann, 1988; F. Dentener & P. Crutzen, 1994) (Table 12).

These results are mainly similar (China, South Korea) or differ insignificantly (Russia, Japan, North Korea). However, the existing national data for South Korea (Park, 1998, Lee and Young, 1998) and Taiwan (Chen et al, 1998; Lin, 1998) deviate significantly, sometimes by 2–3-fold. Since national data are not at present available for the whole East Asian domain, only the international data on deposition were applied.

Riverine discharge of N to the Pacific Ocean

We calculated the river N discharge using national and international data sets (UNESCO, 1978; ESCAP, 1998; Zhu, 1997; Choi, 1998; Shen et al, 1998; Cha et al, 1998; Satake et al, 1998; Neudachin, 1999). The results on N content in various rivers were compared with those published at the beginning of the 1980s (Meybeck, 1982). These results are shown in Table 13.

One can see that during a 20-years period, both content of individual N forms and DIN have been increasing, especially ammonium content. This might be related to

Table 13. Comparison of N content in main East Asian rivers in 1980 and 1996, mgN/L.

River	Cited from M. Meybeck, 1982		Present content	
	$N - NH_4^+$	$N - NO_3^-$	$N - NH_4^+$	$N - NO_3^-$
Yellow	0.01	0.24	1.39	0.87
Yangtze	0.04	1.94	1.30	0.87
Pearl	< 0.01	0.59	1.12	0.75
Mekong	—	0.24	0.07	0.16
Chao Phraya	—	0.63	0.90	0.60

output of untreated wastewater to rivers. In main degree this output of treated water was shown in China rivers and in less degree in Mekong and Chao Phraya. Using these data and water discharge values, the annual fluxes of various N species were calculated for different East Asian rivers (Table 14).

These data were grouped in the scale of various sea basins in Table 15.

The annual total flux of dissolved and suspended N was equal to 5576714 tons in 1994–1997 and DIN was responsible for 65% of N discharge with river water. The amount of discharged nitrogen was mainly related to water discharge values.

Regional N mass balance
The comparison of the input and output data allows us to make the preliminary assessment of N regional budget in East Asia (Table 16).

This regional budget was positive but the values of interregional transformation like crop uptake (N in crop harvest—22510 ktons/yr) were not accounted.

Thus, the following links of the nitrogen biogeochemical cycle accounted for the mass balance calculations in Northeast Asia and the whole East Asian domain: input—deposition, fertilizers, biological N fixation, import of food and products, riverine fluxes and output—crop uptake, denitrification, volatilization, runoff, sedimentation and sea water exchange. All calculations were conducted for 1994–1997 and the mean values were used. (see Table 16).

The values of biogeochemical chains were assembled for the N mass balance calculations in both agroecosystems and the whole area of the Korean peninsula (in the borders of the Republic of Korea). N budget in South Korean agroecosystems was positive (+151 kt/yr or 22% from input) with prevalence of fertilizer N (65%) in input and more similar distribution of output values (crop production—38%, denitrification—19% and ammonium volatilization—21% from input).

Relatively more nitrogen was accumulated in the whole South Korean area (+328 kt/yr or 40% from total input) due to increasing fertilization, deposition and import rates which were not equilibrated by output owing to river discharge, gaseous losses and sea water damping. This gave a rise in N storage in landfill and groundwater as well as gradually increasing riverine N discharge to the Yellow Sea.

Table 14. Assessment of total nitrogen riverine fluxes to the Pacific Ocean from the East Asian domain.

River	Discharge, km³/yr	N species	N content, mg/L	N fluxes, tons/yr
Amur (East sea)	355	NO_3^-	0.43	152650
		NH_4^+	0.03	10650
		NO_2^-	0.46	163300
		N-SPM	0.001	355
		Total		163655
Han	25.0	NO_3^-	1.33	33188
		NH_4^+	1.31	32800
		NO_2^-	0.037	930
		N-SPM	0.064	1600
		Total		68518
Keum	6.4	NO_3^-	1.61	10317
		NH_4^+	1.54	9859
		NO_2^-	0.045	298
		N-SPM	0.056	1040
		Total		21514
Aprock	33.6	NO_3^-	0.20	6920
		NH_4^+	0.41	13840
		N-SPM	0.098	3460
		Total		24220
Liaohe	14.8	NO_3^-	0.20	2960
		NH_4^+	0.39	5920
		N-SPM	0.16	2664
		Total		11544
Haihe	22.8	NO_3^-	0.50	11400
		NH_4^+	0.65	14820
		N-SPM	0.18	4104
		Total		30324
Yellow	59.2	NO_3^-	0.87	51495
		NH_4^+	1.39	76947
		N-SPM	18.3	1083177
		Total		1211619
Yellow & Bohai seas	161.8	DIN		271614
		N-SPM		1096045
		Total		1367959

Table 14. Assessment of total nitrogen riverine fluxes to the Pacific Ocean from the East Asian domain (continued).

River	Discharge, km³/yr	N species	N content, mg/L	N fluxes, tons/yr
Yangtze (East China sea)	924	NO₃⁻	0.87	803880
		NH₄⁺	1.30	1201200
		DIN	2.17	2005080
		N-SPM	0.526	486000
		Total		2491080
Pearl	326	NO₃⁻	0.75	244500
		NH₄⁺	1.12	365120
		DIN	1.87	609620
		N-SPM	0.300	97800
		Total		707420
Langgang	42	NO₃⁻	0.70	29400
		NH₄⁺	1.05	44100
		DIN	1.75	73500
		N-SPM	0.200	8400
		Total		81900
Mekong	510	NO₃⁻	0.16	81600
		NH₄⁺	0.07	39270
		DIN	0.23	122400
		N-SPM	0.50	255000
		Total		377400
Small rivers	210	NO₃⁻	0.06	12600
		NH₄⁺	0.21	44100
		DIN	0.27	56700
		N-SPM	?	?
		Total		> 100000
South China sea	1088	DIN		1249820
		N-SPM		461200
		Total		1711020

Table 15. Nitrogen riverine fluxes in East Asian domain, 1990s.

Sea	River discharge, km^3/yr	N species	Fluxes, tons/yr
East	335	DIN	163300
		SPM	355
		Total	163655
Yellow & Bohai	162	DIN	271614
		SPM	1083177
		Total	1211619
East China	924	DIN	2005080
		SPM	486000
		Total	2491080
South China	878	DIN	1193120
		SPM	361200
		Total	1554320
The whole East Asian domain	2319	DIN	3633114
		SPM	1943600
		Total	5576714

Table 16. Regional budget of nitrogen in East Asian domain, 1994–1997.

Items	Ktons/year	% from input
INPUT		
Deposition	8776	26
Fertilizers	23094	68
N fixation	1947	6
Subtotal	33817	100
OUTPUT		
Riverine runoff	5577	16
Denitrification	7512	22
Volatilization	8652	26
Subtotal	21714	64
BUDGET	+12103	+36

The regional nitrogen balance in the whole East Asian domain was positive (+12103 kt/yr) or +36% from total input of 33816 kt per annum during 1994–1997. These values do not include the N uptake in agroecosystems. The latter makes the regional balance negative.

4. NITRATE BIOGEOCHEMICAL PROVINCES AND CANCER DISEASES IN THE ASIAN REGION

4.1. Nitrogen Toxicology

Nitrogen species, like nitrate and nitrite, have been confirmed to be procarcinogenic. The excessive input of these nitrogen compounds into food and drinking water in the presence of tertiary amines, for instance from medicines, can lead to the formation of carcinogenic N-nitrosoamines.

At present, the extensive application of nitrogen fertilizers in many agricultural regions of the World has led to the formation of agrogenic nitrogen biogeochemical provinces. The characteristic feature of these provinces is the accumulation of nitrate and other nitrogen species in soils and drinking water. This accumulation accompanies an excessive input of nitrate in food webs.

The nitrate (NO_3^-) content in crops is one of the most important indicators of farm production quality. Nitrate content in food is strictly regulated because of its toxicity, especially to young children. The actual toxin is not the nitrate ion itself but rather the nitrite ion (NO_2^-), which is formed when nitrate is reduced by intestinal bacteria, notably *Escherichia coli*. In adults, nitrate is absorbed high in the digestive tract before reduction can take place. In infants, whose stomachs are less acidic, *E. coli* can colonize higher up the digestive tract and therefore reduce the NO_3^- to NO_2^- before it is absorbed.

Nitrite ions are toxic because they can combine with hemoglobin with resulting formation of *methemoglobin*. The association constant for methemoglobin formation is larger than that for oxyhemoglobin complex formation. Thus, the nitrite ion binds with hemoglobin, depriving the tissues of oxygen. Severe cases of disease called *methemoglobinemia* can result in mental retardation of the infant and even a death from asphyxiation.

At stomach pH, nitrite is also converted to $H_2NO_2^+$, which is capable of nitrosating secondary amines and secondary amides. The resulting N-nitrosoamines may be carcinogenic. For example, N-nitrosodimethylamine (or dimethylnitrosoamine) is carcinogenic in many experimental animals, although it is not a confirmed human carcinogen. Dimethylnitrosoamine can also contaminate drinking water supplies, both as a result of industrial activity and also because the compounds may be present in the discharge waters of sewage treatment plants, where it is formed by the microbial degradation of proteinaceous materials.

The average human daily intake of nitrate/nitrite is 95 mg/day in adults. Estimates of the relative contributions of nitrate from drinking water and food to the daily intake vary considerably, depending on how they are calculated. Nevertheless, they show that between 50% and 90% of nitrates in human intake may originate from vegetables, conserved meat (sausages, canned meat, smoked meat, etc.) and even milk products. Vegetables tend to concentrate nitrate ions, especially if they are grown using high rates of nitrogen fertilizers. The concentration of nitrate in vegetables can vary extensively. Lettuce, spinach, cabbage, celery, radish and beetroot can contain as much as 3000–4000 mg/kg of fresh weight. These levels could have potential health effects.

The problem of nitrate accumulation seems to be especially severe in leafy vegetables grown in greenhouses under winter conditions, owing to intensive application of nitrogen fertilizers and low light intensity, which retard nitrate metabolism in crops.

Another source of nitrate and nitrite accumulation in food is their use as food additives. Nitrate and nitrite salts ($NaNO_2$, $NaNO_3$, KNO_2, KNO_3) are added to meats and other food products as a curing salt, color fixative (preventing the meat turning brown) and as food preservative to prevent growth of the dangerous bacterium *Clostridium botulinum*, which produces the highly poisonous botulism toxin.

Cured meats, bacon, ham, smoked sausages, beef, canned meat, pork pies, smoked fish, frozen pizza and some cheeses contain nitrate and nitrite additives, typically at levels of 120 mg/kg. Although without them there would certainly be many deaths due to growth of toxic microbes in meat, excessive intakes of these salts may cause gastroenteritis, vomiting, abdominal pain, vertigo, muscular weakness and an irregular pulse. Long-term exposure to small amounts of nitrates and nitrites may cause anemia and kidney disorders. The level of these additives is strictly controlled (for example, 500 mg/kg as $NaNO_3$ in the UK), and the addition of nitrates and nitrites in baby food is now banned in many countries. The acceptable daily intake (ADI) for $NaNO_3$ is 0–5 mg/kg body weight. For KNO_2 and $NaNO_2$ the ADI is 0 0.2 (temporary), while for KNO_3 it is not specified. The value of ADI is zero for baby food. The WHO sets the ADI at 220 mg an adult.

Since vegetables are a major source of ingested nitrates, the most rational way of reducing the problem is to grow crops with safe levels of nitrates. Most counties do not have actual standards but some kinds of guideline or criteria value based on the ADI. Criteria values of nitrate content exist in the same kind of vegetable consumption and in vegetable production practices. For instance, the nitrate level (mg/kg of wet weight) in spinach in different countries is: USA, 3,600; Switzerland, 3,000; the Netherlands, 4,000; Czech Republic, 730; Russia, 2,100.

Nitrate content is also one of indicators of fodder quality. Numerous cases of cattle poisoning by nitrates present in fodder have been reported in various countries, and many heads of cattle were lost from affected herds. Feed beetroot, cabbage, mustard, sunflower, oat, as well as various types of ensilage used as green fodder may contain potentially toxic levels of nitrate. The toxic level of nitrate for animals is 0.7 mg/kg of body weight. Among farm animals, cattle and young pigs are the most sensitive to nitrates, while sheep are more resistant. Also, consumption of nitrates at sub-toxic levels by cattle has been reported as a cause of reduced milk production and weight gain, vitamin A deficiency, abortions, stillbirths, cystic ovaries, etc. Nitrate-N concentrations of 0.21% in the feed are considered to be toxic to farm animals.

Agriculture is usually the major source of nitrate ions in drinking water. Through manure seepage from feedlots, seepage from the holding tanks used to contain liquid manure from intensive hog production, and excessive use of fertilizers are important. The higher crop yields obtained today compared with 20–40 years ago are largely due to increased use of chemical fertilizers. However, low crop prices combined with high land and machinery cost encourage farmers to cultivate fields up to their margins, thus promoting runoff from fields to water bodies and groundwaters. The accumulation

Table 17. Perennial concentrations of nitrogen species in the food web of the nitrate biogeochemical province of the Desert region of the biosphere, Uzbekistan.

Links of food web	Nitrate	Nitrite
Surface water, ppm	2.03–12.10	0.03–0.42
Drinking water, ppm	3.62–150.80	0–0.09
Vegetables, ppm of wet weight	500–8600	12.4–328.3

of nitrates in groundwater has become an issue of great concern in many European countries and many States of the USA (see Chapter 8).

In 1945, H.H. Comly first estimated the correlation between nitrates in drinking water and the incidence of methemoglobinemia. Research shortly afterwards showed that no cases of methemoglobinemia had been reported in any area of the United States where the water supply contained less than 45 ppm of nitrate ions. This value has become accepted in USA as the upper limit for nitrate concentration in drinking water. At present the WHO limit is also 45 ppm of nitrate but the value of 22 ppm of nitrate has been set for EC countries.

The main reasons of nitrate accumulation are the increasing application of nitrogen fertilizers and increasing generation of nitrogen-containing municipal wastes.

Let us consider in brief the nitrate biogeochemical provinces, where significant growth of cancer incidents has been observed. It is known that the absorption of nitrate and nitrite from water in the intestinal tract is two times as much as from food products. This means that the accumulation of nitrate in drinking water sources should be of special concern during consideration of food webs in any relevant biogeochemical province.

4.2. Nitrogen Biogeochemical Provinces in Central Asia

As an example, we will consider the formation of nitrate biogeochemical provinces in the Desert region of the biosphere in the Zerafshan river watershed that occupies most of the Samarkand administrative region of Uzbekistan. This is an agricultural area with irrigation production of cotton and some other crops like cereals, vegetables and fruits. The natural biogeochemical provinces are characterized by iodine, copper, and zinc deficiency and lithium excess.

The concentration of nitrogen species in various links of biogeochemical food webs of this province is shown in Table 17.

We can see that the concentrations of both nitrate and nitrate in drinking water and vegetable are excessive. The accumulation of nitrate in the food web of this biogeochemical province was accompanied by a high content of residual pesticides in soils of agroecosystems (Table 18).

The accumulation of pesticides is connected with their long-term application at heavy rates, 15 kg/ha on average during the 1970–1980s. The elevated concentrations of PCB are related to irrigation by polluted waters from the Zerafshan River (Galiulin

Table 18. The residual content of persistent chourorganic compounds in soils of nitrate biogeochemical province of the Desert region of the biosphere, Uzbekistan, ppb.

Xeno-biotic	Cotton		Tobacco		Orchards	
	Mean	Limits	mean	limits	Mean	Limits
DDT	8.3	tr–47.5	27.2	tr–157.7	24.7	tr–70.9
$\sum$ DDT*	113.4	tr–970.0	282.7	tr–1715.0	19.9	tr–432.5
$\sum$ PCB	28.0	2.3–100.6	121.1	28.4–425.8	24.9	9.2–67.5

Note: $\sum$ DDT—DDT + DDE + DDD; the corresponding standard for DDT content in soil is 100 ppb, and temporary criterion for PCB is 60 ppb.

Table 19. Growth of liver and stomach cancer rates in the Samarkand nitrate biogeochemical province of the Desert region of the biosphere, Uzbekistan, cases per 100,000 individuals.

Period	Liver cancer	Stomach cancer
1950–1955	1.4	12.5
1956–1960	1.5	17.8
1961–1965	2.4	18.9
1966–1970	2.9	23.2
1971–1975	3.9	25.3
1980–1985	4.7	27.8
1985–1990	5.3	32.6
1990–1995	6.0	34.6

and Bashkin, 1996). The mean content of residual DDT and its metabolites DDE and DDD is higher than the corresponding standard in soils of Cotton and Tobacco agroecosystems. The mean content of total amount of PCB is also higher than the standard in soils of Tobacco agroecosystems.

Thus, the combination of natural biogeochemical features of the food web (deficiency of J, Zn, and Co with excess of Li) with anthropogenic pollution has led to increased cancer rates (Table 19).

FURTHER READING

1. Bashkin V. N., 1987. *Nitrogen Agrogeochemistry*. ONTI Publishing House, Pushchino, 267 pp.

2. Butcher S. S., Charlson R. J., Orians G. H. and Wolfe G. V. (Eds.), 1992. *Global Biogeochemical Cycles*. Academic Press, London et al, Chapter 15.

3. Moldan B. and Cherny J. (Eds.), 1994. *Biogeochemistry of Small Catchments*, John Wiley and Sons, Chapter 13.

4. Dobrovolsky V. V., 1994. *Biogeochemistry of the World's Land*. Mir Publishers, Moscow and CRC Press, Boca Raton, Ann Arbor, Tokyo, London, 206–211.

5. Howarth R. W. (Ed.), 1996. *Nitrogen Cycling in the North Atlantic Ocean and its Watersheds*, Kluwer Academic Publishers, 304 pp.

6. Radojevic M. and Bashkin V., 1999. *Practical Environmental Analysis*. Royal Society of Chemistry, UK, Chapter 5.13.

7. Boyer E. W. and Howarth R. W. (Eds.), 2002. *The nitrogen cycle at regional to global scale*. Kluwer Academic Publishers, 519 pp.

8. Bashkin V. and Howarth R., 2002. *Modern Biogeochemistry*. Kluwer Academic Publishers, 604 pp.

WEBSITES OF INTEREST

1. http//:www.escap.un.org

QUESTIONS AND PROBLEMS

1. Present a definition of environmental biogeochemistry and define the process of environmental pollution from the biogeochemical point of view.

2. Explain the choice of nitrogen as a main element for consideration in environmental biogeochemistry. Give relevant examples.

3. Consider the disturbance of the biogeochemical cycle of nitrogen and explain why leaching plays the most important role in environmental biogeochemistry of this element.

4. Discuss the biogeochemical cycle of nitrogen in mixed ecosystems. Focus your attention on wetlands and upland landscapes.

5. Explain the different approaches to a consideration of the role of various airborne nitrogen species during regional mass balance calculations.

6. Discuss the accumulation of nitrate nitrogen in groundwaters of different regions. Why may these waters be considered as a biogeochemical barrier in the environmental biogeochemical cycle of nitrogen?

7. Present an example of regional biogeochemical mass balance of nitrogen. Discuss the ratio between input and output items for different regions.

8. Consider the environmental biogeochemistry of nitrogen in East Asia. What are the similarities and differences between the North Atlantic and East Asian domains regarding the disturbance of natural N cycling?

9. Describe the regional biogeochemical mass balance of nitrogen in the example of South Korea. Focus your attention on characteristic features of the nitrogen cycle in this country.

10. Discuss the regional mass budget of N in the Bohai-Yellow Seas. Consider the role of riverine discharge in nitrogen fluxes and pools of this marine system.

11. What are the main reasons for N accumulation in the East Asian regions?

CHAPTER 14

ORGANIC XENOBIOTICS

1. INTRODUCTION

The toxic persistent organic pollutants (POP) are of great environmental concern in the Asian region. The famous insecticide DDT is only one of the dangerous substances that belong to POPs. POPs are long-lived organic compounds that become more concentrated as they move up the food chain, i.e., they have high BCF values. Moreover, these compounds can travel with air mass thousands of kilometers from their points of release. Although POPs cover a broad range of chemical classes, the concern and much of research activity both in the World and the Asian region, focus on about a dozen chemicals, like various pesticides, polychlorinated biphenyls, dioxins, furans, chlorophenols, etc. Most of these chemicals have now been phased out in developed countries but they are still in production and use in most Asian countries.

In this chapter, we will consider various questions regarding various classes of organic xenobiotics, their production, toxicology, and environmental behavior in polluted areas.

2. PESTICIDES IN THE ASIAN COUNTRIES

2.1. Trends in Pesticide Application

Definition

The term *pesticide* is used to indicate any substance, preparation, or organism used for destroying pests. This broad definition covers substances used for many purposes, including insecticides, herbicides, fungicides, nematocides, acaricides, lumbricides, growth regulators, and insect repellents. The latter two groups are usually included as pesticides, even though they do not kill the target organisms. An enormous spectrum of chemical types of pesticides is covered by the definition of pesticides. According to their chemical nature, a first rough classification distinguishes between organic and inorganic pesticides. Actually, organic chemical pesticides receive virtually all of the regulatory attention and public concern, and these are the subject of this section. The most important classes of pesticides are organochlorines, organophosphates, carbamates, triazines, phenoxyacids, phenylureas and sulfonylureas, acetoanilides, benzimidazoles, and pyrethroides. In some cases, compounds with different chemical

Table 1. Selected indicators of pesticide consumption in the Asian region* (ESCAP, 2000).

Item	1983	1988	1993	1998	2003 (planned)
Consumption, tons	885	806	784	819	870
Cost, US$/kg	6.29	7.48	8.69	10.22	11.94
Consumption, US$ millions	5,571	6,025	6,814	8,370	10,390
Herbicides	1,750	1,970	2,180	2,600	3,150
Insecticides	2,318	2,470	2,790	3,400	4,200
Fungicides	990	1,100	1,260	1,580	2,000
Other	513	485	584	790	1,040

* including the Oceania region.

natures that are used similarly are classified as a single group; for example, the fumigants, which include compounds that have an elevated vapor pressure and acts as vapors.

Application
In comparison with developed countries, some Asian countries have emerged as large users of pesticides, less so in the case of fungicides and herbicides. Total sale of pesticides in the Asian region has grown from ≈ US$5.0 billion in 1983 to almost US$8.0 billion in 1998 and the amount is projected to reach US$10.3 billion in 2003 (ESCAP, 2000). The decrease in pesticide application at the end of the 1980s has been charged with constant growth in the 1990s (Table 1).

In recent times, Asia has accounted for 10% of global sales of pesticides (World Resources Institute, 1998). It is estimated that agriculture accounts for 85% of total pesticide use. Unlike the developed countries, pesticides in developing countries are mostly insecticides with a higher level of acute and chronic toxicity. As in the case of fertilizers, pesticide sale and application have been accelerated with the spread of the so-called green revolution. Use of these chemicals started in a big way with the introduction of new high yielding varieties of wheat, rice and other crops mostly in the 1970s–1980s. Pesticide use has been associated with pesticide resistance, health damage, loss of biodiversity and environmental problems. With increasing use of larger amounts and stronger pesticides, an ever-increasing number of species of insects (over 900 now against 182 in 1965), pathogens and weeds are becoming resistant to these chemicals.

2.2. Major Environmental Pathways
The major environmental pathways of pesticides into streams and into the food chains are presented in Figure 1.

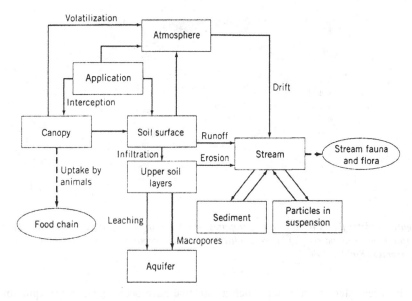

Figure 1. Major environmental pathways for the distribution of pesticides in the biosphere (Richter, 1999).

Transport in soil and groundwater

The soil pays a major role for both transport and biodegradation of a pesticide. The upper soil layer above the plough horizon is the most active zone for microorganisms. From experimental results obtained from artificial soil columns, it is unlikely that many pesticides in significant amounts can pass this zone. Most of them are absorbed in a solid organo-mineral mass of soils. The possible exception is irrigated sandy soil with heavy water drainage (Galiulin & Bashkin, 2001). However, the pollution of groundwater is well documented in many Asian countries. One possible reason is that the random structure of real soils with cracks and micropores facilitates fast transition to deeper layers (Figure 2).

During the dry season, the formation of soil cracks, macro- and micropores is common in the Asian countries with monsoon climate. This process is especially important in irrigation fields where water fluxes will stimulate the pesticide migration downward in the soil profile.

Drift, volatilization and runoff

Drift and volatilization are important factors for large-scale transport from the treated plots. Whereas the range of drift of pesticide in droplets is limited to adjacent ecosystems, volatilization (i.e., vaporization) may lead to a large-scale aerial transport. The physics of spray drift involves Stokes' law for the derivation of sedimentation velocities of droplets (see Chapter 3).

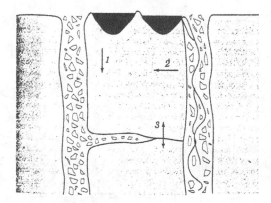

Figure 2. Water fluxes in soil with micropores. The following processes are relevant: (1) infiltration into soil matrix, (2) lateral infiltration from macropores, and (3) exchange between aggregates (Richter, 1999).

In recent studies, pesticides such as atrazine have been found in precipitation. Therefore volatilization and subsequent transport in the gaseous phase is an important environmental pathway. Vaporization rates of pesticides deposited on the surface of soil and plant leaves depend on the physical-chemical properties of the substance. A useful physicochemical criterion is Henry's constant, K_H, which is defined as the equilibrium air-to-water partial pressure ratio of the substance (see Chapter 7).

As an example, we can compare the solubility of lindane and 2,4-D pesticides with respect to their major loss pathways. The K_H values of these pesticides are 1.33×10^{-4} and 5.5×10^{-9} mol L^{-1}, respectively. In addition, the retardation coefficient or degree of sorption of lindane is much higher than that of 2,4-D. Therefore, 2,4-D is more likely to be leached, whereas lindane is more likely to remain near the soil surface from which it can vaporize. Volatilization is thus the major pathway of lindane and degradation and leaching are the major loss pathways of 2,4-D in the Asian environments.

The volatilization rate at the surface is further influenced by temperature and by the thickness of the stagnant air layer over the surface. Furthermore, random micrometeorological conditions of the crop stand are also very important (turbulence, thickness of boundary layer, humidity, and wind velocity). Special models are available to calculate the volatilization and vaporization rates of pesticides (Richter, 1999).

In the Asian region with prevalent monsoon climate and heavy rain events during rainy seasons, the surface runoff plays an important role in the environmental pathways of many pesticides. The major factors influencing the amount of pesticides carried away by runoff are as follows (Richter, 1999):

- the time course of the rainfall event characterized by the total precipitation and peak values;

- the timing of spaying with respect to an event;

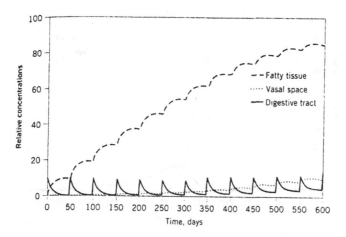

Figure 3. Simulation of the accumulation of an organochlorine in fatty tissue under a long-term intake scheme (Richter, 1999).

- the degree of water saturation of the soil prior to the event;

- soil structure and texture;

- surface crusting and compaction;

- hill slope;

- vegetation cover;

- management practices, for example, vegetative buffer strips near a river with high resistance to flow or contour ploughing.

Some models like CREAMS (Chemicals, Runoff and Erosion from Agricultural Management systems) are available for simulation of pesticide surface runoff (Knisel, 1980).

Transport in the food chain

Because organochlorines dissolve in fats and oil, bioaccumulation in the food chain may occur (see Chapter 11). Figure 3 shows the simulation of the accumulation of a substance in fatty tissues under a regular intake.

We can see that increasing accumulation of an organochlorine compound occurs in fatty tissues, whereas in vasal space and the digestive tract this accumulation is much less evident. However, when the intake is decreasing, then the concentration of any organochlorine compound in fatty tissues will reduce with a relative increase in vasal space and the digestive tract (Figure 4).

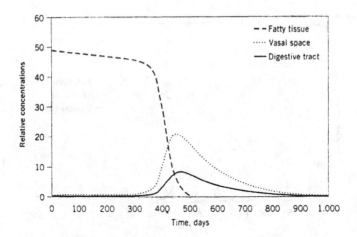

Figure 4. Simulated effect of reduction of fatty tissue concentration of organochlorine with relative increase in other tissues (Richter, 1999).

This leads to a release of the pesticide from fatty tissue into other compartments of the body. Since milk contains fat, mother's milk also constitutes a storage medium for the pesticide (see below the example of DDT and TCDD).

Let us consider now the food chain. Chemicals which accumulate in one organism at one trophic level in the food chain (a prey) owing to tissue binding are further concentrated at the subsequent trophic level. The highest tissue will occur at the top of the trophic pyramid. This holds especially true for human beings who are indeed at the top of different food chains. The mechanism of bioaccumulation can be best understood by a simple ecosystem model (Spain, 1982). Figure 5 shows the time course of concentration of a non-metabolized pesticide at each trophic level following the constant influx of a pesticide into the environmental compartments.

The influx is stopped at time $t = 500$ weeks. One can see from this simulation that the pesticide is concentrated from level to level and that high levels are maintained even after release of the pesticide is stopped. This model is the most fitting for the non-metabolized or very slow metabolized organochlorines like dioxin or DDT. We can see that this is an additional step in understanding the dioxin bioaccumulation in ecosystems to the example shown in Chapter 11.

2.3. DDT

Chemistry of DDT

DDT is an abbreviation for DichloroDiphenylTrichloroethane, which is incorrect name for 2,2-bis-(*p*-chlorphenyl)-1,1,1-trichloethane (Figure 6).

Preparation of DDT, which is both an alkyl chloride and aryl chloride, was first synthesized in 1874. DDT is synthesized from trichloacetaldehyde and chlorobenzene by a modified Friedel-Crafts reaction (Figure 7).

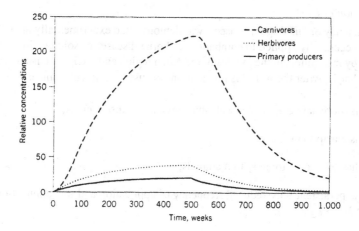

Figure 5. Simulated bioaccumulation of fat-soluble pesticide in the food chain (Richter, 1999, RAM, f.8, 1229).

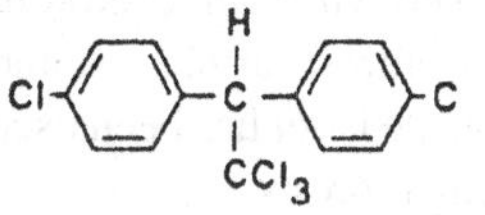

Figure 6. Structural formula of DDT.

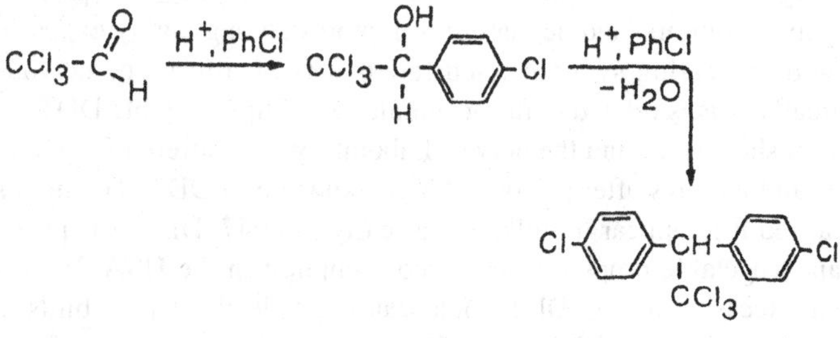

Figure 7. Formulation of DDT.

Since chlorobenzene substitutes electrophilically at the o- and p- positions, the isomer shown (p,p'-DDT) is accompanied by small amounts of the o,p- and o,o'-isomers. The first isomer, p,p'-DDT undergoes metabolic dehydrochlorination and also photochemical dehydrochlorination in the environment, giving 1,1-chloro-2,2-bis(p-chlophenyl)-ethene, also known as DichloroDichlorophenylEthylene, DDE. DDE is separable from DDT by gas chromatography, but since DDE always accompanies DDT, the sum of the two is usually reported.

Application of DDT

The efficiency of DDT against insects was demonstrated experimentally in 1942, and it was successfully applied to combat insect-borne diseases of soldiers, like malaria, carried by mosquitos, in the jungles of the Asian region and the Pacific islands during World War II. After the war, this preparation was used worldwide, since:

- it is more active against insects than any known insecticides;

- it is cheap to produce;

- it has low acute toxicity to mammals;

- it is persistent with reduced frequency of application and low application rates $(0.2–0.3\,\mathrm{kg\,ha^{-1}})$;

- it can be sprayed by aircraft and pests could be eradicated even in locations inaccessible from the ground.

Earlier targets of DDT were insects which attacked cotton, but it was soon successfully tried against almost any pest in forest farmlands, and even suburbs as a method of mosquito control. For example, the latter has been used in North Asian oil exploration regions to combat insects in taiga forests.

Ecotoxicology of DDT and its metabolites

Early investigators of DDT toxicology were aware of its negative properties, such as aching joints, tremors, and depression symptoms among highly exposed soldiers. These slow disappearing symptoms indicate that DDT affects the nervous system. Toxicologically, this is related to the accumulation of lipid-soluble DDT in the insulating myelin sheaths around the nerves. Laboratory rats suffered fatty degeneration of the liver and kidneys after prolonged high exposure to DDT. The lipid solubility of DDT caused it to appear in milk, and as early as 1947, DDT was phased out for pastures and vegetable crops for human consumption in the USA. By 1946 it was known that excessive use of DDT could cause the death of fish, birds, and other wildlife and that residues of DDT were detected in the adipose tissue of a wide range of wildlife, especially carnivores. This is an earlier example of DDT biomagnification, see Chapter 11. Carnivorous birds appeared to fare particularly badly; reduced hatchability of eggs and physical deformities in the chicks were linked statistically to high DDT levels in the parents.

Humans are also at the top of food webs, and by the late 1960s citizens of several countries were horrified to learn that the levels of DDT in mothers' milk ranged up to 130 ppb, which would have been classified as unfit for human consumption.

Until recently, DDT isomers were thought to be of no significance; however, current evidence suggests that o,p-DDT can act in the environment as an estrogen mimic, meaning that it can interfere with the normal signal processing of the

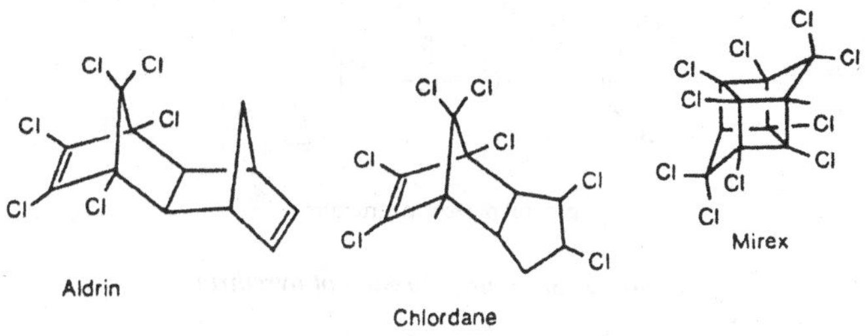

Figure 8. Structural formulas of chlorinated insecticides.

female sex hormone estradiol (Bunce, 1994). In humans, concern about exposure to environmental estrogens, including o,p-DDT, exists over their role in the development of estrogen-dependent breast, uterine and ovarian cancers in females, and demasculinization effects in males.

Societal concern on DDT application

Rachel Carson's book *Silent Spring* that was published in 1962 was the first popular work to bring the uncontrolled environmental contamination by pesticides to public attention. Well-publicized and well-organized campaigns were mounted in several countries to prohibit the use of DDT and other persistent chlorinated insecticides such as Aldrin and heptachlor. Governments in many developed countries like USA, UK, Italy, USSR, Sweden, Norway, Denmark and Finland, proclaimed bans on DDT or severely restricted its use around 1960–1970.

Despite problems with its use, DDT has saved countless lives in regions where malaria is endemic. Tropical Asian countries, which have discontinued its use, have seen malaria incidence increase; India and some other countries still cling to use of DDT because it is cheap and effective. Tissue levels in wildlife of the regions where DDT was banned began to decline about 10 years after DDT was banned; however, DDT and its metabolites may routinely be detected in soil and bottom sediments even today (Galiulin & Bashkin, 1996). DDT longtime residence permits atmospheric migration, for example from India to the Asian polar zone, and the contamination of local ecosystems.

2.4. Other Pesticides

Chlorinated insecticides

Many other chlorinated insecticides were introduced during the 1950s. Some characteristic examples are shown in Figure 8.

All these compounds are highly chlorinated, persistent, lipophilic, and neurotoxic. In addition, they are rather unselective; that is, the safety factor between the doses of DDT and other highly chlorinated insecticides that will kill target insects, and those which will harm non-target organisms, is relatively small.

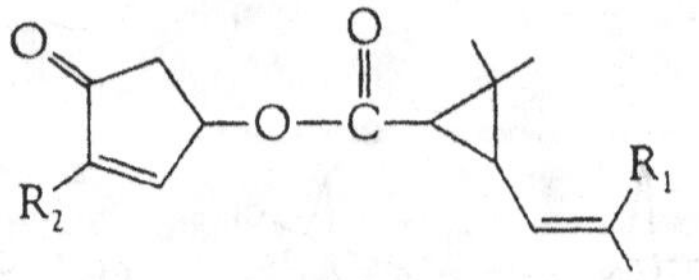

general pyrethrin structure

Figure 9. Structural formula of pyrethrin.

Pyrethroids

Pyrethroids represent a new class of highly selective insecticide. These preparations are almost completely non-toxic towards mammals. Pyrethroids are related to natural insecticide pyrethrin, which is obtained from *Pyrethrum*, a daisy-like flower grown in large quantities in Kenya, Africa (Figure 9).

Although we have just emphasized the thrust in pesticide research towards the development of environmentally non-resistant species, natural pyrethrin is so labile, especially in sunlight, that cattle barns, for example, would need to be sprayed by pyrethrin several times a day for fly control. The modern derivatives of pyrethrin retain high selectivity towards insects and low mammalian toxicity but they are resistant during a few days. Some examples of other pesticides were considered in Chapter 11.

3. POLYCHLORINATED BIPHENYLS, PCBs

3.1. PCBs as a Class of Compounds

Production

Polychlorinated biphenyls, PCBs, are among the most hazardous human-made substances, distributed so widely that they have been classified as global POPs. They comprise a group of 209 structurally similar compounds, so called congeners, with diverse teratogenic, reproductive, immunotoxic, neurotoxic and carcinogenic effects.

PCBs have been in extremely wide use in numerous industrial applications from the 1930s. The US EPA estimated that roughly 600,000 tons of PCBs were purchased by US industry from the time they were first manufactured in 1929–1930 and the mid-1970s when this production was restricted in many developed countries due to environmental concerns. Globally, more than one million tons were produced and these compounds are still under production in many Asian countries and worldwide (Table 2).

Generic structure of PCBs

The generic structure of PCBs is shown in Figure 10a.

The hydrogen atoms on the biphenyl nucleus are replaced by from one to ten chlorines, for a total of 209 possible congeners. We omit here the explanation of the PCBs classification. We can say only that the congeners are numbered by a scheme

Table 2. Trade names of some major commercial PCB formulations.

Name (Producer)	Uses
Arochlor (USA)	Transformers
Pyranol (USA)	Transformers, capacitors
Pyrochlor (USA)	Transformers
Phenochlor (France)	Transformers, capacitors
Pyralene (France)	Transformers, capacitors
Clophen (France)	Transformers, capacitors
Elaol (Germany)	Transformers, capacitors
Kanechlor (Japan)	Transformers, capacitors
Santothem FR (Japan)	Transformers, capacitors
Fenchlor (Italy)	Transformers, capacitors
Apirolio (Italy)	Transformers, capacitors
Sovol (Russia)	Transformers, capacitors

originally suggested by K. Ballschmitter and D. Zell in 1980 and later adopted by the International Union of Pure and Applied Chemistry (IUPAC).

The commercial formulations of PCBs are made by controlled chlorination of biphenyl. Each formulation contains various percentages of many of the 209 individual congeners, and there are significant variations among individual production lots. Hazardous waste samples may contain mixtures of two or more commercial formulations.

Physical and chemical properties

PCBs are extremely stable to heat, chemical, and biological decomposition. They are excellent insulating and cooling fluids, extensively used for many years in manufacture of transformers and capacitors. PCBs are also used in hydraulic fluids, lubricating oils, paints, adhesive resins, inks, fire retardants, wax extenders, and numerous other products. The chemical and physical properties of PCBs make the remediation of polluted sites difficult. They resist degradation and absorb into soils and colloidal materials in water. Some persist with half-lives of 8–15 years in the environmental compartments. This stability contributes to their dispersion in the environment and long-range air pollution. Because they are lipophilic, these species are stored in fatty tissues and accumulate in the food webs (see Section 2.2).

The physical and toxicological properties and molecular structures of PCBs are related to their degree of chlorination. Substitution of the electronegative chlorines at one or more *ortho*-positions leads to rotation about the bond between the phenyl rings,

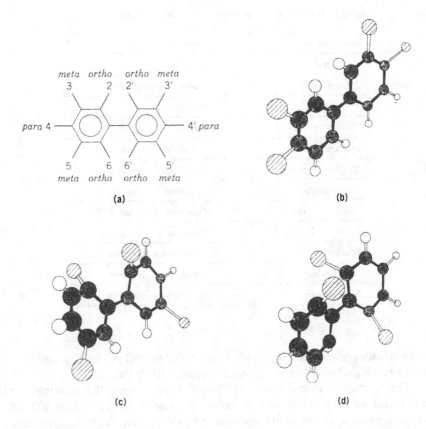

Figure 10. Generic structure and structural formulas of PCBs (Karu et al, 1999) (a) Generic structure of polychlorinated biphenyls. There are 209 possible combinations of hydrogen and chlorine atoms on the rings; (b) energy-minimized gas-phase structures of 3,4,3',4'-tetra-chlorobiphenyl (PCB 77); (c) 2,5,2',5'-tetrachlorobiphenyl (PCB 52) and (d) 2,6,2',6'-tetra-chlorobiphenyl (PCB 54), showing the effect of 0,2, and 4 ortho-chlorines on ring coplanarity. The dihedral angle between the two phenyl rings is 0°, 7.31° and 19.27°, respectively. Much larger dihedral angles have been observed in solution (white circle, hydrogen; gray circle, chlorine; black circle, carbon).

as shown in Figure 10c–d (Karu et al, 1999). The three most toxic congeners are 3,4,3',4'-tetrachlorobiphenyl (PCB 77), 3,4,3',4',5'-pentachlorobiphenyl (PCB 126), and 3,4,5,3',4',5'-hexachlorobiphenyl (PCB 169). Because these have no chlorines at the *ortho*-positions, the biphenyl rings can be coplanar and structurally resemble dioxin, which is the most toxic organic compound (see below). The coplanar congeners exert their toxic effects by binding to the aryl hydrocarbon (Ah) receptor, which leads to induction of cytochrome P-450-associated enzyme activities (see Chapter 11). The nonplanar, *ortho*-chlorinated congeners are much more abundant in commercial

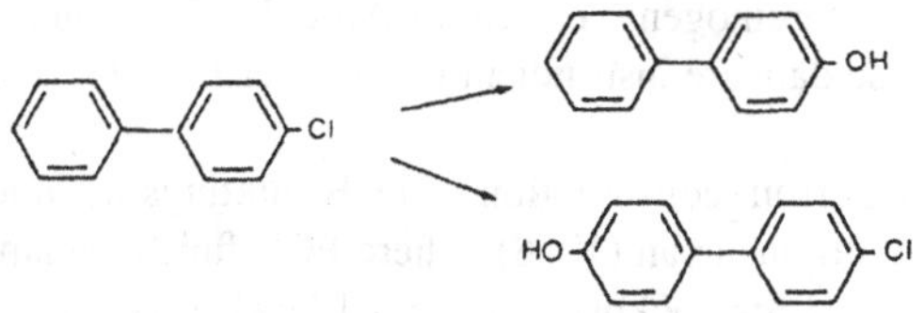

Figure 11. Microbial anaerobic dechlorination of PCB.

PCBs and hazardous waste, and they have different mechanisms of toxicity that are still not well understood.

Degradation of PCBs

Photolysis and vapor-phase reaction with hydroxyl radicals formed by sunlight play a role in environmental breakdown of PCBs. These reactions in atmosphere are relatively slow, as is evident from the transport of PCBs over long distances through the atmosphere, as discussed earlier.

Direct photolysis is a minor pathway for most congeners because of the poor overlap between PCB (Ar – Cl) absorption and the tropospheric solar spectrum. Following excitation, cleavage of the C – Cl bond can occur, but the quantum yield is often low; the aryl radicals thus formed suffer immediate oxidation by atmospheric oxygen (reaction (1))

$$Ar - Cl \longrightarrow Ar^\bullet + Cl^\bullet. \tag{1}$$

In the environmental compartments, PCBs are degraded mainly by microorganisms. Mono-, di-, and trichlorobiphenyls are broken down relatively quickly, whereas the more highly chlorinated PCBs are much more persistent. Highly chlorinated PCBs are reductively dechlorinated anaerobically with replacement of Cl by H, and the less-chlorinated products are subject to aerobic breakdown processes. The formation of *ortho*-diols is thought to involve arene epoxides (Figure 11).

Toxicology of PCBs

The toxicity of PCBs is complicated by the presence of a large number of congeners, each with its own toxicity, and the presence in PCB formulations of impurities such as chlorinated dibenzofurans (PCDFs), some of which are more toxic than the PCBs themselves.

The acute toxicity of PCB mixtures is relatively low; oral LD_{50} values in rodents are near 1 g per kg of body weight that is similar to those for aspirin. The acne is one of the most characteristic toxic responses to chlorinated compounds in humans. The condition, which is a disfiguring acne-like rash affecting principally the face and upper back, can last months to years. Other symptoms of PCB exposure include abnormal skin pigmentation, generalized feelings of fatigue, headache, and joint pain.

There are data on carcinogenic effects of Arochlor 1242 and Arochlor 1254 to animals, but still these data are insufficient to consider most PCBs as proved human carcinogens.

Human poisonings from consumption of PCB mixtures include the well-known Yusho rice oil poisoning in Japan (1968), where PCB fluids became mixed inadvertently with rice oil used for cooking. The actual level of contamination of the rice oil was low, about 0.2%. About 2000 people were poisoned in 1978 in a similar incident in Yu-Cheng, Taiwan. The principal symptoms in both cases were related to headache. Adipose tissue from Yusho patients was found to contain up to 75 ppm of PCBs. Cancer development was not statistically connected with people poisoning by PCBs. Later investigations have shown that the toxicity of these PCB-contaminated rice oils was probably not primarily due to the PCBs themselves, but to the traces of polychlorinated dibenzofurans (PCDFs), which they contained. We will consider the formation of furans and dioxins in the following section.

Mammalian and human metabolism involves hydroxylation, conjugation to thiols, and other water-soluble moieties including protein. The major mammalian metabolites are chlorinated mono- and dihydrodiols, phenols, and phenol acetates, methyl esters, and methylthio- and methylsulfonyl compounds.

PCB analysis

The most frequently used detection method for PCBs is capillary gas chromatography (GC) and high resolution mass spectrometry (MS) or GC/MS. However, GC/MS cannot resolve the most toxic coplanar congeners from *ortho*-substituted congeners that have similar elution properties. More recently, high-resolution capillary GC has been used as a more sensitive and reliable method, especially to identify coplanar congeners. GC/MS and capillary GC remain expensive and require highly trained analysts to perform very careful sample preparation and data interpretation. Instrumental analyses depend on GC peak or pattern matching of the test sample with reference standards of commercial formulation and individual congeners. Sample preparation and instrumental analysis are done mostly in centralized laboratories. The turnaround time may be from several days to a few months. Although new extraction methods and automation are bringing the cost down, instrumental PCB analyses may cost anywhere from US$50 to 500 per sample.

4. POLYCHLORINATED DIBENZO-*p*-DIOXINS (PCDDs) AND POLYCHLORINATED DIBENZOFURANS (PCDFs)

These compounds have been much in the news, mainly because of the extreme toxicity of the congener 2,3,7,8-tetrachlorodibenzo-*p*-dioxin (TCDD), usually called "dioxin" in the news media. Like the PCBs, PCDDs and PCDFs are families with many congeners: there are 75 PCDDs and 135 PCDFs (Figure 12).

Most are colorless solids of moderate to high melting point and low volatility. Their low environmental concentrations and complex congener pattern make their analysis extremely challenging.

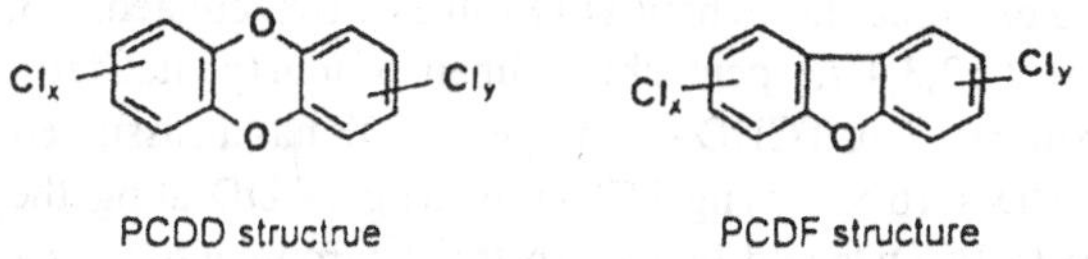

PCDD structrue   PCDF structure

Figure 12. General structural formulas of PCDD and PCDF.

Table 3. Toxicity Equivalency factors for PCDDs and PCDFs (Bunce, 1994).

Congener	PCDD series	PCDF series
2,3,7,8	1(by definition)	0.1
1,2,3,7,8	0.5	0.05
2,3,4,7,8		0.5
1,2,3,4,7,8	0.1[a]	0.1[b]
1,2,3,4,6,7,8	0.01	0.01[c]
octachloro	0.001	0.001

Notes: [a] same values for 1,2,3,6,7,8- and 1,2,3,7,8,9-congeners,
  [b] same values for 1,2,3,6,7,8-, 1,2,3,7,8,9-, and 2,3,4,6,7,8-congeners,
  [c] same values for 1,2,3,4,7,8,9-congeners.

Analysis of dioxin-like compounds

The standard chemical method for analyzing dioxin and dioxin-like compounds is experimentally demanding, tedious, and therefore expensive. The procedure includes the spiking with an isotopically labeled dioxin surrogate, such as $[^{13}C_{12}]$-TCDD as a recovery standard. After that, the conventional analysis involves solvent extraction, multiple chromatographic procedures to isolate the polychlorinated dibenzodioxin/polychlorinated dibenzofuran (PCDD/PCDF) fractions, and quantitative determination of the individual PCDD and PCDF congeners by capillary gas chromatography (GC)/high resolution mass spectrometry (MS). This yields the concentrations of each congeners in units such as nanagram (ng) congener per gram sample, ppt or even ppq level (Bunce and Petrulis, 1999).

Toxic Equivalency Factors, TEFs

Usually in the context of environmental remediation, human potential hazard, or/and comparison of dioxin-like compounds, DLC, an estimation of the toxic potency of the whole sample is desired, rather than the concentrations of the sample's components. One such measure, the toxic equivalents, TEQ, is obtained from the analytical data by using Toxic Equivalency Factors, TEFs, relating the potency of DLC "X" to the potency of TCDD, taken as a reference toxicant (TEF = 1). In this TEF scheme, the toxicity of any PCDD or PCDF congener is related, using animal data, to the amount of TCDD that would have equal toxicity (Table 3).

The following example shows how TEQ values are calculated. 2,3,7,8-TCDD has TEF = 1 (as defined); 2,3,4,7,8-pentachlorodibenzofuran (pentaCDF) has TEF = 0.5. Therefore, a mixture of 2 ng TCDD + 6 ng pentaCDF has an assumed toxicity equivalent to 2 ng TCDD + (6 × 0.5) ng TCDD, or 5 ng TCDD altogether. Estimates of human exposure to PCDDs and PCDFs in the environments are ca. 0.1 ng TCDD TEQ per person per day, mainly from food.

Selected dioxin sources

Dioxin emissions inventories have been constructed for a number of countries, mainly for developed countries of Europe and North America, as well as for the world as a whole. On average, municipal and hospital waste incineration was estimated to be the largest source. The next largest sources were estimated to be cement kiln and industrial boilers burning hazardous wastes and other materials; biomass combustion of all kinds, including industrial wood burning, residential wood burning, and forest fires; and secondary copper (recycling) smelters, which may have chlorinated compounds as contaminants in the feedstock. Some dioxin emission factors are shown in Table 4.

At present the inventory of dioxin sources is absent for the whole Asian region. However, expert estimates show that the prevalent sources of dioxin and dioxin-like compounds, include hospital incinerators, open garbage burning, uncontrolled tire fires, forest and agricultural burning, residential wood burning, used motor oil burning, PCB wood preservatives, 2,4-D and 2,4,5-T herbicides, cigarettes smoking, etc.

Hospital incineration yields predominantly octachlorodibenzo-*p*-dioxin, and the very toxic TCDD amounts to only a few percent of the mixture (Bunce, 1994). More dioxin emission with high TEQ values is connected with open garbage burning that is often practiced in the Asian cities and uncontrolled landfills.

Contamination from the 2,4,5-T family of herbicides (2,4,5-trichlorophenoxyacetic acid and its salts and eaters) which are manufactured from 2,4,5-trichlorophenol, affords principally 2,3,7,8-TCCD.

TCDD arise in 2,4,5-T formulations during the manufacture of 2,4,5-trichlorophenol from 1, 2, 4, 5-tetrachlorobenzene. The reaction is carried out at temperatures in the range 140–170 °C, in a polar, high-boiling solvent such as ethylene glycol (Figure 13).

In the 1970s, formulations of 2,4,5-T typically contained 10–100 ppm of TCDD. The concentration of TCDD in the product can be limited to about 0.1 ppm by careful temperature control of the exothermal reaction. However, the manufacture of the herbicide seems inevitably to produce traces of TCDD, and the registration of 2,4,5-T was ultimately withdrawn in North America due to public pressure. At present, this formulation is still manufactured in some Asian countries.

Toxicology of PCDDs and PCDFs

The first attention was paid to 2,3,7,8-tetrachlorodibenzo-*p*-dioxin (TCDD) in the 1970s, when 2,4,5-T was suspected of being a teratogenic agent. At that time, the formulation of 2,4,5-T was employed in very large quantities as the chemical defoliant

Table 4. Dioxin emission factors and TEQ for various sources (Thomas and Spiro, 1999).

Emission sources	Emission factors, $\mu g\ kg^{-1}$ feed	
	Total	TEQ
Consumer waste		
MSW incinerators	10	0.2
Hospital incinerators	20	0.35
Apartments incinerators	60	1
Open garbage burning	60	1
Sewage sludge incinerators	1	0.02
Industrial waste		
Hazardous waste incineration	3	0.06
Copper recycling	20	0.04
Steel recycling	0.1	0.004
Steel drum reconditioning furnaces	30/drum	0.5/drum
Used motor oil burners	0.04	0.01
Bleached pulp production	0.01	0.0002
Tire fires(uncontrolled)	0.09	0.004
Carbon regeneration	0.06	0.01
Tired incineration (controlled)	0.009	0.0004
Biomass combustion, etc.		
Forest and agricultural burning	0.4	0.004
Residential wood burning	0.4	0.004
Industrial wood combustion	0.05	0.001
Structural fires	0.4	0.004
PCB-treated wood combustion	8	0.1
PCB fires	1000	20
Cigarettes	0.1	0.002
Fossil fuel		
Oil combustion (except gasoline)	0.003	0.00005
Leaded gasoline	0.03	0.0005
Unleaded gasoline	0.003	0.00005
Coal combustion	0.001	0.00002
Dioxin-contaminated chemicals		
PCP wood preservative (to air)	2,000,000	20,000
2, 4-D herbicide (to soils)	200	0.2
Tetrachloroethylene (to air)	10	0.1

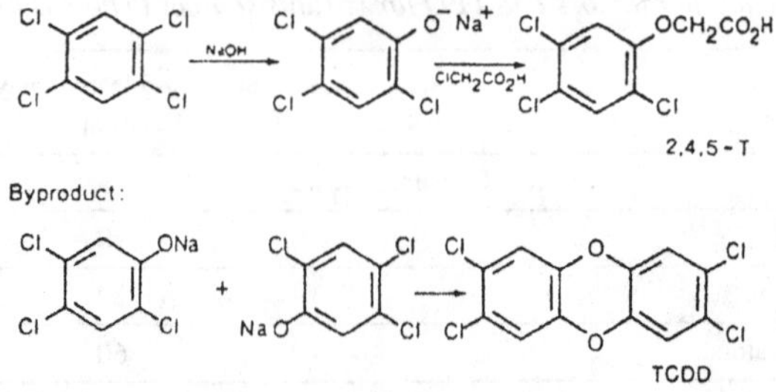

Figure 13. Formation of TCDD during 2, 4, 5-T herbicide formulation (Bunce, 1994).

Table 5. Oral LD_{50} values for TCDD in laboratory animals (after Bunce, 1994).

Species	Sex	LD_{50}, $\mu g\ kg^{-1}$
Guinea pig	Male	0.6
	Female	2.1
Rat	Male	22.0
	Female	50–500
Rabbit	Male & Female	115
Monkey	Female	70
Hamster	Male & Female	1100
Frog	Male & Female	1000

Agent Orange by the US Army in the Vietnam War. There is a statistical relationship between the development of risk for some diseases and exposure to this chemical of ex-servicemen. The US Government has a policy of making disability payments to these exposed ex-soldiers suffering from certain illnesses claimed to be Agent orange-related. Similar relationships have been shown for ex-soldiers for the Vietnam Army.

Toxicity of TCDD for experimental animals has been confirmed in many trials. Unusual facets of TCDD acute toxicity are that its LD_{50} is very species-dependent (Table 5).

Sub-lethal effects of TCDD include teratogenicity, carcinogenicity, reproductive complications, suppression of the immune systems, skin lesions and porphyria. Most of these effects have been studied in laboratory animals.

By comparison with 2,3,7,8-TCDD, the toxicity of other PCDD and PCDF congeners are reduced if chlorine atoms are removed from the lateral (2,3,7, or 8 positions) and/or are present at positions 1,4,6, or 9. These structural changes reduce greatly the toxicity of considered chemicals (Table 6).

Humans appear to be less sensitive than animals. The most common syndrome is chloracne. Other effects, including carcinogenicity, are rather uncertain.

Table 6. LD$_{50}$ values for administration of a single dose of PCDD congeners to guinea pig and mouse (after Bunce, 1994).

Congener	LD$_{50}$, μgkg^{-1}	
	Guinea pig	Mouse
Unsubstituted		> 50, 000
2,8-Cl$_2$	> 300, 000	> 800, 000, 000
2,3,7-Cl$_2$	29,000	> 3, 000
1,3,6,8-Cl$_4$	> 15, 000, 000	> 3, 000, 000
2,3,7,8-Cl$_4$	1	200
1,2,3,7,8-Cl$_5$	3	300
1,2,4,7,8-Cl$_5$	1,100	> 5, 000
1,2,3,4,7,8-Cl$_6$	73	830
Cl$_7$	> 600	
Cl$_8$		> 4, 000, 000

OCDDs and PCDFs are present in human tissues and human breast milk. Meat and dairy products have been shown to be the major routs of exposure of these lipophilic xenobiotics in most people. Nursing infants are more highly exposed per kg of body weight than adults because of the relatively high concentrations of these compounds in mother's milk. The transport of organochlorines into the milk is substantially depleting the fatty reservoirs of the mother's body (see Figure 4).

5. CHLORINATED PHENOLS

Chlorinated phenols were first synthesized in the 19th century. The most characteristic properties of these compounds are related to their antiseptic features. At present, 19 chlorinated phenols are known. For industrial use, the most important congeners are 2,4-dichlorophenol, 2,4,5-trichlorophenol, and pentachlorophenol, PCP. The first of these continues to be commercially important in the production of chlorinated phenoxyacetic acid herbicides and 2,4,5-trichlorophenol is also used as the precursor of the antiseptic hexachlorophene through condensation with formaldehyde.

Non-point sources of 2,4-dichlorophenol and 2,4,5-trichlorophenol are mainly agricultural, since the phenoxy herbicides are hydrolyzed back to the phenols with a lifetime of about a week near 20°C. A minor local source of chlorophenols is chlorination of raw drinking water, which is contaminated with phenol (see Chapter 10). The most important chlorinated phenol is the pentachloro congener.

Pentachlorophenol, PCP, is a highly effective biocide developed in 1935 by the Monsanto Chemical Company and used today in the Asian region primarily as a wood preservative.

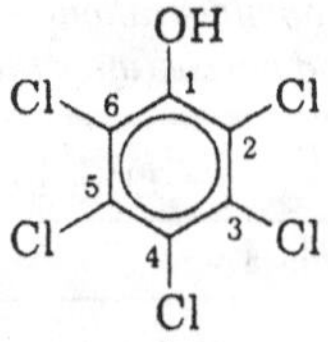

Figure 14. Structure of pentachlorophenol. Positions 2 and 6 are the ortho-positions, 3 and 5, the meta-positions, and 4, the para-position (Diehl and Borazjani, 1999).

5.1. Chemical Structure and Properties of PCP

PCP (Figure 14) is an organic compound (Cl_5C_6OH) manufactured by the catalytic chlorination of phenol.

It is quite stable, only slightly soluble in water, soluble in most organic acids, and moderately volatile. Only the hydroxyl group participates in nucleophilic reactions. Some physical and chemical properties of PCP are shown in Table 7.

Pentachlorophenol is unusually acidic with pK_a 5.01 at 25 °C. This means that extraction of PCP from environmental samples such as soil or bottom sediments, and tissues, requires prior adjustment of pH to less than 5. We can note that the increase in the acidity of PCP (from pK_a 10 for the parent phenol) is due to five electron-withdrawing chloro substituents. In the natural surface waters with pH of 6–8, PCP will be completely dissociated.

Technical grade PCP, the commercial form, is a tan or gray crystalline solid, containing 85% of PCP, 4–8% of tetrachlorophenol and 2–6% of other phenols, and trace quantities of chlorinated dibenzo-p-dioxin and dibenzofuran. These chlorinated dibenzo-p-dioxin and dibenzofuran contaminants contain six to eight chlorines. None of the highly toxic TCDDs are usually found in PCP formulations. Because of the presence of the other contaminants, PCP has been restricted or banned in several countries, including Sweden, Finland, Germany and Japan.

5.2. Toxicology of PCP

PCP is acutely toxic to animals. PCP appears to uncouple oxidative phosphorylation, which allows transport of electrons down the respiratory chain to oxygen but prevents synthesis of ATP. Ingestion of PCP occurs by inhalation, oral ingestion, and dermal absorption. Symptoms include dermatitis, irritation to the eyes, nose, and throat, elevated respiratory rates, hyperglycemia, elevated blood pressure, and cardiovascular distress. The no observable effects level (NOEL) of PCP, which is the amount that can be ingested without any symptoms or effects, for fetotoxicity is 5.8 mg kg^{-1} of body weight per day. For chronic toxicity the NOEL is 500 mg kg^{-1}day^{-1}. For the chlorinated dioxin contaminants, the fetotoxicity NOEL is 1 μg kg^{-1}day^{-1}. PCP does not exhibit mutagenicity but is highly embryolethal and embryotoxic to rats. In rats, purified PCP has LD$_{50}$ of about 150 mg kg^{-1} of body weight per day. Reduced reproductive capacity has been observed upon long-term administration of PCP to

Table 7. Physical and chemical properties of PCP (Diehl and Borazjani, 1999).

Physical and chemical properties	Values
Molecular weight, g	266.35
Melting point, °C	190.2
Boiling point, °C	300.6
Density, g cm^{-3}	1.85
Vapor pressure, torr, mm Hg	
0 °C	1.7×10^{-5}
20 °C	1.7×10^{-4}
50 °C	3.1×10^{-3}
Solubility in water, g L^{-1}	
0 °C	0.005
20 °C	0.014
30 °C	0.020
50 °C	0.035
Solubility in organic solvents, g L^{-1}, at 25 °C	
Methanol	180
Acetone	50
Benzene	15
pK$_a$, at 25 °C	4.70
Log K$_{ow}$, at 25 °C	5.01

female rats at 30 mg kg^{-1} per day. In most animal species, PCP is rapidly metabolized and eliminated from the body. This is related to the fact that at physiological pH, PCP will be almost completely in the anionic form, thus favoring excretion rather than bioaccumulation.

5.3. PCP contamination problems and sources

In the 1980s, total worldwide production of PCP was estimated to be about 50,000 tons per year. Past uses, which led to PCP's widespread occurrence and contamination problems in the Asian region, included a herbicide for paddy and upland rice fields in many Asian countries, a molluscicide in water systems in the far East, as a coating on wicker products in Hong Kong, and as biocide coating on burlap, cotton, rope, twine and leather, an insecticide in seed plots and greenhouses, an additive to paints for mildew prevention, and as a biocide for reducing slime in industrial cooling waters. This is still used as crop fungicide in some Asian countries.

As a wood preservative, PCP is a very effective biocide and resists leaching from the wood into the soils, but because of its low water solubility it must first be dissolved in a petroleum oil before being impregnated into wood. Concentrations of PCP in the oil solvent range from 5–10%. PCP in heavy petroleum oil is used to treat utility poles and sawn timber, in light petroleum oil with insecticide for joinery, and in light oil for remedial treatment of fungal decay in buildings. A water soluble salt form of PCP, sodium pentachlorophenate (NaPCP) is applied as an aqueous spray or dip to green lumber or freshly peeled poles for control of sap stain.

FURTHER READING

1. Richter O., 1999. Pesticides: environmental fate. In: Meyers R. A. (Ed.), *Encyclopedia of Environmental Pollution and Cleanup*, John Wiley, NY, 1224–1231.

2. Galiulin R. V., Bashkin V. N. and Galiulina R. R., 2001. Behavior of 2, 4-D herbicide in coastal area of Oka River, Russia. *Water, Air and Soil Pollution*, 129: 1–12.

3. Knisel W. G. (Ed.), 1980. *CREAMS: a flied-scale model for chemical runoff and erosion.* USDA-SEA Conservation Research Report, 26, US Governmental Printing Office, Washington, DC.

4. Spain J. D., 1982. *Basic computer models in biology*. Advanced book program. Addison-Wesley Publishing Co., Reading, Mass.

5. Karu A. E., Lesnik B. and Haas R. A., 1999. Polychlorinated biphenyls: detection by immunoassay. In: Meyers R. A. (Ed.), *Encyclopedia of Environmental Pollution and Cleanup*, John Wiley, NY, 1231–1238.

6. Thomas V. M. and Spiro T. G., 1999. Dioxin emission inventory. In: Meyers R. A. (Ed.), *Encyclopedia of Environmental Pollution and Cleanup*, John Wiley, NY, 436–440.

7. Bunce N. J. and Petrulis J. R., 1999. Dioxin-like compounds: screening assays. In: Meyers R. A. (Ed.), *Encyclopedia of Environmental Pollution and Cleanup*, John Wiley, NY, 440–449.

8. Diehl S. V. and Borazjani A., 1999. Pentachlorophenol-contaminated soils, bioremediation. In: Meyers R. A. (Ed.), *Encyclopedia of Environmental Pollution and Cleanup*, John Wiley, NY, 1216–1221.

WEBSITES OF INTEREST

1. http//:www.escap.un.org

QUESTIONS AND PROBLEMS

1. Characterize the typical parameters of substances that are known as being organic xenobiotics.

2. Describe the application of pesticides in the Asian region and note the driving forces for permanent increasing of these agrochemicals use.

3. Discuss the environmental behavior of organic xenobiotics in various compartments and indicate the principal pathways for migration of these compounds in terrestrial and freshwater ecosystems.

4. Discuss the bioaccumulation of organic xenobiotics in the body and focus your attention on the redistribution processes between various biochemical compartments.

5. Characterize the chemistry of DDT formulation and its principal physical and chemical parameters that enhance the worldwide application of this preparation.

6. Characterize the toxicological and ecotoxicological behavior of DDT and its principal metabolites in the environment and food chains.

7. Discuss the previous and present, if any, application of DDT in your country and the causes that enforced this application.

8. Characterize the past and present classes of pesticides; emphasize the role of resistance of these species in the environment.

9. Present the general characteristics of chlorinated biphenyls and explain the structural formulas of any PCB congeners.

10. What are the main applications of PCBs in your countries? Have these chemicals been restricted or banned in your country and why?

11. Discuss the toxicological parameters of PCBs in animal and human organisms and explain the differences in PCB metabolism in various living species.

12. Discuss the modern chemical methods of PCB analysis in environmental samples and explain the relevant difficulties.

13. Describe the general structural formulas of PCDD and PCDF congeners. Stress the role of chlorine substitution in toxicity of these compounds.

14. Characterize the application of TEQ values and present corresponding example for calculating TEFs for mixtures of different dioxin-like compounds.

15. Characterize the sources of dioxin-like compounds in the Asian region and in your country. Try to calculate the DLC emissions using emission factors shown in Table 4.

16. Discuss toxicology of PCDDs and PCDFs and give some examples of LD_{50} for different animals.

17. Discuss the physical and chemical properties of pentachlorophenols. Focus your attention on the solubility of PCP in natural water using the values of pK_a for this compound.

18. Describe the toxicology of PCP for animals and human beings with explanation of the differences.

19. Characterize the application of PCP in the whole Asian region and in your country.

PART IV

ENVIRONMENTAL CHEMISTRY FOR SUSTAINABLE DEVELOPMENT IN ASIA

CHAPTER 15

CRITICAL LOADS

1. INTRODUCTION

There is agreement both nationally and internationally that long-range transboundary air pollution may span continents: pollutants are transferred from Europe to North America and Asia as well as in the opposite directions (Sofiev, 1998). Consequently, the calculation and mapping of critical loads as indicators of ecosystem sensitivity to acid deposition in regions outside Europe and North America are of great scientific and political interest. Some researches have been made to calculate the acidification loading for Asia (Dianwu et al, 1994; Shindo et al, 1995; World Bank, 1994; Kuylenstierna et al, 1995; Kozlov et al, 1997; Bashkin and Kozlov, 1999).

It has been argued (Bashkin et al, 1996a) that the best approach to the calculation and mapping of critical loads on ecosystems in Asia is to use various combinations of expert approaches and geoinformation systems, including different modern methods of expert modeling and environmental risk assessment. These systems can operate using databases and knowledge bases relative to the areas with great spatial data uncertainty. As a rule, the given systems include an analysis of the cycles of various elements in the key plots, a choice of algorithms describing these cycles, and corresponding interpretation of the data. This approach requires numerous cartographic data, such as vegetation, soil, geochemical and biogeochemical maps, information on pollution and buffering capacity of soil, water and atmosphere. This approach is the most appropriate for Russia as well as for other Asian countries such as China, India, and Thailand where, at present, adequate information on the great spatial variability of natural and anthropogenic factors is either limited or absent.

The applicability of these approaches for the assessment of acidification loading on the terrestrial ecosystems in Asia is demonstrated here using the examples of Asian domain (Asian part of Russia, China, Japan, Taiwan, Korea and Thailand). In spite of the great differences in climate, soil and vegetation conditions, these regions can serve as a good test of the proposed methodology.

2. DEFINITION OF CRITICAL LOAD

It is well known that biogeochemical cycling is a universal feature of the biosphere, which provides its sustainability against anthropogenic loads, including acid forming

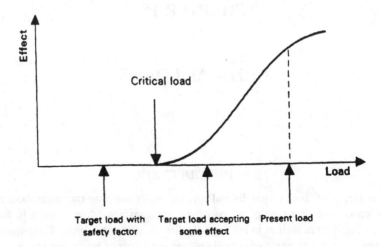

Figure 1. Illustration of critical load and target load concepts.

compounds. Using biogeochemical principles, the concept of *critical loads* (CL) has been developed in order to calculate the deposition levels at which effects of acidifying air pollutants start to occur. A UN/ECE (United Nations/Economic Committee of Europe) working Group on Sulfur and Nitrogen Oxides has defined the critical load on an ecosystem as: *"A quantitative estimate of an exposure to one or more pollutants below which significant harmful effects on specified sensitive elements of the environment do not occur according to present knowledge"* (Nilsson & Grennfelt, 1988). These critical load values may be also characterized as *"the maximum input of pollutants (sulfur, nitrogen, heavy metals, POPs, etc.), which will not introduce harmful alterations in biogeochemical structure and function of ecosystems in the long-term, i.e., 50–100 years"* (Bashkin and Park, 1998).

The term *critical load* refers only to the deposition of pollutants. Threshold gaseous concentration exposures are termed *critical levels* and are defined as *"concentrations in the atmosphere above which direct adverse effects on receptors such as plants, ecosystems or materials, may occur according to present knowledge"*. Correspondingly, regional assessments of critical loads are of concern for optimizing abatement strategy for emission of both N and S compounds and their transboundary transport (Figure 1).

The critical load concept is intended to achieve the maximum economic benefit from the reduction of pollutant emissions, since it takes into account the estimates of differing sensitivity of various ecosystems to acid deposition. Thus, this concept is considered to be an alternative to the more expensive BAT (Best Available Technologies) concept (Posch et al, 1996). Critical load calculations and mapping allow the creation of ecological-economic optimization models with a corresponding assessment of minimum financial investments for achieving maximum environmental protection.

3. THE CRITICAL LOAD CONCEPT

In accordance with the above-mentioned definition, a critical load is an indicator for sustainability of an ecosystem, in that it provides a value for the maximum permissible load of a pollutant at which risk of damage to the biogeochemical cycling and structure of ecosystem is reduced. By measuring or estimating certain links of biogeochemical cycles of sulfur, nitrogen, base cations and some other relevant elements, sensitivity to both biogeochemical cycling and ecosystem structure as a whole to acidic deposition and/or eutrophication deposition can be calculated, and *a critical load of acidity*, or the level of acidic deposition, which affects the sustainability of biogeochemical cycling in the ecosystem, can be identified, as well as *critical nutrient load*, which affects the biodiversity of species within ecosystems. According to the political and economic requirements of the protocol for reduction of N and S emissions and deposition, as well as the parameters of subsequent optimizing models, the definitions of critical loads are given separately for sulfur, nitrogen and for total acidity, which is induced by both sulfur and nitrogen compounds. Hence, critical loads for acidity can be determined as the maximum input of S and N before significant harmful acidifying effects occur. When assessing the individual influences of sulfur and nitrogen, it is necessary to take into account the acidifying effects induced by these elements and the eutrophication effect caused only by nitrogen. In this case, critical load for nitrogen can be determined as the maximum input of nitrogen into the ecosystem, below which neither significant harmful eutrophication effects nor acidifying effects together with sulfur occur during long-term periods.

Both ratio of the base cations to aluminum and the aluminum concentrations are used as indicators for steady-state geochemical and biogeochemical processes. By assigning established critical loads to these indicators (for example, the concentrations of aluminum in soil solution should not exceed 0.2 meq/L and the base cations to aluminum ratio should not be less than 1), it is possible to compute the allowable acidification for each ecosystem. An extensive overview of critical values for the ratio of base cations to aluminum for a large variety of plants and trees can be found in H. Sverdrup's papers (Sverdrup H. and de Vries W., 1994).

4. CRITICAL LOAD VALUES OF ACID FORMING COMPOUNDS ON ECOSYSTEMS OF ASIA

The interest in acidic deposition has resulted in the development of intensive biogeo-chemical investigations of a large number of ecosystems in North America, Europe, Asia and South America (Moldan & Cherny 1994; Bashkin & Park 1998). The bio-geochemical cycling concept is designed to summarize the cycling process within various components of ecosystems such as soil, surface water and groundwater, bottom sediments, biota and atmosphere. Ecosystem and soil maps can serve as a basis for biogeochemical mapping (see Chapter 2).

The experimental data obtained in various countries of East Asia allow us to consider the applicability of methodology of critical loads related to an assessment of ecosystem sensitivity to acid rains. The critical load (CL) and Environmental Risk

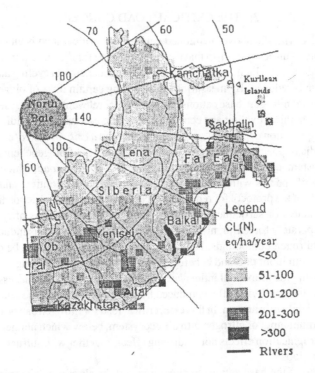

Figure 2. Critical loads of nitrogen on the ecosystems in the Russian Northern Asia (After Bashkin & Kozlov, 1999).

Assessment (ERA) approaches were used for the evaluation of ecosystem sustainability to acid deposition in East Asia. Calculations of critical loads for assessment of the sensitivity of the ecosystem to acidic deposition have been made using biogeochemical approaches including the intensity of biogeochemical cycling and period of active temperature duration. On the basis of these coefficients the soil-biogeochemical regionalization is carried out for the whole area of Asia and the values of critical loads for acid-forming compounds are calculated using modified steady-state mass balance (SSMB) equations.

4.1. Critical Loads of Sulfur and Nitrogen on North Asian Ecosystems

The models for calculation of critical loads in North Asian ecosystems are shown in Box 1. In the ecosystems of the Asian part of Russia these values of critical loads for N, CL(N), and S, CL(S), compounds are shown to be less than in Europe due to many peculiarities of climate regime (long winter with accumulation of pollutants in snow cover) and depressed biogeochemical cycling of elements (see Section 2, Chapter 7). The minimum values of both CL(N) and CL(S) are < 50 eq/ha/yr and the maximum ones are > 300 eq/ha/yr (Figures 2 and 3).

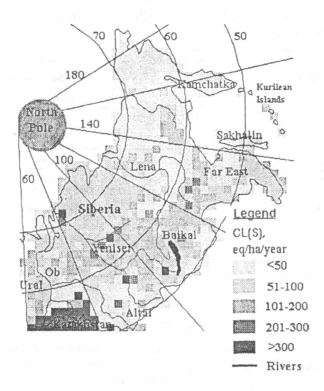

Figure 3. Critical loads of sulfur on the ecosystems in the Russian Northern Asia (After Bashkin & Kozlov, 1999).

At current rates of atmospheric deposition, 88.0% and 72.7% of ecosystems in North Asia have no or small (< 50 eq/ha/yr) exceedances of CL(N) and CL(S), respectively (Table 1).

However, atmospheric deposition in excess of calculated critical loads of N and S are exceeded for 10% and 20%, respectively, of the studied ecosystems in this region.

A comparison of these critical load values with those calculated for corresponding ecosystems in Europe reveals lower values in the Asian areas (Bashkin & Kozlov, 1999). This can be explained by the more prolonged winter period causing an accumulation of pollutants in the snow layer and their influence on biogeochemical cycling of nutrients during the short spring and summer period. The values of active temperature coefficient, C_t are within the limits of 0.15–0.57 for the majority of the North Asian ecosystems, whereas in the European part of Russia, for example, the corresponding values are 0.25–0.87. This points to the shorter but very active periods of biogeochemical turnover in almost all of North Asian ecosystems accelerating the effects of the acidification loading on the ecosystems.

Table 1. Distribution of critical load values of sulfur, nitrogen and their exceedances for the North Asian ecosystems.

Values range, eq/ha/yr	Percentage of area under different critical load values		Percentage of area under different critical load values	
	For nitrogen	For sulfur	For nitrogen	For sulfur
< 50	8.3	40.5	88.0	72.7
50–100	40.8	32.4	6.9	14.3
101–200	41.3	18.2	3.9	8.3
201–300	8.0	1.5	1.1	2.6
> 300	1.6	7.4	0.1	2.1

Box 1. Models for calculation of S and N critical loads at North Asian Ecosystems (After Bashkin and Kozlov, 1999)

Critical loads of nitrogen

$$CL(N) = {}^*N_u + {}^*N_i + {}^*N_{de} + {}^*N_{l(crit)},$$

where * means that each of the terms refers to the values at the actual total atmospheric deposition at a side. N_u, N_i, N_{de} and $N_{l(crit)}$ are permissible nitrogen uptake, soil immobilization, denitrification and leaching, correspondingly.

Permissible atmospheric nitrogen uptake ($^*N_u$) was given as:

$$^*N_u = N_{upt} - N_u,$$

where N_{upt} is the annual accumulation of nitrogen in biomass and N_u is the annual uptake of N from the soil.

N_{upt} was calculated accounting for the coefficients of biogeochemical turnover (see Chapter 7). Annual N_u from soil was calculated on a basis of nitrogen mineralizing capacity (NMC) of soils, which was determined experimentally or calculated using regression equations (Bashkin, 1997). So,

$$N_u = (NMC - N_i - N_{de})C_t,$$

where

$N_i = 0.15NMC$, if $C : N < 10$, $N_i = 0.25NMC$, if $10 < C : N < 14$,

$N_i = 0.30NMC$, if $14 < C : N < 20$, $N_i = 0.35NMC$, if $C : N > 20$,

$N_{de} = 0.145NMC + 6.447$, if $NMC > 60\,kg/ha/yr$,

$N_{de} = 0.145NMC + 0.900$, if $NMC < 10\,kg/ha/yr$,

$N_{de} = 0.145NMC + 2.605$, if $10 < NMC > 60\,kg/ha/yr$.

Permissible immobilization of atmospheric deposition N ($^*N_i$) was found as:

$$N_i = [(0.20NH_y + 0.10NO_x)/C_b]C_t, \quad \text{if } C:N < 10,$$
$$N_i = [(0.30NH_y + 0.20NO_x)/C_b]C_t, \quad \text{if } 10 < C:N < 14,$$
$$N_i = [(0.35NH_y + 0.25NO_x)/C_b]C_t, \quad \text{if } 14 < C:N < 20,$$
$$N_i = [(0.40NH_y + 0.30NO_x)/C_b]C_t, \quad \text{if } C:N > 20,$$

where NO_x and NH_y are the oxidized and reduced N wet and dry deposition. Permissible denitrification from atmospheric deposition N ($^*N_{de}$) was found as:

$$^*N_{de} = (N_{de}/NMC) N_{td} C_t,$$

where N_{de}/NMC is the denitrification fraction, which depends on many features of soils and calculated on a basis of experimental data and N_{td} is the total N deposition.

Finally, permissible critical leaching of atmospheric nitrogen ($^*N_{l(crit)}$) was given as

$$^*N_{l(crit)} = Q C_{Ncrit},$$

where Q is the annual surplus of precipitation (runoff) and C_{Ncrit} is the permissible nitrogen concentration in surface water.

Critical loads of sulfur

Since for the majority of ecosystems in North-Eastern Asia the ratio of precipitation to potential evapotranspiration (P : PE) is equal to ≤ 1.00 or slightly exceeds the 1.00 (except of ecosystems with some cambisols, histosols and andosols) the values of runoff can be neglected in calculation of critical loads of acidity, CL(Ac), and they were found as

$$CL(Ac) = (BC_wC_t)/C_b$$

and critical loads of sulfur were calculated as

$$CL(S) = S_fCL(Ac),$$

where BC_w is the weathering of base cations and S_f is the sulfur fraction in total sum of sulfur and nitrogen deposition.

The values of BC_w were determined on a basis of FAO soil nomenclature, soil parent material and soil texture according to UBA (1996) and values of C_b and C_t were applied to accounting biogeochemical cycling intensity and duration of active temperature period (see Section 2, Chapter 7). The root zone was assumed to be equal to 0.5 m.

Exceedances of critical loads

The values of exceedances were calculated as follows:

$$Ex(N) = N_{td} - CL(N),$$
$$Ex(S) = S_{td} - CL(S),$$

where Ex(N), Ex(S) are the values of excessive input of nitrogen and sulfur compounds above the calculated critical loads, and N_{td}, S_{td} are nitrogen and sulfur deposition.

Regarding the uncertainty analysis of critical loads of acid forming compounds in different terrestrial ecosystems of the Asian part of Russia, one can see the following. The strongest influential factor for the values of CL(N) is the parameter nitrogen uptake by plant biomass, N_u. For the biggest part of Siberia and Far East territory this parameter has the first rank and only in the case of Humid Cambisols and Cryic Gleysols ecosystems is its rank second. Among others, nitrogen immobilization, N_i and nitrogen denitrification, N_{de}, parameters, which are closely inter-correlated, are the weakest ones. The given parameters do not have any practical influence on the values of CL(N), except Dyctric Cambisol ecosystems. The influence of nitrogen leaching, N_l, values is decreasing in the following ecosystem-forming soils: Regosols > Humid Cambisols = Cryic Gleysols = Gelic Podzols > Andosols > Dystric Cambisols > Eutric Cambisols > Luvic Phaeozems = Chernozems > Kashtanozems. These results reflect, in significant extent, the geographical change of ecosystems from north to south and correspondingly an alteration of relationship of temperature and moisture constituents in hydrothermic coefficient. So, the main impact to the assessment of the influence of inputting parameters (N_u, N_i, N_{de}, N_l) on both uncertainty and sensitivity of outputting values of CL(N) belongs to N_u. It is connected firstly with a deficit of nitrogen as the main nutrient in all North Asian ecosystems as well as an existing spatial and temporal variability of this parameter that relates to a significance and correctness of experimental and computed values of N_u. In accordance with relatively better knowledge of the hydrological picture and relatively homogenous values of critical concentration of nitrogen in surface waters, C_{Ncrit}, included in the calculation of critical nitrogen leaching, $N_{l(crit)}$, values, the input of the given parameter into the uncertainty of CL(N) is expressed in a lesser degree. Furthermore, the runoff processes are practically not significant for ecosystems of Luvic Phaeozems, Chernozems and Kashtanozems due to low P:PE ratio. During the calculations of CL(N) for ecosystems of North Eastern Asia, the values of critical immobilization and denitrification from N depositions both in relative and absolute meanings played a subordinate role that obviously reflects their minor contribution to uncertainty and sensitivity analysis of the computed output values of ecosystem sensitivity to acidic deposition.

Thus, the ERA estimates shown in Table 2 characterize the significance of such endpoints as nitrogen content in plant issues and surface waters for many ecosystems of North Eastern Asia. These endpoints have to be taken into account during risk management steps of the ERA flowchart for emission abatement strategy development.

4.2. Critical Loads of Sulfur in China

Based on the mineralogy controlling weathering and soil development, sensitivity of ecosystems to acid deposition is assessed with the comprehensive consideration on the effect of temperature, soil texture, land use and precipitation. The results show

Table 2. Percentage of various endpoints contribution to total environmental risk assessment of ecosystem sensitivity to acid deposition in Northern Asia (Bashkin and Park, 1998).

Ecosystem forming soils	Endpoints assessment method	Endpoint			
		N content in plant issues, N_u	N content in surface waters, N_l	Denitrified N, N_{de}	N enrichment of soil, N_i
Regosols and Lithosols	SRC	50	41	2	7
	RTU	30	27	21	22
Cryic Gleysols and Humic Cambisols	SRC	37	56	2	5
	RTU	22	37	21	20
Gelic Podzols	SRC	47	45	3	5
	RTU	40	39	10	11
Andosols	SRC	40	39	10	11
	RTU	81	17	1	1
Eutric Cambisols	SRC	41	9	25	25
	RTU	63	10	14	13
Distric Cambisols	SRC	96	2	1	1
	RTU	38	32	16	14
Luvic Phaerozems and Chernozems	SRC	96	1	1	2
	RTN	65	6	12	17
Kashtanozems	SRC	95	1	4	5
	RTU	61	5	19	15

Note: SRC—Standard Regression Coefficient;
      RTU—RooT of the Uncertainty.

that the most sensitive area to acid deposition in China is Podzolic soil zone in the Northeast, then followed by Latosol, Dark Brown Forest soil and Black soil zones. The less sensitive area is Ferralsol and Yellow-Brown Earth zone in the Southeast, and the least sensitive areas are mainly referred to as Xerosol zone in the Northwest, Alpine soil zone in the Tibet Plateau, and Dark Loessial soil and Chernozem soil zone in central China. These regional different soil sensitivities to acid deposition can be attributed to the differences in temperature, humidity and soil texture (Hao et al, 1998). It has been shown that the assessment of ecosystem sensitivity to acidic loading depends strongly on the calculation of chemical weathering of soil base cations due to an input of protons with depositions. The critical loads of acid deposition have been mapped for the Chinese ecosystems, as shown in Figure 4.

It can be seen that the areas most sensitive to acid deposition are in Southeast China, and that the insensitive is in the Northwest. The sensitive areas, including the catchment of the Changjiang (Yangtze) River and the wide areas to the southward, are

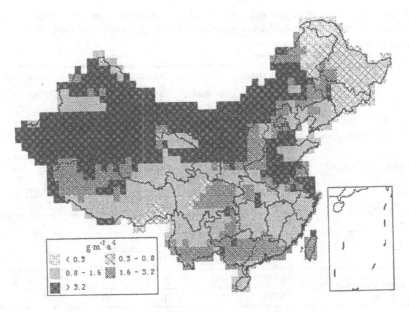

Figure 4. Critical loads of sulfur at terrestrial ecosystems of China (Hao et al, 1998).

warm and rain-abundant. The natural vegetation is the Tropic Rain Forest, Seasonal Rain Forest, and Subtropical Evergreen Forest ecosystems. The dominant soils in these areas are acid Ferralsols, with obvious accumulation of iron and aluminum. These soils can tolerate approximately 0.8–1.6 g/m2/yr of sulfur from acid deposition and belong to the intermediate sensitivity class 3. In Northwest China, the climate is semiarid or arid. The predominant ecosystems are Dry Steppe and agroecosystems, with spots of Deciduous Forest or Mixed Forest ecosystems. The soils represented by Xerosol and Alpine types are carbonate-rich and saline. Consequently, they are resistant to acidification.

The critical load class 1, the most sensitive class of ecosystems, chiefly refers to Podzol soils of Coniferous Forest ecosystems. This class occupies the small areas in the northeast of China, mostly on the north part of Da-xing-an-ling Mountain, which is covered by coniferous forest, with annual precipitation 400–500 mm and annual mean temperature −4.9–0 °C. Podzol soils were derived from acid granite or quartz rocks and their formation was often influenced by leaching and settling of organic acid complex compounds. Hence, the soils show acid reactions and the base saturation is very low. The soil clay minerals are composed of hydrous mica and small amounts of other unweatherable minerals such as montmorillonite, kaolinite, roseite and chlorite. The low temperature and the coarse texture of soil constitute the importance to the low weathering rate of soil minerals there. Therefore, these areas must be paid great attention, even if acid deposition has not yet appeared.

Class 2 is found in Dark Brown Forest soil and Black soil zones in the Northeast, and Latosol soil zone in the south of the Taiwan province, in the north of Hainan province and near Hekou in the Yunnan province. The Dark Brown soil zone, with annual mean temperature -1–$5\,°C$, is covered by coniferous forest. The coarse soil was derived from granite parent rocks and the chief clay mineral is hydrous mica. As in the Dark Brown soil zone, the temperature in Black soil areas is very low. The Black soil contains clay minerals such as hydrous mica, fulonite, gibbsite and kaolinite. It is obvious that the climate conditions and the mineral composition favor the chemical weathering of soil minerals. Contrarily, Latosol soil zone is under high temperature rain abundant conditions where the annual mean precipitation is 1900 mm and the annual mean temperature is $12\,°C$. The texture of Latosol is fine and the fraction of clay in the soil is high, which is advantageous to weathering. However, the clay minerals in this soil are dominated by kaolinite and gibbsite. Anorthite has completely decomposed and K-feldspar is rare. The weathering rate is still low as a result of the lack of weatherable minerals.

Soils of class 3 include Lateritic Red Earth in the areas southward from the Nanling Mountain, Red Earth and Yellow Earth between the Nanling Mountain and the Changjiang (Yangtzee) River, Yellow-Brown Earth in the lower reaches of the Changjiang(Yangtzee) River, Subalpine Meadow soil and Alpine Meadow soil on the Plateau of Tibet.

Class 4 is found in Paddy soil zones sporadically distributed throughout China and in the Purplish soil zone in the Sichuan River Basin.

Class 5 (the least sensitive) soil includes Kashtanozem, Brown soil and Sierozem soil zones in the Plateau of Inner Mongolia and the Loess Plateau, Desert soil zones in He-xi-zou-lang and the Talimu River Basin, Subalpine Steppe soil, Alpine Steppe soil and Alpine Desert soil in the Plateau of Tibet. These kinds of soils, belonging to the soil class of Xerosol or Alpine soil, consist of easy weathering minerals such as carbonate. They show alkaline reactions, with weak leaching and sparse vegetation. Those kinds of soils are insensitive to acid deposition.

Figure 5 illustrates the percentage of areas shared by each critical load class. As we can see, the most sensitive soil (class 1) shares only 2% of the whole area in China, and the sensitive soil (class 2) is no more than 8.7%. The intermediate sensitivity and the least sensitive soils are the most extensive, which account for 35.5% and 42.4%, respectively. Class 4 covers 11.4%.

To summarize, the sensitivity of an area to acid deposition and the critical load values depend on soil and vegetation types and meteorological conditions. In China, acid deposition often occurs in the Allite areas in the southeast of China, where soils are of intermediate sensitivity, except Latosol. The annual mean temperature in these tropic or subtropic areas is 16–$25\,°C$, the accumulated temperature of $\geq 10\,°C$ is 5000–9000 $°C$, and precipitation more than 1500 mm. The characteristics of temperature-high, rain-abundant, and wet-warm-in-same-season, promote reactions of soil minerals and cause the rapid weathering of soil minerals and the rapid circulation of biological materials. As a result, the weathering rates are sufficient enough for high values of corresponding critical loads. On the other hand, the areas of Podzol soil, Dark

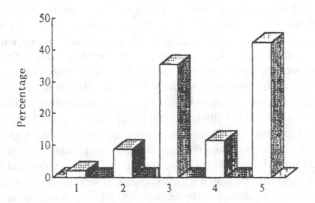

Figure 5. Percentage of areas shared by each critical load class in China (Hao et al, 1998).

Brown Forest soil and Black soil are in frigid temperate conditions. The accumulated temperature of $\geq$ 10 °C is about 2000, the annual mean temperature is lower than 5 °C, and the precipitation is low. These soils contain large sand and small clay fractions. Thus, the chemical weathering rates are low and critical loads are also low.

It can be concluded that high temperature, high humidity, wet-warm-in-same-season and high concentrations of clay and loam, are important reasons why soils and water bodies have not been found acidified in heavy acid rain areas such as the Chongqing, Guiyang and provinces of Guangdong and Guangxi. However, it should be noted that the weatherable minerals in soils in south China are scarce, and these soils are potentially at risk of acidification. Acid deposition must be strongly controlled in the Northeast area of China to protect terrestrial and aquatic ecosystems from acidification and decrease of forest production like those in Central and North Europe and North America. Comparison of the calculated critical loads with the sulfur deposition of 1995 in China (Hao et al, 1998) led to the critical load exceedance map of sulfur deposition (Figure 6).

As we can see from this map, sulfur deposition exceeds critical load in a wide land area that amounts for 25% of total Chinese ecosystems, which mainly refers to the southeast of China. Among these areas, the exceedances are especially serious in the lower reaches of Changjiang (Yangtzee) River, in the Sichuan River Basin, and in the Delta of Zhujiang River.

4.3. Critical Loads of Sulfur in South Korea

Air pollution is still of crucial importance in South Korea. Emission of sulfur dioxide, nitrogen oxides and ammonia that have been rising in the last decades are responsible for air pollution in many parts of the country. The amount of sulfur dioxide emitted in the area of South Korea was 1,494,000 tons in 1995 and was gradually decreasing to 1,490,000 tons in 1997.

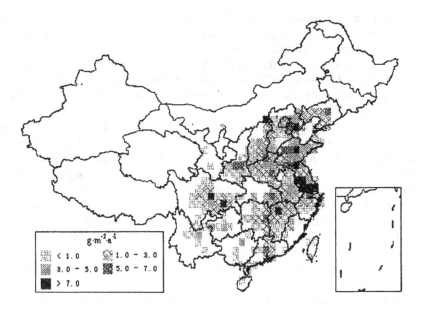

Figure 6. Sulfur critical load exceedance map of China (Hao et al, 1998).

However, it is well known that various pollutants including sulfur compounds can be transported by air from country to country in the whole Asian domain and especially in Northeast Asia. Thus, the model calculations have shown that in 1991–1994 about 35% of oxidized sulfur species deposited in South Korea was transported from somewhere, mainly from China (Sofiev, 1998). Accordingly, in spite of national reduction in SO_2 emission, the sulfur depositions are still very significant.

The preliminary research has shown a high sensitivity of most part of South Korean ecosystems to input of acid forming sulfur compounds (World Bank, 1994). This is generally related to acid geological rocks (predominantly, gneisses and granites acting as a soil parent material), acid soils with predominant light texture and low buffering capacity and sensitive Coniferous Forest (mainly pine species) ecosystems.

It has been argued that the best approach to the calculation and mapping of critical loads on ecosystems of East Asia is to use various combinations of expert approaches and geoinformation systems (EM GIS), including different modern techniques of expert modeling. However, this approach is difficult in practice due to a scarcity of data and insufficient understanding of basic controls over major pathways in the biogeochemical structure of many Asian ecosystems and their alteration under the influence of different air pollutants. Under the given conditions, the EM GIS can operate using databases and knowledge basis relative to the areas with different spatial data uncertainty. As a rule the given systems include an analysis of the cycles of various elements in the key plots, a choice of algorithms describing these cycles, and corresponding interpretation of the results. This approach requires cartographic data, such as vegetation, soil and geological maps, information on buffering capacities of soils, precipitation and runoff patterns.

Indeed this approach is shown to be the most appropriate for the Asian part of Russia, China, Thailand, and Taiwan, where, at present, adequate information on the great spatial variability of natural and anthropogenic factors is either limited or absent (Bashkin & Park, 1998).

During recent decades, significant efforts have been applied for monitoring of main environmental compartments in South Korea. The soil, geological, and vegetation maps were created in different scales, from 1:25,000 to 1,000,000. The monitoring networks of precipitation and their chemical composition were enlarged to cover the whole area of the country. However, till now there are significant gaps in precise understanding of biogeochemical structure of different terrestrial and aquatic ecosystems that governs the sustainability of these ecosystems to acidity loading.

The Comprehensive Acid Deposition Model (CADM) has been created for calculation of dry and wet deposition of sulfur species over South Korea (Park et al, 1997). This model presents the quantitative assessment of the acidity loading and alterations in deposition rates.

South Korea is in the southern part of the mountainous Korean peninsula between 126 °E–130 °E and 33 °N–38 °N. Geographically this is the northern temperate zone of the Eastern Hemisphere. The overall area of Republic of Korea (ROK) comprises 99,766 km^2.

In accordance with international approaches the following information has been collected for calculation of sulfur critical loads for all Korean ecosystems:

a. general information: soil map, scale 1:1,000,000; land use map, scale 1 × 1 km; geological map, scale 1:1,000,000;

b. soil chemical and physical parameters: soil texture; soil layer height; moisture content; soil bulk density; Mg + Ca + K content; log K gibbsite; pH; soil temperature; soil cation exchange capacity;

c. vegetation type parameters: annual net productivity growth; net uptake of nutrients (N, Ca, Mg, K, Na, S);

d. geochemical analysis: Ca, Mg, K, Na, Al, trace elements; base cation chemical weathering from soils and underlying geological parent material.

The calculation of critical loads of maximum sulfur was carried out in scale of 11 × 14 km cells. There are 665 cells on the area of South Korea.

The calculations were carried out for the 1994–1997 period that allows us to account for the temporal variations of temperature and precipitation as well as the dynamic pattern of sulfur and base cation depositions.

Using the above-mentioned constituents, the critical loads of sulfur were calculated for the natural terrestrial ecosystems over Korea. Geographical distribution of CLmaxS values is shown in Figure 7.

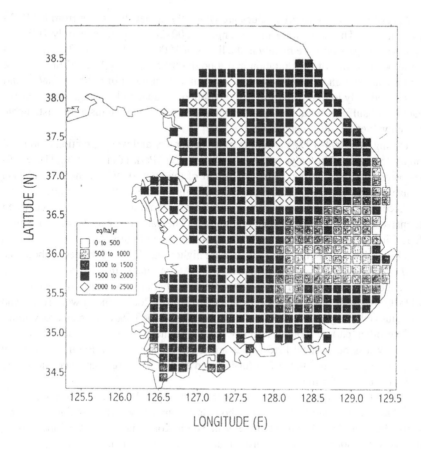

Figure 7. Critical loads of sulfur at terrestrial ecosystems of South Korea (Park and Bashkin, 2001).

We can see that about 71% of Korean ecosystems have the rank of 1000–2000 eq/ha/yr Very low values of critical loads (< 500 eq/ha/yr) are characteristic for insignificant part of ecosystems (1.0%), low values (500–1000 eq/ha/yr) are for 15.5% of ecosystems and on the contrary high values (> 2000 eq/ha/yr) are shown for 12.1% of considered ecosystem types.

These values are significantly different from those that have been earlier calculated by RAINS–ASIA model (World Bank, 1994). These differences might be related to the much more detailed and comprehensive national data sets on geological, soil, climate (precipitation, temperature, evapotranspiration, runoff, etc.) and vegetation mapping, physico-chemical properties of soils and geological rocks. This allows the authors to calculate more precise values of all constituents used for maximum sulfur critical load calculation and mapping.

Accordingly, our CLmaxS values are generally higher than those from RAINS–ASIA model. The latter were in the range of 200–2000 eq/ha/yr, mainly 200–500 eq/ha/yr, whereas ours are mainly in the limits of 1000–2000 eq/ha/yr. This is related to the peculiarities of precipitation and topography patterns in South Korea. These in turn lead to high values of surface runoff of annual mean of 6,200 m^3/ha/yr, and correspondingly to high values of acid neutralizing capacity leaching. The intensive leaching of sulfur from the soil profiles makes the local ecosystems more sustainable to high values of sulfur deposition.

During the 1994–1997 period the mean sulfur dry and wet deposition amounted totally to about 47 kg/ha/yr or about 3000 eq/ha/yr (Park et al, 1999b). These values were maximum in the southeastern part of the country, where the Pusan-Ulsan industrial agglomeration takes place and minimum in the northeastern part.

Accordingly, a significant part of the Korean ecosystems was subjected to an intensive input of S acid forming compounds. The values of exceedances of sulfur deposition over sulfur critical loads (ExS) are shown in Figure 8.

During 1994–1997 the Sdep values were higher than CLmaxS values at about one third of terrestrial Korean ecosystems (38%). Among them, the ExS values were in the range 176–500 eq/ha/yr for 16.1% of total number of ecosystems, in the range of 500–1000 eq/ha/yr were for 7.9%, in the range of 1000–2000 eq/ha/yr were 10.7% and the values even higher than 2000 eq/ha/yr were found for 3.5% of Korean ecosystems.

The other part of Korean territory (61.8%), where the sulfur depositions were relatively less but critical load values are relatively higher (see Figure 8), was not subjected to excessive input of sulfur-induced acidity. This area can be considered as sustainable to sulfur input.

As we have mentioned above, during the 1990s up to 30–35% of sulfur deposition was due to emission of SO$_2$ by transboundary sources, occurring mainly in China. Thus, the emission abatement strategy in South Korea has to be developed taking into account both local and transboundary emission reduction in the whole East Asian domain. The values of CL and their mapping present a good possibility for the creation of ecological optimization models. At present, these CL values and corresponding mapping have been carried out by national research teams in almost all the East Asian countries, such as China, Japan, South Korea, Asian part of Russia and Taiwan (Bashkin and Park, 1998). Accordingly, this national-based mapping can be considered as a scientific basis for decreasing local and regional air pollution in the East Asian domain.

4.4. Critical Loads for Southeast Asia

Taiwan Forest ecosystems. A RAINS-ASIA impact module is used to assess ecosystem sensitivity to acid deposition and to calculate critical load of sulfur to six forest ecosystems in Taiwan (Lin, 1998). Results indicate that forest ecosystems in Taiwan are very sensitive to acid deposition due to their low soil pH (< 5.5). Lowland subtropical forest ecosystems in Taiwan have low or moderate low critical loads for S suggesting that they are vulnerable to acid deposition (Table 3).

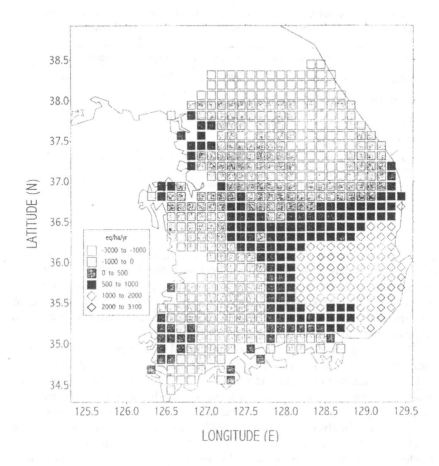

Figure 8. Exceedances of critical loads of sulfur over South Korea (Park and Bashkin, 2001).

Yet, many forest ecosystems are exposed to acid deposition far exceeding their critical loads. Although these forest ecosystems appear healthy, there may be a sudden detrimental change once the current buffering capacity is depleted. Cation leaching both from the forest canopy and forest soils is observed in some forest ecosystems. Continuous exposure to high levels of acid deposition can lead the forest to be in nutrient imbalance and thereby undermine forest health.

North Thailand Forest ecosystems. The input of < 1 meq/100 g soil to the forest soil in the northern part of Thailand, with the organic content of 1.33%, has no changes in pH value due to existing hydrogen buffering capacity. Simulated acid rain at pH 2 led to Al leaching from this soil naturally enriched by aluminum and iron, although there is no significant change of pH in the soil. The acid depositions cause also the reduction of bacteria, actinomyces and ammonifying microbes in the upper

Table 3. Data input for the calculation of critical loads (eq/ha/yr), the calculated critical loads and current deposition rates for S at six forest ecosystems in Taiwan.

Site	Fu-shan	Pin-lin	Lien-hua-chi	Pi-lu-chi	San-pan	Tai-ma-li
ANC_w (eq/ha/yr)	70	70	140	70	70	240
Q^1 (mm/yr)	2000	1600	1200	1300	2000	1250
K_{gibb}	150	150	150	300	150	150
CL (eq/ha/yr)	365	343	560	300	365	890
DR^2 (eq/ha/yr)	2800	2700	1400	650	1750	800

[1] Q: runoff, the difference between rainfall and evaportranspiration;
[2] DR: deposition rate.

part of the soil. These could lead to low nutrient cycling rate in the ecosystem as these microorganisms play a significant role in organic matter decomposition in the soil. Owing to certain buffering ability of soils to acid depositions the forest ecosystem can be sustainable in short-term periods, however some negative chemical and biological changes occur in soil. This will gradually decrease the ecosystem sustainability to acidic loading (Kozlov and Towprayoon, 1998).

This and other researches carried out in North Thailand provides the results of sensitivity assessment in term of maps for the study area, and to determine and identify the sensitive receptors and locations where abatement strategies would be implemented to reduce environmental impacts of acidification on forest and agriculture. Sensitivity of ecosystems to acid deposition was carried out using a two step procedure. Firstly, the sensitivity mapping according to revised methodology of Stockholm Environment Institute has been conducted for Thailand conditions (see Kuylenstierna et al, 1995 for details). The purpose of such sensitivity mapping is to define the distribution of ecosystems with the same relative level of reactions to the given rate of acidic deposition. On the second stage the sensitivity of Thailand ecosystems to acidic deposition has been described by means of Critical Loads (CL) and exceedances, using a modified Steady-State Mass Balance model with simplified expert-modeling approach (Figure 9).

The maps for CL of sulfur derived from this methodology have been overlaid with current (or projected future) deposition maps in order to show areas where the CL of sulfur is (or will be) exceeded. In order to manipulate the numerous maps and data a geographical information system (GIS) was used (Kozlov and Towprayoon, 1998). As a result of both the high sensitivity of ecosystems and level of exceedances across Northern Thailand, more than 75% of the ecosystems across about 50% of this territory is at significant risk from acid deposition.

Thus, the acidity of precipitation has increased during the latest decades in many developed and developing countries of Asia. There exist clearly transboundary air pollution problems. The harmful single and synergetic effects of acidity, SO_2, NO_x

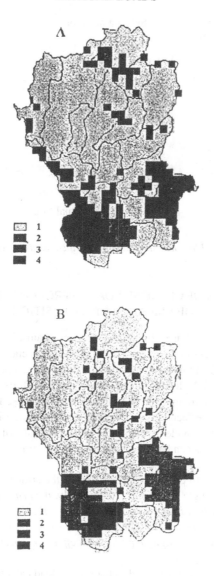

Figure 9. Critical loads of acidity on the ecosystems in Southeastern Asia (Northern Thailand).

and O_3 are experimentally shown for various natural and agricultural ecosystems. The critical load concept is shown to be a good guide for assessing ecosystem sensitivity to acid deposition in many East Asian countries. The calculated critical loads are exceeded at present acidity loading for many ecosystems in East Asian countries (Russia, China, Thailand, Taiwan, Japan, South Korea).

Select a receptor
↓
Determine the critical limits
↓
Select a computation model
↓
Collect input data
↓
Calculate critical loads
↓
Compare with the actual load

Figure 10. Flowchart for calculating critical loads of heavy metals.

5. APPLICATION OF CRITICAL LOAD APPROACH FOR CALCULATION OF HMs AND POPs DEPOSITION

We have shown already that anthropogenic loading changed significantly the natural biogeochemical cycles of many heavy metals (HM), especially those like mercury or lead (see Chapter 12). In order to make adequate environmental policies concerning the reduction of the HM loading to the environment, efforts for the assessment of the effects of such pollutants and the identification of those loads of heavy metals below which harmful effects no longer occur, the so-called critical loads, is needed. Accordingly, we will provide here only the idea that these critical loads at terrestrial and aquatic ecosystems can be also calculated using similar approaches as for S and N acid.

On the contrary with critical loads for acidity, the critical loads for heavy metals and POPs refers to single metal only. Accordingly, *the critical loads equal the load causing a concentration in a compartment (soil, soil solution, groundwater, plant, animal and human organisms, etc.), that does not exceed the critical limits set for HM and POPs, thus preventing significant harmful effects on specified sensitive elements of the biogeochemical food web.*

Discussing the problems, related to the critical load calculation, attention should be paid to (i) selection of a receptor of concern, (ii) critical limits, (iii) possible calculation methods, (iv) the necessary input data and (v) the various sources of error and uncertainty (De Vries and Bakker, 1998).

The following flowchart is useful for calculation of critical loads of heavy metals and persistent organic pollutants for both terrestrial and aquatic ecosystems (Figure 10).

The application of critical load approach for HMs and POPs is discussed in more detail in (De Vries and Bakker, 1998; Bashkin and Howarth, 2002).

FURTHER READING

1. Posch M., de Smet P. A. M., Hettelingh J.-P. and Downing R. J., 1999. *Calculation and Mapping of Critical Thresholds in Europe*, Status Report 1999. CCE/RIVM, Bilhoven, the Netherlands, 166 pp.

2. De Vries W. and Bakker D. J., 1998. *Manual for calculating critical loads of heavy metals for terrestrial ecosystems.* DLO Winand Staring Centre, Report 166, The Netherlands, 144 pp.

3. De Vries W., Bakker D. J. and Sverdrup H. U., 1998. *Manual for calculating critical loads of heavy metals for aquatic ecosystems.* DLO Winand Staring Centre, Report 165, The Netherlands, 91 pp.

4. Bashkin V. and Park S.-U. (Eds.), (1998) *Acid Deposition and Ecosystem Sensitivity in East Asia*, Nova Science Publisher, 427 pp.

5. Sverdrup H. and de Vries W., 1994. *Critical loads of acidity to Swedish forest soils.* Report 5:1995, Lund University, Sweden, 104 pp.

6. Bashkin V. and Howarth R., 2002. *Modern Biogeochemistry.* Kluwer Academic Publishers, Chapter 10.

WEBSITES OF INTEREST

1. Critical loads calculation and mapping,
   http://www.rivm.nl/cce

QUESTIONS AND PROBLEMS

1. Discuss the applicability of critical load concept in various countries of Asia. Present relevant examples of CL calculations and mapping.

2. Characterize the role of China in transboundary pollution in East Asia. Describe the CL and Ex maps on the area of China. Indicate the reasons of differentiated sensitivity of Chinese ecosystems to acid deposition.

3. Discuss the acid rain problems in South-East Asia. Focus your attention on acid deposition effects on biogeochemical cycling in Tropical Rain Forest ecosystems.

4. Discuss the interactions between acid deposition and accelerating migration of heavy metals in biogeochemical food webs. Present examples.

5. Outline the applicability of critical load concept for setting environmental standards for heavy metals.

6. Describe the flowchart for CL calculation for heavy metals. Compare the various approaches.

7. What data are of crucial importance for calculating CL in countries like Russia or China? Discuss the application of GIS technologies for CL calculation and mapping.

8. Discuss the calculation and mapping of critical loads of any pollutants in your country. Outline the existing problems of such calculations.

9. Discuss the political and economic limitations of applying CL concepts in your country. What alternative approaches do you know? Discuss the pros and cons of various approaches to emission abatement strategy in your country or region.

10. Compare the biogeochemical approaches to setting environmental standards with other methods like ecotoxicological experimental testing. Discuss the advantages and drawbacks of different approaches.

CHAPTER 16

WASTE TREATMENT

1. INTRODUCTION

This chapter consists of four main parts: (i) sources of wastes and wastewater in the Asian region; (ii) treatment of both solid waste and wastewater; (iii) recycling as an emerging area of waste minimization; and (iv) remediation of polluted areas.

With expanding economic activity and consumption of consumer goods, quantities of wastes are increasing rapidly in the Asian region. Waste is an unwelcome and often non-avoidable side effect of rapid development. At present, the Asian region houses about 60% of the global 6 billion people. Uncontrolled urbanization, rapid industrialization and extensive agricultural practice are generating huge quantities of various wastes. Furthermore, the ecological jurisdiction in many Asian countries is still much more diffuse in comparison with developed countries, thus stimulating different companies to construct branches of waste-generating facilities in the Asian countries without corresponding industrial waste and wastewater treatment capacities. These factors make the waste problem of crucial environmental concern in almost all the Asian countries.

We can not here discuss the whole spectrum of waste production and treatment in the Asian region. Only some fundamental principles will be discussed with rather arbitrary selection of waste generation system and waste treatment facilities. You can find more information from additional recommended books and websites of interest at the end of this chapter.

2. TYPE AND SOURCES OF WASTES IN THE ASIAN REGION

Primary sources of wastes are municipal, agricultural, commercial and industrial, hospitals and other health-care sectors. Accordingly, major categories of waste are agricultural waste and residues, municipal waste, industrial waste and hazardous waste. Major components of waste are organic and inorganic constituents, nutrients, toxicants and different species of micro-organisms. Wastes are presented in solid, semi-solid (sludge) and liquid forms with one or more of the following characteristics: corrosive, ignitable, reactive, toxic, radioactive, infectious, bioaccumulative, carcinogenic, mutagenic and teratogenic (Figure 1).

Figure 1. Types and characteristics of waste (ESCAP, 2000).

2.1. Municipal Solid Waste

Types of municipal solid waste

In the Asian cities, municipal solid waste (MSW) is generated from residences, commerce, institutions, construction and demolition, cleaning services and treatment plants. Generation rates of MSW vary from city to city and season to season. In most of the Asian cities, these rates range from less than 0.5 to 0.8 kg per capita per day. Some cities have higher generation rates of more than 1.0 kg per capita per day.

Generation of MSW depends usually on the level of economic development of the country. The MSW production in high-income countries (Japan, Hong Kong, South Korea, and Singapore) is between 1.5–2.5 kg per capita per day, in middle-income countries (China, Thailand, Malaysia etc.), 0.7–0.9 kg per capita per day and in low-income countries (like Bangladesh, Sri Lanka, Vietnam, Myanmar etc.), 0.4–0.6 kg per capita per day. The trends in the total MSW production for different groups of countries are shown in Figure 2.

Gross annual generation of MSW in some selected Asian countries is depicted in Figure 3.

Composition of municipal solid waste

The composition of MSW varies significantly from city to city. The compostable waste fraction represents typically more than 25% in most of the Asian cities, followed by non-compostable waste, paper and cardboard, and plastics, as shown in Figure 4.

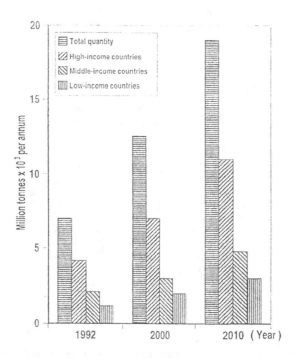

Figure 2. Municipal solid waste generation in different groups of countries in the Asian region (ESCAP, 2000).

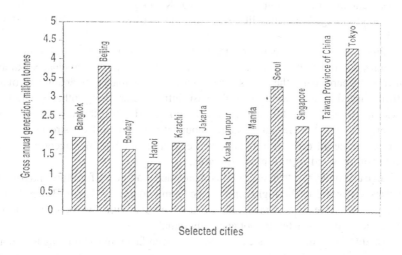

Figure 3. Gross annual generation of MSW in some selected Asian countries (ESCAP, 2000).

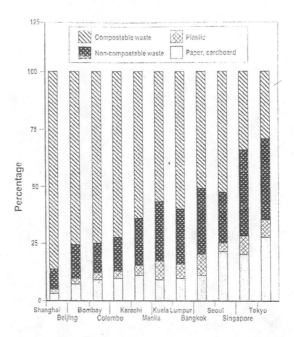

Figure 4. Approximate composition of municipal solid waste in selected cities of the Asian region (ESCAP, 2000).

The moisture content of MSW is generally between 35 and 55% and the caloric values are also variable, from typical 500–700 Kcal kg^{-1} to much higher values of 800–1500 Kcal kg^{-1} in cities like Bangkok, Jakarta, Manila and Tokyo (ESCAP, 2000).

Climate and fuel use are also related to the composition of municipal solid waste. The cities where heating is needed in winter, such as Beijing, Shanghai, Seoul and Tokyo and where coal is the main source of energy, have much greater amounts of ash in their waste during the wintertime. The basic infrastructure brings other variations in cities and towns, such as Calcutta, Dhaka (India), and Hanoi (Vietnam) with unpaved or poorly paved streets that have a large amount of dust and dirt from street sweeping. There are large differences in organic waste fraction among cities according to the number of trees and shrubs in public places.

Large and bulky waste items, such as abandoned motor cars, furniture and packing residues are found in the higher-income economy countries (Japan, Brunei Darussalam, South Korea, and Singapore), on the contrary with low-income countries, where such type of wastes are not generally found.

The amount of human waste in the MSW is significant in squatter areas of many Asian cities, like Hanoi, Calcutta, Dhaka, etc., where "wrap and throw" sanitation is common or bucket latrines are emptied into waste containers.

Table 1. Wastewater production in various Asian countries (ESCAP, 2000).

Country	Wastewater, million m^3 per day
Singapore	1,0
South Korea	15.0
Malaysia	0.0025
China	0.145
India	12.0

Municipal wastewater

Municipal wastewater is a combination of liquid and water-carried waste from residences, commercial buildings and institutions. This is actually the wastewater arising from human sanitary water use and a combination of water coming from kitchen, toilets, bathrooms, laundries, etc., where it may have been used for washing, cooking, food preparation and cleaning, showers, toilet flushing, etc.

The important parameters for proper management of municipal wastewater are the quantity and the chemical composition of wastewater. The quantity can be estimated using a number of factors, such as population, water supply data, housing type and likely presence of washing machines and dish-washing machines, presence of hotels, restaurants, hospitals, offices, schools, colleges and universities. Using this information, approximate wastewater quantities can be calculated to design a proper wastewater treatment plant.

Various countries of the Asian region produce different quantities of municipal wastewater and sewage (Table 1).

We can see that developed countries, like South Korea, produce much more wastewater in spite of much less population (44 million in 2000) than multi-million China.

The composition of municipal wastewater is quite variable, being highly dependent on the local conditions. In addition to any materials present in the home water supply, it will contain different contaminants, like feces, urine, paper, soap, oil and grease, dirt and soil, food waste, and home used chemicals. Typical composition of an untreated municipal wastewater includes dissolved solids, 140–1200 mg L^{-1}, organic nitrogen, 8–35 mg L^{-1}, ammonia, 2–50 mg L^{-1}, organic phosphorus, 1–5 mg L^{-1}, inorganic phosphorus, 3–10 mg L^{-1}, chloride, 30–100 mg L^{-1}, oil and grease, 50–150 mg L^{-1}. These waters contain also a significant number of various microorganisms, including some poisoning species (ESCAP, 2000).

2.2. Industrial Waste

We will consider briefly sources of both industrial solid waste and industrial wastewater in the Asian region.

Industrial solid waste

Various industries in the Asian region can be grouped into following broad categories:

1. iron and steel industries;
2. chloralkali industries;
3. sulfuric acid and nitric acids plants;
4. hydrochloric acid and fluorochemical plants;
5. petroleum oil refineries;
6. fertilizer industries;
7. mining industries;
8. metal fabricating industries;
9. surface coating industries;
10. pesticide industries;
11. pharmaceutical industries;
12. natural rubber industries;
13. palm oil mills;
14. dyes and intermediates industries;
15. textile industries;
16. leather tanneries;
17. fermentation industries;
18. sugar industries;
19. food processing industries, and
20. jute and coir industries.

As a result of relevant industrial activities, huge quantities of solid waste are generated in the Asian region with a very wide spectrum of waste materials of different environmental toxicity. Due to the absence of a regularly up-dated and systematic statistical database on solid waste industrial generation from different groups of industry, the exact rates of generation are very uncertain for individual industries. However, we can conclude that the typical waste products include rubbish, paper, packaging materials, waste from food processing, oils, solvents, resins, paints and sludge, glass, ceramics, stones, metals, plastics, rubber, leather, wood, cloth, straw, abrasives, etc.

Ratio between municipal and industrial solid waste generation varies significantly from country to country. For example, in China this ratio is three to one, whereas in Bangladesh, Sri Lanka and Pakistan, much less. In developed countries, like Japan, this ratio is equal to 1 : 8. Based on an average ratio for the Asian region, the industrial solid waste generation is approximately 1.5 billion ton per annum (ESCAP, 2000). These data are depicted in Figure 5.

The amount is expected to increase significantly with planned economic growth in the region, doubling the current generation rates. This incremental growth will pose very serious environmental problems because existing industrial solid waste collection, processing and disposal systems are grossly inadequate and the required improvement of solid waste treatment is unlikely in the near future.

Industrial wastewater

Industrial wastewater is produced by all industries, which use water in different technological processes. The generation of this wastewater varies depending upon nature of industries, their production processes and water usage. The present estimate shows

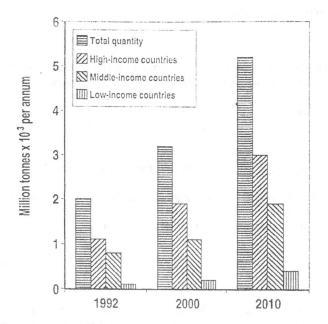

Figure 5. Estimated generation of industrial solid waste in different income-groups of countries in the Asian region (ESCAP, 2000).

that huge quantities of wastewater are generated both in developed and developing countries of the Asian region.

However, the generation rates are very different. China, for example, generates over 2 kg of wastewater for an industrial production equivalent of one US dollar, followed by Japan, 1 kg, South Korea, 0.5 kg, Hong Kong, 0.3 kg and Singapore, 0.2 kg (ESCAP, 2000).

The composition of industrial wastewater is variable as flow rates with type and size of industry. The typical chemicals of environmental concern in wastewater are shown below (the list is incomplete):

- acrylonitrile, arsenic, and aldrin/dieldrin;

- benzene, benzidine, benz(a)pyrene, beryllium, bis(2-chloroethyl) ether, bromodi-chloromethane, bromoethane, and bromoform;

- cadmium, carbon tetrachloride, chlordane, chloroform, chloromethane, chromi-um, copper, cyanides;

- DDT, dibromochloromethane, dibutyl phtalate, 1,2-dichloroethane, 1,1-dichlro-diethyl phtalate, di-*n*-butylphtalate, 2,4-dinitrophenol, 1,2-diphenylhydrazine;

- endosulphan, endrin, ethylbenzene;

- fluoride;

- heptachlor, heptachloroepoxide, hexachlorobutadiene, α-,β-hexachlorocyclohexane, hexachlorocyclopentadiene, hexachloroethane, hydrogen cyanide;

- iron, isophorone;

- lead, lindane;

- malathion, mercury, methylene chloride, molybdenum, mirex;

- nickel, nitrobenzene, N-nitrosodimethylamine, N-nitrosodi-n-propylamine;

- pentachlorophenol, phenol, PCBs, PAHs, photomirex;

- total hydrocarbons;

- VOCs;

- 2,3,7,8-TCDD, 1,1,2,2-tetrachloroethane, trichloroethylene, toluene, toxaphene, 1,2,3-trichlorobenzene, 1,1,1- and 1,1,2-trichloroethylene, 2,4,6-trichloroenol, etc.

We can add here, that this typical chemical composition is complicated by high concentrations of both dissolved and suspended solids, nutrients, acids, oil and greases.

2.3. Agricultural Waste

Agricultural waste and residues

The expanded production of a range of agricultural crops has naturally resulted in the generation of annually increased quantities of waste (including manure and animal dung) and residues in the form of livestock waste, agricultural crop residues and agro-industrial by-products. The approximate estimate shows that among the Asian countries, China produces the largest quantities of both agricultural waste and crop residues, followed by India (Table 2).

Plastic waste

Another form of agricultural waste produced in some countries of the Asian region is plastic. On-farm plastic use appears to be rapidly increasing in the form of low-density polyethylene (LDPE) agricultural plastic film. The use of this film is extensively practiced in South Korea, China and Japan as a means to maintain soil temperature favorable to plant growth, retain moisture and reduce weeds. LDPE agricultural plastic film is an increasingly popular alternative to more traditional methods of storage and cover. It is used in agricultural operations for silo bunker cover, silage bags, bale

*Table 2. Approximate estimate of annual production of agricultural
waste and residues in selected Asian countries (ESCAP, 2000).*

Country	Annual production, million tons		
	Agricultural waste	Crop residues	Total
Bangladesh	15	30	45
China	255	587	842
India	240	320	560
Indonesia	32	90	122
Malaysia	12	30	42
Myanmar	28	4	32
Nepal	4	12	16
Pakistan	16	68	84
Philippines	20	12	32
Sri Lanka	6	3	9
South Korea	15	10	25
Thailand	25	47	72

wrap, greenhouse cover, haulage cover, row cover, and mulch film. Crops for which
LDPE is used include cotton, maize, tobacco, juvenile rice and fruit tree seedlings,
melon, vegetables and groundnuts. In China alone, the use of plastic in agricultural
practices has increased almost 70 times since 1990. Similar numbers have been shown
for South Korea (ESCAP, 2000).

Aquaculture waste

In the Asian region, marine and freshwater aquaculture farms are producing almost
90% of the World's farm-bred seafood. China accounts for about half of the region
output, followed by India, Thailand, Vietnam and Japan. Huge quantities of wastewater
are produced from freshwater aquaculture. At present, the exact values are not
available, but it is expected that these figures will be doubled by 2010.

2.4. Hazardous Waste

With rapid development of industry, agriculture, hospital and health-care facilities,
the Asian region is consuming significant quantities of toxic chemicals and producing
a large amount of hazardous waste. Each year, another 1000 new chemicals are added
to the market for industrial and other uses.

Most hazardous waste is the by-product of a broad spectrum of industrial, agricultural and manufacturing processes, nuclear power plants, and hospitals. Primarily,

Table 3. Estimate of annual production of hazardous waste in some Asian countries (after ESCAP, 2000).

Country	Estimated annual production, thousand tons		
	1993	2000	2010
Bangladesh	109	275	514
China	50,000	130,000	250,000
Hong Kong	35	88	165
India	39,000	82,000	156,000
Indonesia	5,000	12,000	23,000
Japan	82	220	415
Malaysia	377	940	1750
Pakistan	786	1735	3100
Philippines	115	285	530
Singapore	28	72	135
Sri Lanka	114	250	460
South Korea	269	670	1265
Taiwan	3,000	6,700	12,000
Thailand	882	2,215	4,120
Vietnam	460	910	1,560

high-volume generators of industrial hazardous waste are the chemical, petrochemical, petroleum, metals, wood treatment, pulp and paper, leather, textiles and energy production plants, especially coal- and lignite-burning plants. Municipal waste also contains hazardous acids and solvents, dry cleaners and pesticide residues, photochemical waste, plastics, medical waste and paint sludge.

Among the prominent groups of hazardous waste that are generated in the Asian region are waste solvents, chlorine bearing waste, and pesticide formulation waste and residues. Chlorine is the primary ingredient of PVC formulation. In the Asian countries, chemical industries are using about 20% of the global production of chlorine and caustic soda, 10 and 12 million tons respectively for the formulation of around 11,000 organochlorine species.

The estimated quantities of hazardous waste in some selected Asian countries are shown in Table 3.

We can see a significant increase of hazardous waste production in planning for China, India, Indonesia, Philippines and Thailand by year 2010.

Huge amounts of petroleum waste are produced annually in some countries of the region, like Brunei Darussalam, China, India, Indonesia, Malaysia, Japan and South Korea. Two latter countries and Taiwan produce significant quantities of radioactive waste.

Hazardous waste could be solid, semi-solid (sludge), liquid and gaseous and possesses variable characteristic properties such as toxicity, reactivity, flammability, explosivity, corrosivity and radioactivity.

2.5. Transboundary Movement of Hazardous Waste to the Asian Region

The Asian region has become the favorite dumping ground for hazardous waste exporters. From 1994 to 1997, industrialized countries have sent totally 3.5 million tons of hazardous waste or harmful trash to the Asian countries.

For example, waste export to China has been documented from United State, Japan, South Korea and Taiwan (ESCAP, 2000). Another example is related to India. Cheap labor, poor environmental standards, a sieve-like import regime and a growing market for cheap raw materials are the main driving forces. Ignoring its law courts, India is thus helping developed countries to beat the International Basel Convention on the ban on the damping of toxic industrial waste in developing countries. Thousands of tons of toxic waste are being illegally shipped to India for recycling or dumping, including 73,000 tons of toxic lead and zinc residues from 49 countries. In 1995, Australia alone exported more than 1,450 tons of hazardous waste such as scrap lead batteries, zinc and copper ash to India.

Huge amounts of PVC waste produced in the Netherlands are regularly shipped to Pakistan and the Philippines. Since 1990, more than 100,000 tons have been exported for local recycling into various products of poor quality, which end up in burning dumpsites in a few years.

Similarly, other countries of the region, such as Bangladesh, Pakistan, Indonesia, and Thailand have become dumping grounds of huge quantities of hazardous waste for the exporters of industrialized countries both within and outside the Asian region (ESCAP, 2000).

2.6. Environmental Impacts of Waste

The adverse environmental impacts of waste have been well documented in the form of water (riverine, coastal and marine, and ground) pollution, air pollution and land pollution (see Chapters 7–9). Many cities like Bangkok, Beijing, Bombay, Calcutta, Dhaka, Hanoi, Jakarta, Kuala Lampur, Manila and Shanghai are plagued by garbage problems. Bangladesh, India, Indonesia, Malaysia, Pakistan, the Philippines and Thailand are faced with growing problems of both domestic and industrial hazardous waste. Rivers are choked by untreated domestic and industrial wastewater in many Asian countries.

Some industrial waste, biomedical and nuclear waste constitute the most obvious categories of hazardous products, which are detrimental for human health and environment. Agricultural waste includes residues of fertilizers and pesticides with known toxicity (see Chapter 11).

Various incidents of pollution have been reported from industrial waste, effluents from sewage treatment plants, food processing plants along with biocides and toxic effluents from sawmills and timber processing areas. These effects are often enhanced by disposal of hazardous wastes in municipal landfills.

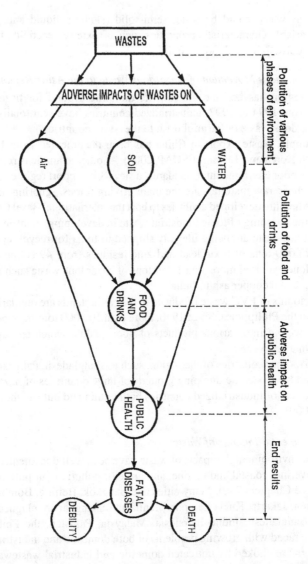

Figure 6. Various adverse environmental impacts of wastes and their end results (ESCAP, 2000).

The general scheme of combined impacts of hazardous waste on the human being and environment is shown in Figure 6.

Among these are direct transmission of disease and the spread of epidemics, loss of healthy amenable rural and urban environment. The potential spread of AIDS and other infectious diseases through the discharge of biomedical and health-care wastes into general municipal waste streams is a growing threat in the Asian region. Waste containing oils, pesticides and heavy metals is a major contributor to soil pollution and land contamination in many Asian countries. Other environmental impacts include the adverse effects of windblown and floating debris and plastics on crops, birds and aquatic biota.

3. WASTE TREATMENT IN ASIAN COUNTRIES

Here we will consider municipal wastewater treatment, land disposal and incineration of solid municipal waste.

3.1. Municipal Wastewater Treatment

Municipal wastewater collection, treatment and disposal is a basic problem in most Asian countries, especially in urban areas. Wastewater is generally collected by sewer systems from residential, commercial, recreational, and institutional areas and is transported to a central sewage treatment plant, where sewage undergoes physical, chemical and biological treatment. Such systems exist in Seoul, Tokyo, Singapore, Ankara and other cities. However, in many cities of the region, the existing sewarage systems are old. Maintenance has often been poor due to lack of funds and leakages into ground and surface water are common. In a few cases, where sewers and water mains have been placed in close proximity, cracks in both lines have led to sewage contamination of water supplies, particularly, during periods of peak flow.

Typical municipal sewage contains oxygen-demanding materials, sediments, grease, oil, scum, pathogenic bacteria, viruses, salts, algae nutrients, pesticides, refractory organic materials, heavy metals, and an astonishing variety of flotsam.

Current processes for the treatment of wastewater may be divided into three main categories of primary treatment, secondary treatment, and tertiary or advanced treatment (Figure 7).

Primary treatment

The sewage enters a lagoon, or clarifier, whose capacity is large enough to allow a residence time of several hours for wastewater in this lagoon. A coarse screen at the entrance to the clarifier removes large objects, such as pieces of wood, tree branches, and similar debris. A grid tank ahead of the main clarifier allows the deposition of grit, sand, and similar materials. Typically, the sewage enters and leaves at opposite ends of the lagoon, and moves through slowly enough for settling down of any solid substances. Some greasy material may float to the surface, where it is removed by a skimming device. The effluent from the primary settler is almost transparent. However, it has typically very high BOD values, of the order of 500–1500 ppm. The major sources of BOD are the tiny particles, which fail to separate during primary settling. These particles can be coagulated and settled upon the addition of filter alum or ferric chloride. This may reduce BOD significantly, up to the level of 50–150 ppm. In some cities, this coagulation stage is applied as an alternative to the secondary treatment.

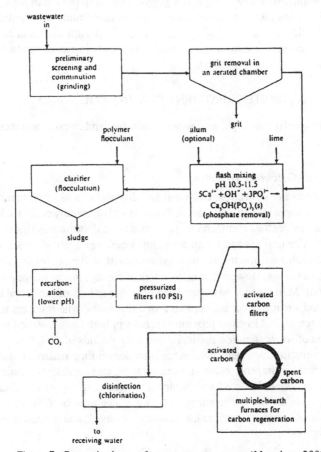

Figure 7. General scheme of wastewater treatment (Manahan, 2000).

Solids removed at the primary stage are collected and deposited at landfill.

Secondary treatment

The main objective of the wastewater secondary treatment is to reduce the BOD values. The successful reduction may achieve 90% from the initial level. We will consider two common approaches that are in use in the Asian region, the trickling filter and the activated sludge reactor (see Figure 8).

The trickling filter is a large round bed of sand and gravel, with coarse gravel on the top and successively finer layers beneath. A rotating boom sprinkles the bed with water from the primary settler. The gravel bed quickly becomes colonized with micro-organisms, which use the carbon compounds in the water as the energy materials. A well-maintained trickling bed may remain in operation for several decades (Figure 8).

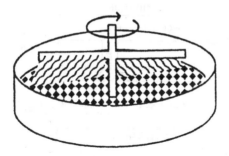

Figure 8. Trickling filter for secondary wastewater treatment (Manahan, 1994).

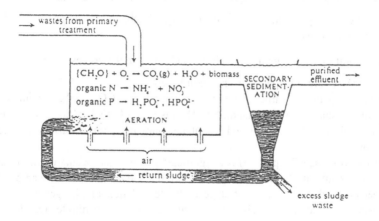

Figure 9. Activated sludge process (Manahan, 1994).

An alternative method of secondary treatment is the activated sludge reactor. The reactor is a large tank, in which the wastewater is agitated and aerated to provide the oxygen required by the micro-organisms. Wastewater aeration is the principal difference with the trickling filter, where oxygen diffuses naturally from the air. In order to keep the concentration of micro-organisms high and in the maximal growth phase, a portion of the sludge of the micro-organisms is removed from the exit stream of the reactor and recycled into the influent stream (Figure 9).

The advantages of the activated sludge reactor in comparison with the trickling filter are related to the maintenance at the optimum temperature for biological activity and relatively small operational space. The first point is important for countries with moderate climate, like South Korea, Japan and partly China, whereas the second one, for many Asian cities (Hong Kong, Singapore, Seoul, etc.) where land is expensive.

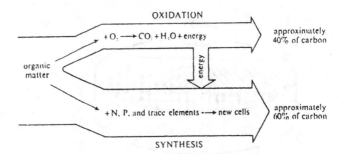

Figure 10. Pathways for the removal of BOD in biological wastewater treatment (Manahan, 1994).

The activated sludge process provides two pathways for BOD reducing:

- oxidation of organic matter to provide energy for the metabolic processes of the micro-organisms;

- incorporation of the organic matter into cell mass in the process of biosynthesis.

These processes are schematically shown in Figure 10.

In the first pathway, carbon is removed as the gaseous CO_2, in the second, as the solid biomass that needs to be landfilled. The disposal of this biomass is a problem, since it is only about 1% solids and it contains many undesirable hazardous species (see below).

Nitrification (see Chapter 13) is a significant process that occurs during biological waste treatment. Ammonium ions are normally the principal inorganic nitrogen species produced at the first stage of biodegradation of nitrogenous organic compounds. Ammonia is oxidized first to nitrite and then to nitrate (reactions (1) and (2))

$$2\,NH_4^+ + 3\,O_2 \longrightarrow 4\,H^+ + 2\,NO_2^- + 2\,H_2O, \tag{1}$$

$$2\,NO_2^- + O_2 \longrightarrow 2\,NO_3^-. \tag{2}$$

Both these reactions occur in the aeration tank of the activated sludge plant and are favored in general by long retention time, low organic loading, large amounts of suspended solids, and high temperatures. The subsequent process, denitrification, removes nitrate from wastewater and occurs in the anaerobic condition (reaction (3))

$$4\,NO_3^- + 5\,CH_2O + 4\,H^+ \longrightarrow 2\,N_2(g) + 5\,CO_2(g) + 7\,H_2O. \tag{3}$$

Sewage sludge

Both primary and secondary sewage treatments involve settling of particulate matter, and thus produce sludge. The term sludge refers to a material having a high water content (95–99%), even when is has been air-dried. Dewatering is aided by heating (digesting) the sludge, which causes the small particles to coagulate. Digestion is anaerobic, and is accompanied by the release of reduced gases, such as methane. The dewatered sludge with much coarser particles can be air-dried to a material that is really solid. This material is disposed of by landfilling, by incineration, by ocean dumping, or by spreading it on the land for use as a fertilizer. These methods are applied in the Asian region in various extents depending on the country.

Sewage sludge is rich in organic matter, nitrogen and phosphorus and may be used as fertilizer and soil-conditioner. It may be applied for the remediation of disturbed soil cover, for instance, in the areas of mining. The typical rates are of hundreds of tons per hectare. In these areas, artificial landscapes can be made up by planting trees and grasses.

The actual restriction for application of sludge is high transportation cost and high content of water (up to 60–65%) even in the dewatered material.

A potential drawback to the use of sewage sludge as fertilizer in agricultural fields is the presence of both organic and inorganic toxic compounds. The former compounds are oxidation-resistant organic substances, such as organochlorine species, which become bound in the organic matrix of the sludge. The inorganic toxicants are represented by heavy metals, mainly arsenic, cadmium, lead, mercury and zinc. These metals can be taken up by crops and introduced in the food chains or leached to the groundwater.

The following considerations govern the safe use of sewage sludge as a fertilizer in the Asian region:

- the amount of sludge which may be safely applied varies with the soil type, being higher in clay soils and lower in the sand soils;

- agricultural soils that are fertilized with sewage sludge should be analyzed regularly for content and accumulation of toxic metals in plant-available forms;

- sludge should never be applied to a growing crop, otherwise the crop will uptake any toxic species before they have been immobilized by the soil mass.

Tertiary wastewater treatment

Tertiary wastewater treatment or advanced waste treatment is a term used to describe a variety of processes performed on the effluent from secondary waste treatment. The contaminants removed by tertiary waste treatment typically are suspended solids, dissolved organic compounds and dissolved inorganic materials, including nitrogen and phosphorus as the important algae nutrients. Each of these categories presents its own problems with regards to water quality. Suspended solids are primarily responsible for residual biological oxygen demand in secondary sewage effluent waters. The dissolved organics and heavy metals are the most hazardous toxicants in the effluent.

In addition to these chemical pollutants, secondary sewage effluent often contains a number of pathogens (disease-causing micro-organisms), requiring disinfection. Among the bacteria that may be found in the secondary waste water effluent are organisms causing tuberculosis, dysenteric bacteria (*Bacillus dysenteriae, Shigella dysenteriae, Shigella paradysenteriae, Proteus vulgaris*), cholera bacteria (*Vibrio cholerae*), bacteria causing typhoid fever (*Salmonella typhosa, Salmonella paratyphi*), and bacteria causing mud fever (*Leptospira icterohemorrhagiae*). Various viruses, causing diarrhea, eye infections, infectious hepatitis, and polio are also encountered in many Asian countries.

Tertiary wastewater treatment includes different physical-chemical processes:

- removal of scum and tiny solid objects;

- clarification, generally with addition of a coagulant, and frequently with the addition of other chemicals, for example, lime for phosphorus removal;

- activated carbon absorption;

- disinfection.

Two methods are in use to reduce the BOD below 50 ppm O_2, which is typical for secondary sewage effluent, microstraining and coagulation. Microstraining involves forcing the water through very fine (μm) screens made from stainless steel. The particles are trapped on the screens, and have to be removed by back-flushing whenever the pressure needed to filter the water becomes excessive.

Coagulation process is very similar to that discussed in the advanced primary treatment and drinking water purification (see Chapter 10).

Superchlorination of wastewater effluent by high dose of chlorine may be applied to kill any remaining pathogens. Some by-products of such superchlorination, like $CHCl_3$ and organochlorines, are unavoidable and this can restrict the application of superchlorination in the wide scale. However, in the Asian region the threat of microbiological pollution is so much more pronounced that by-product formation and superchlorination must be applied in many treatment plants.

3.2. *Traditional Methods of MSW Processing*

Disposal of municipal, agricultural, industrial and biomedical wastes is given relatively low priority in many countries of the Asian regions despite increasing loads. It has been estimated that only 70% of the municipal waste in the urban areas are collected, of which only 5–10% is properly disposed. There exist limited actions in reducing, reusing and recycling of wastes in the region. Wastes are disposed in a variety of ways. Apart from reuse and recycling, these include animal feeding; indiscriminate dumping on land and in mangroves, creeks, lagoons, coastal waters, and on drains and roadsides; land burying, landfilling, composting, open burning and incineration. In general, the problems of indiscriminate dumping are much more common near major urban centers. In rural areas and suburbs, open burning, composting and land spreading and burying are common practices.

Various disposal methods of MSW in the Asian region include animal feeding, composting, vermi-composting, dumping, landfilling, open burning and incineration.

Animal feeding

A large portion of the household organic waste in China, India, Indonesia, Malaysia, the Philippines, Thailand and Vietnam is eaten by farm animals or fed to pets. Pig and poultry farmers routinely collect food waste from households, hotel and restaurants for animal feed.

Composting

Backyard composting of the organic portion of MSW is a longstanding tradition in countries like Japan, China, Indonesia and the Philippines. Organic municipal waste is composted in many countries of the region to reduce waste for landfilling. However, most of the composting plants in the developing countries are not functioning well and are not operating at full capacity owing to

(a) high operating and maintenance cost compared with open dumping or landfilling;

(b) cost of compost is higher than commercial fertilizers;

(c) incomplete separation of materials such as plastic and glass, making the compost poor for agriculture applications; and

(d) poor operation and maintenance of the facilities.

Co-composting of organic MSW with agricultural waste and sludge from municipal sewage treatment plants is gaining acceptance in Bangladesh, China, India, Philippines and Thailand, where land is available. High operation cost and maintenance and transportation cost, along with incomplete separation of waste materials are major constraints for the adoption of this process.

Vermi-composting

Vermi-composting is a process which uses worms and micro-organisms to convert organic materials into nutrients-rich humus to use as fertilizer. Vermi-composting of organic MSW is in practical application in China, India, Indonesia and the Philippines.

Open dumping

In the Asian region, among various methods open dumping is the most common method of solid waste disposal. Dumping sites are especially uncontrolled disposal sites. In some cities, solid waste disposal combines open dumping and burning and sea disposal. In many coastal cities solid waste is disposed on open dumps and then burnt and some waste is pushed off the cliff into the sea. Very often, the dumping sites are swamp lands and ravines and the waste is used for reclamation. Filling of wetlands and ravines with municipal solid waste is being practiced for land development in South Korea, Japan, India, Sri Lanka and the Philippines.

3.3. Landfilling

Landfill historically has been the most common way of disposing of solid waste, including municipal solid waste. Landfill involves disposal that is at least partially underground in excavated cells, quarries, or natural depressions. Usually fill is continued above ground to utilize space most efficiently and provide a grade for drainage of precipitation.

Landfill design

The greatest environmental concern with landfilling of municipal solid waste that unavoidably contains hazardous compounds, is the generation of leachate from infiltrating surface water and groundwater with resultant contamination of groundwater suppliers. Modern waste landfills provide elaborate systems to contain, collect, and control such leachate (Figure 11).

In modern containment landfills, a liner is constructed at the base to minimize percolation of leachate into the underlying groundwater. Depending on the landforms where a waste is disposed, there are three types of containment landfills: (1) below-grade landfills (see Figure 11); (2) at-grade landfills; and (3) canyon landfills.

Environmental impacts of landfill

An improperly designed and operated landfill can cause discharge of leachate to groundwater aquifer and groundwater pollution. Table 4 presents the limits of concentrations for various contaminants in landfill leachate.

From a disposal standpoint, the primary item of interest is the chemical composition of the leachate generated from a waste. We can see from Table 4 that leachate quality from waste type is not unique. Both the local climate and operational practices heavily influence leachate chemical composition. This also varies with time. The time-quality curve is somewhat bell-shaped.

In addition to groundwater pollution, an improper landfill operation can impact several elements of the physical environment like surface water, air, forest lands, and wetlands. The problems of odor and fire hazard owed to gas migration, breeding and harboring of disease vectors are also of special concern.

Landfilling in the Asian region

Landfilling operation of solid waste in the Asian region is generally unsatisfactory except in countries such as Japan and Singapore, though it is an attractive disposal option. Landfill sites vary from uncontrolled tipping/dumping, controlled dumping, basic sanitary landfill and full sanitary landfill. Many countries such as China, Japan, South Korea, Malaysia and Thailand have adopted controlled tipping or sanitary landfilling for solid waste disposal. Some cities including Tokyo and Seoul do have well-designed and reasonably operated sanitary landfills. Kuala Lumpur has used old tin mines for MSW landfills around the city. The high percentage of organics, combined with much plastic, which forms layers when compacted, contributes to the

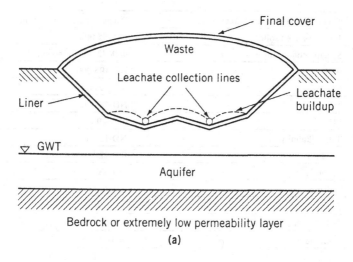

(a)

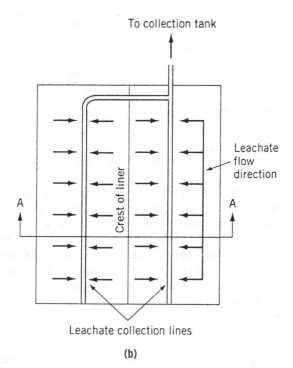

(b)

Figure 11. Schematic presentation of modern containment landfill: (a) cross section A-A and (b) plan view (Bagchi, 1999).

Table 4. Limits of contaminant concentrations in landfill leachate (Bagchi, 1999).

Chemical and physical parameters	Range of concentration, mg L^{-1}
Total dissolved solids	584–55,000
Specific conductance	480–72,000 μmho cm^{-1}
BOD	ND-195,000
COD	6.6–99,000
TOC	ND-44,000
PH	3.7–8.9
Total alkalinity	ND-15,050
Hardness	0.1–225,000
Chloride	2–11,375
Calcium	3.0–2,500
Sodium	12–6,010
$NH_4^+ - N$	ND-1,200
Aluminum	ND-85
Zinc	ND-371
Manganese	ND-400
Nickel	ND-7.5
Lead	ND-14.2
Arsenic	ND-70.2
Mercury	ND-3.0
$NO_3 - N$	ND-250
$NO_2 - N$	ND-1.46

Note: ND—not determined.

build-up of landfill gas (methane) at landfill sites; fires often break out and workers are made ill by the gases.

In large cities in the Asian region, the availability of land for landfill siting is a major issue, especially in urbanized and heavily populated areas. For example, in Hong Kong, Singapore and Taiwan, severe land constraints have led to the selection of landfill sites in coastal areas, offshore islands and mountainous terrain rendering complex engineering infrastructure facilities necessary to ensure proper operational and maintenance control. In Singapore, there are two controlled landfill sites; their capacities are almost exhausted and offshore landfill is now in operation. A similar situation is in the Bangkok metropolitan area.

3.4. Incineration

Many chemical species of MSW, including polymeric or carbonaceous solids, are degraded by high temperature. In the presence of oxygen, any carbon, hydrogen, and sulfur are oxidized to CO_2, H_2O, and SO_2, respectively. The range of incineration

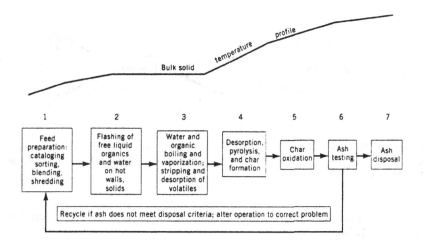

Figure 12. The schematic view of incineration process (Diemer et al, 1999).

increases rapidly with temperature. A range of 700–760 °C is generally required for combustion, and most general-purpose incinerators operate between 760 and 1100 °C. For an incinerator to operate with auxiliary fuel or air preheating, the waste feed or refuse must contain less than 50% moisture or 60% ash, and have more than 25% combustibles.

It is extremely difficult to burn and recover useful energy from unsorted municipal waste because of its heterogeneity in size, shape, chemical composition, and heating value. However, pretreatment of the waste before thermal treatment facilitates burning. Such pretreatment contributes to the front-end cost but reduces furnace cost. The waste is upgraded by separation of the nonorganic fraction and drying, shredding, and densifying solids.

The various types of incinerators include the moving grate incinerators and the multichamber incinerators. The rotary kilns are used to incinerate a large variety of liquid and solid industrial wastes. Any liquid capable of being atomized by steam or air can be incinerated, as well as heavy tars, sludges, pallets, and filter cakes.

Thermal destruction of solid waste

The generalized process flow chart for the thermal treatment of municipal solid waste is shown in Figure 12.

The analysis of the evolution and/or destruction of hydrocarbons during the incineration of MSW and hazardous waste involves heat transfer, mass transfer, and reaction kinetics. The key phenomena include the flashing of liquid hydrocarbons; the vaporization, desorption, and stripping of hydrocarbons; the pyrolysis and charring of hydrocarbons; and the oxidation of char. To a certain extent these processes occur in parallel (steps 2, 3, 4, and 5) and are common to most thermal treatment processes.

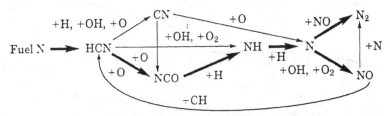

Figure 13. Schematic showing reaction pathways by which fuel nitrogen is converted to NO *and* N_2. *The bold lines indicate the key pathway. Thermal* NO *is formed from* $N_2 : N_2 + O \rightarrow NO + O$ *(Diemer et al, 1999).*

The key variables affecting the rate of destruction of solid waste are temperature, time, and gas-solid contact.

Pollutant emission from solid waste incinerators

NO_x *and* SO_2 *emissions.* Oxides of nitrogen (NO and NO2) and sulfur are emitted from most combustion systems including MSW and hazardous waste incinerators. The two principle mechanisms are shown in Figure 13.

The thermal NO_x pathway is important in any high temperature process containing molecular nitrogen and oxygen (see Chapter 3). The fuel NO pathway is also important if the fuel or waste contains nitrogen. The level of nitrogen oxides is difficult to estimate quantitatively since rate of formation of NO by both pathways is determined by complex kinetics and gas-phase mixing (Diemer et al, 1999). Sulfur oxide production in incinerators is generally easy to predict because nearly all organic sulfur species are completely converted to SO_2.

Partitioning of heavy metals. Metals entering a solid waste incinerator can leave the system with the bottom ash, the captured fly ash, or the exhaust gases. The fraction leaving with the exhaust gases can include metal vapors such as mercury and very small particles that escape capture in the stack. Metals entering the incinerators as liquid streams are usually carried out of the reactor in the fly ash, which is typically enriched with heavy metals, especially lead and cadmium, relative to the entering solid waste. The enrichment of small particles results from the condensation of vaporized metals. Metal concentrations of particles also typically increase with decreasing particle diameter.

Dioxin and furan emissions. PCDD and PCDF congeners are emitted mostly as absorbates on fly ash, rather than free gaseous species. Although the most toxic PCDD and PCDF congeners are only minor components of the total, their presence raises considerable public opposition to incineration of MSW. Surprisingly, hospital incinerators are among the more significant sources of these pollutants, mainly because of the presence of chlorinated plastics such as polyvinyl chloride in the feed. The proposed mechanism by which chlorinated dioxins and furans form has shifted from one of incomplete destruction of the waste to one of low temperature, downstream formation on fly ash particles. Two mechanisms are proposed: a *de novo* synthesis, in which PCDD and PCDF are formed from organic carbon sources and Cl in the presence of metal catalysts, and a more direct synthesis from chlorinated organic precursors, again involving heterogeneous catalysis. Bench-scale tests suggest that the optimum temperature for PCDD and PCDF formation in the presence of fly ash is roughly 300 °C. Chlorine may be formed in reaction (4) at temperatures below 900 °C

$$2\,HCl + 1/2\,O_2 \longleftrightarrow Cl_2 + H_2O. \tag{4}$$

Both Cl_2 and HCl have been shown to chlorinate hydrocarbons on fly ash particles. Pilot-scale data involving the injection of fly ash from municipal waste combustion show that intermediate oxygen concentration (4–7%) produce the highest levels of both PCDDs and PCDFs. The data also show significant reductions in emission of these contaminants with the upstream injection of $Ca(OH)_2$ at about 800 °C (Diemer et al, 1999).

Incineration in the Asian countries

Incineration process of MSW disposal is capital-intensive and skilled manpower is required for operation and maintenance. Up-to-date, full-scale incinerators are currently in service only in Japan, Singapore, China, Hong Kong, South Korean, Indonesia, and Taiwan. High capital investment, high operation and maintenance costs, and stringent air pollution regulations have severely limited the application of this technology for the disposal of MSW in the region.

We will give some examples from various Asian countries. Singapore operates three incinerator plants, burning more than 75% of the daily 6,700 tons of MSW collected. No sorting of waste is carried out before the MSW is fed to the incinerators, with exception of crushing bulky waste. The waste is mixed and burned using rotating roller grates. Auxiliary oil burners are used to start up the combustion process. Combustion is self-sustaining in some case, while at other times wood is added. In general, the combustion fraction of MSW is high and in some instances has been raised by moisture-reducing compaction at transfer stations. Total electrical energy recovery from the plants is about 60 megawatts or 250–300 kWh per ton of MSW incinerated. This energy is used partially to run the incinerators operations, and the balance is sold to the national electricity grid. The new plant, costing US$700 million and constructed in 2000, is generating 80 megawatts of electricity from the waste heat, of which 20 megawatts will be consumed by the plant and the remainder sold to Singapore Power (ESCAP, 2000).

Incineration plants in other Asian countries like Japan, South Korea and Taiwan are of similar design to those of Singapore. In South Korea, the rise in incinerated waste was shown from 3% in 1994 to 20% in 1999.

There are 1900 general waste incinerators across Japan, of which 1584 incinerators are operated by local governments and the other 316, by private companies. China has two incinerators and five are now in construction for MSW disposal with foreign assistance.

4. WASTE RECYCLING

4.1. Recycling as the Best Solution of Waste Problems

The disposal of waste is a monumental problem on a global scale, and in particular of the Asian society, because of the enormous amounts of wastes produced in the region. Although industry receives much bad press over the waste disposal issue, a great deal of the waste generated by industry is actually recycled: defective products, manufacturing overruns, machined scrap, etc. There is no particular altruism in this, since it is generally cheaper to recycle scrap than to make new product. For instance, production of new aluminum cans from recycled cans is 5% cheaper than from raw materials. Today, major corporations are very conscious of the adverse corporate image of poor waste disposal practices.

As a slogan Recycling sounds excellent, and indeed should be practiced wherever possible. However, this is not the only option for waste minimization. Recycling is one of the 3-Rs: waste Reduction, Reuse, and Recycling. Advantages of recycling are related to saving of raw materials, saving of time and energy, lower treatment and disposal cost, less risk of legal liability, and less employee exposure to hazardous materials. Central to the program is a Waste Audit, to establish the identities and quantities of the wastes being generated. In some cases, Waste Exchanges have been set up to allow companies to make use of another's wastes. The following example is characteristic for Waste Exchange. Company A, having waste acid to neutralize may be able to use waste alkali from Company B, rather than using virgin base.

In order to estimate the role of waste recycling in the waste disposal system, let us consider a hierarchy of options for waste disposal:

1.   disposal into environment;

2.   placed in a permitted landfill;

3.   burned within a permitted waste-to-energy incinerator or energy recycling;

4.   put to a low value use, as in conversion of waste plastic into park benches;

5.   put to a high value use, as in recycling aluminum cans by melting them down;

6.   rebuilt, as in repair of a car;

7.   reused, as in refilling beverage containers.

At first glance, it would seem that the higher the category, the more environmentally friendly the option. This will not always be true, if total energy and resource cost are computed. For example, at some point the cost of repair of the car (option 6) exceeds the costs of recycling the components and starting afresh (option 5). An unforeseen problem that has arisen with respect to municipal recycling is that its success is being undermined by insufficient capacity to treat and sell the material collected, especially newsprint. In many countries, virgin product is cheaper than the recycled material, because of collection and transportation costs (Bunce, 1994). As a result, recycling programs in the developed countries tend to be heavily dependent upon government subsidy. However, this is not a case in developing Asian countries with low labor cost.

4.2. Plastic Waste Recycling

Plastic wastes are now contributing 7–8% by weight of North American municipal garbage, and even more, up to 10–15% in the Asian countries. About half of this amount represents discarded packaging material (altogether, packaging forms nearly one-third of all municipal garbage). The littering of roadsides, the countryside, and ocean beaches by plastic rubbish calls attention to the near-indestructibility of synthetic plastics such as polyethylene, polystrene, and polyvinyl chloride. As with most other pollutants, many of the virtues of plastics (light weight, inertness to chemical and microbial attack, low cost) become their weaknesses when they are discarded. At present, the proportion of plastic recycling in the various countries is very small, mostly polyethylene terephthalate from soft drink bottles and polyethylene from plastic milk jugs (see details below). This requires manual sorting of the plastics waste stream to produce a homogeneous feed for recycling, since chemically different plastics are not necessary compatible with each other.

Plastic life cycle analysis

Once the plastic product has been used in its application, it has arrived at the end of its life cycle, as shown in Figure 14.

At this stage, there are three general options for plastics: recycling, incineration, and/or dumping. Although incineration is often seen as thermal or energy recycling, it is mentioned here as a separate option because it does not linger on in the life cycle. The material, burned only once, stimulates the greenhouse effect, leads to air pollution, and to a waste of resources with its (contaminated) ash residue.

In the case of plastic recycling, energy recycling is often not a permitted option, on account of the negative image of incineration and PCDD/PCDF formation. An intermediate solution, which is under development, is feedstock recycling, in which the plastic is depolymerized and the monomers recycled as chemical feedstocks. A recent development is the recycling of mixed plastic waste into items such as fence posts, park benches and other structural units, which have the advantages over wood and steel that they do not rot. Even incompatible plastics can be blended together to form articles having the strength needed for these application.

Degradable plastic materials

Two approaches to degrading plastics are now available: photodegradation and biodegradation. However, in most Asian countries, these product forms represent only a very

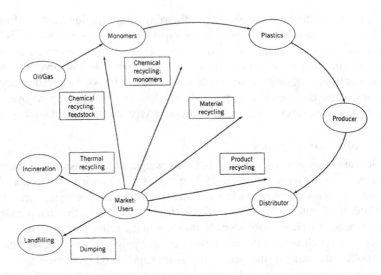

Figure 14. Schematic overview of the plastic life cycle and corresponding recycling options (van der Broek et al, 1999).

tiny fraction of all plastics. Furthermore, they do not degrade away literally to nothing. Plastics buried in the landfills do not photodegrade because they are not exposed to sunlight, and even biodegradation is extremely slow under these conditions.

Photodegradable plastics are copolymers of conventional polymers such as polyethylene with a small amount of carbon monoxide. The presence of carbon monoxide in the feed affords a polymer containing a small proportion of randomly located carbonyl groups, the key feature of which is their ability to undergo chain scission upon absorption of light in the 300–330 nm range. This is a well-known reaction in organic photochemistry, the mechanism of which creates breaks in the polymer chain to loss of structural integrity (Bunce, 1994), as shown in Figure 15.

The structural integrity of polymers is maintained by van der Waals attractions of neighboring chains. Breaks in the polymer chains lead to the plastic disintegrating physically, although in the chemical sense little of the material has already decomposed. The chief advantage of photodegradable plastic is that unsightly litter disappears to powder upon sunlight exposure.

We can add one more advantage of this plastic. The carbonyl chromophers present in photodegradable plastics absorb nearly 300 nm, in the UV-B range of tropospheric solar radiation. Artificial lighting, both incandescent tungsten bulbs and fluorescent lamps, produces very little UV-B, and window glass filters out this radiation from sunlight. Thus, the plastics will be degraded only when are discarded to landfill.

Biodegradable plastics are also copolymers of conventional polymers such as polyethylene with a small amount of starch molecules. The presence of starch allows micro-organisms to eat this organic component and disintegrate the polymer chains.

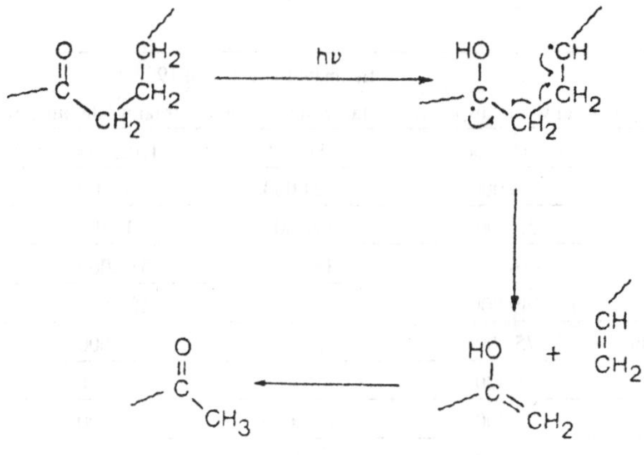

Figure 15. Photodegradation of plastics modified by including carbon monoxide in polymer chain (Bunce, 1994).

Another example of biodegradable plastics is related to the application of polyhydroxyalkanoate (PHAs), which may be biosynthesized from activated sludge of municipal wastewater (Satoh et al, 1998).

Recycling in the Asian region

The rate of recovery of recyclables from the MSW stream in many countries of the region such as Japan, South Korea and Singapore has increased dramatically in recent years (Hara, 1997, ESCAP, 2000). Overall resource recovery grew from less than 10% of all MSW in 1988 to 30% in 1998 in the region. Most of the increase is attributable to greater rates of recovery of paper and paperboard, plastics, glass and metals. The paper and paperboard category dominates in terms of total tons of materials recovered (almost 60%), which tends to mask the importance of the total mass of recovered materials, but in terms of its total economic value, it far exceeds the paper products category.

Among the Asian countries, Japan recycles huge quantities of materials from MSW stream. Almost half of the waste paper is retrieved or recycled in Japan and the retrieval rate increased from about 48% in 1990 to about 56% in 1997. Similarly, recycling rates of aluminum cans and steel cans increased from 40% to 60% and 45% to 70% from 1990 to 1997, respectively. Glass recycling rate also increased from 48% in 1990 to 57% in 1997. South Korea and Singapore recycle huge amounts of paper and cardboard, plastics, glass and metals. The example of Singapore is shown in Table 5.

In the Asian region, recycling of plastics derived from MSW is thriving as a booming business. The plastic recycling boom is due to a combination of various factors such as unemployment, low labor cost and a wider market. Plastic recycling

Table 5. Various categories of materials recycled from MSW in Singapore in 1997 (ESCAP, 2000).

Waste type	Estimated quantity in 1997, tons			
	Total waste disposed	Total waste recycled	Total waste out	Recycling rate, %
Food waste	1,085,000	24,700	1,109,700	2.2
Paper/cardboard	576,000	324,000	900,000	36
Plastics	162,000	35,300	197,300	17.9
Construction debris	126,000	188,000	314,000	59.9
Wood/timber	249,000	34,800	283,800	12.3
Horticultural waste	75,400	67,700	143,000	47.3
Earth spoils	75,400	—	75,400	—
Ferrous metals	75,400	893,000	968,400	92.2
Non-ferrous metals	14,000	76,000	90,000	84.4
Used slag	120,000	135,000	255,000	52.9
Sludge	50,200	—	50,200	—
Glass	30,800	4,600	35,400	13
Textile/leather	25,200	—	25,200	—
Scrap tires	5,600	5,700	11,300	50.4
Others	126,000	1,300	127,300	1
Total	2,976,000	1,790,000	4,586,000	39

has become very popular in countries such as Bangladesh, China, India and Thailand. In these countries, paper and paperboard recycling is also enhanced.

Recycling in developing countries relies largely on informal collection of materials from waste carried out by human scavengers or wastepickers. In the Asian cities, it has been estimated that up to 2% of the population survives by recovering materials from waste to sell for reuse or recycling or for their own consumption.

5. SITE REMEDIATION AFTER WASTE DISPOSAL

We have considered already (see Chapter 6) soil remediation after release and accumulation of some types of pollutant like excessive salt accumulation and accidental release of oil products. In this subchapter we will further discuss the problems of site remediation, including bioremediation, after accumulation and release of heavy metals and persistent organic pollutants.

5.1. Heavy Metal Remediation

This term refers to cleaning up the soils and conjugated compartment like undersurface and groundwater at a former manufacturing facility, waste site or site after accidental release. These sites may be strongly contaminated by different heavy metals; lead, cadmium, chromium, mercury, and arsenic are only a few from the list. Soil remediation after pollution of different heavy metals is typically a very expensive process on account of the large volumes of soil involved and the relevant remediation of groundwater is much more expensive. Here we must refer to the known paradigm that prevention of pollution is cheaper than control and remediation.

Let us consider some technologies for soil remediation from heavy metals.

Electrokinetic remediation

Electrokinetic remediation involves passing a low level (mA cm^{-2}) DC electric current having potential difference a few volts through the soil, either *in situ* with electrodes placed into the soils, or in an external reactor. It has been shown that heavy metals such as zinc, copper, cadmium, chromium, lead, arsenic, mercury, nickel and iron can be efficiently removed from soils by electrokinetics.

An externally supplied fluid is used as the conducting medium. When the current is applied to a soil-water system, electrolytic reactions generate an acid-front at the anode and a base-front at the cathode. The acid-front moves towards the cathode due to the electrical, chemical, and hydraulic potential gradients that are established. Simultaneously, the base-front moves towards the anode due to electrical and chemical gradients. Consequently, the pH profile changes in the soil. The acid-front eventually flushes across the soil and neutralizes the base-front developed at the cathode. Ionic migration and diffusion, and any electro-osmotic advecton enhance the transport of the acid-front, generated at the anode. After the current is applied for several hours, the pH drops to 2 at the anode and rises to 12 at the anode and pH profile in the soil is slowly changed by the production of H^+ and OH^- at each electrode. The change of the soil pH affects the removal efficiency of heavy metals by affecting the pore fluid composition in the soil, and the surface charge density of soil particles.

The acid-front of the soil facilitates desorption of heavy metals from the soil surface and dissolution of hydroxyl complexes of heavy metals. As a result, this increases the heavy metals fractions present in the liquid phase, their mobilities, and the removal efficiency of heavy metal. In contrast, the base-front in the cathode zone can immobilize heavy metals by forming their hydroxides and heavy metal precipitation occurs in the soils close to cathode, causing low removal efficiency.

The movement of acid-front to the cathode is due to migration (electric potentials), diffusion (chemical potentials), and advection (hydraulic potential) and causes desorption of heavy metals from clay surfaces and transports them into the pore fluid. Electro-osmotic flow and its associated phenomena constitute the mechanisms for removing heavy metals from soils.

If traces of electrolysable solute exist in the pore fluid, it is then possible to carry out water electrolysis by inserting electrodes in a saturated soil mass in the field site. The primary electrode reactions are as follows:

$$2 H_2O - 4 e^- \longrightarrow O_2 + 4 H^+ \quad \text{(anode)}, \qquad (5)$$

$$4 H_2O + 4 e^- \longrightarrow 2 H_2 + 4 OH^- \quad \text{(cathode)}. \qquad (6)$$

Table 6. *Condition of soil remediation experiment (Yang, 1999).*

Experimental conditions	Current (mA)	concentration	Initial Pb^{2+} concent-ration, ppm
System I	20	5	201
System II	20	5	205
System III	20	5,10	203, 200
System IV	20	5,10	202, 203

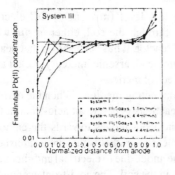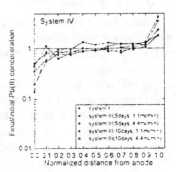

Figure 16. *Lead concentration in soil slides after remediation (Yang, 1999).*

Secondary reactions may exist depending on the concentration of the available species as reactions (7) and (8)

$$2\,H^+ + e^- \longrightarrow H_2 \quad \text{(anode)}, \tag{7}$$

$$Pb^{2+} + 2\,e^- \longrightarrow Pb \quad \text{(anode)}. \tag{8}$$

The production of H^+ ions at the anode reduces the pH according to the reaction (5) and the production of OH^- ions at the cathode increase the pH according to the reaction (6). As a result of migration of anions and cations in the pore fluid of the soil under the electrical field and reactions (5) and (6), two supplemental ionic species (H^+ and OH^-) can significantly influence the removal of heavy metals from the soil.

For example, we can consider the results of soil remediation in South Korea from lead pollution (Yang, 1999). The initial concentration of lead was 200–205 ppm (Table 6)

The content of lead, taken as a ratio between the final and initial concentrations in ten different slides of soil after remediation process, is shown in Figure 16.

Lead was partially removed along the anode section, and it was accumulated in the cathode section for all of the experiments. It was found that the lowest lead concentration occurred in the soil slides nearest to the anode because lead in the elecrokinetic process becomes cationic and moves towards the cathode, most probably as free Pb^{2+} ions in the pore fluid. Generally, the higher removal efficiency of lead occurred in system III at a 1.1 ml/min flow rate.

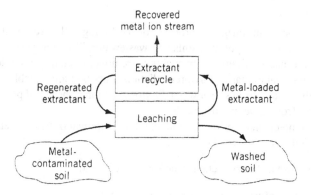

Figure 17. General concept for using soil washing to remove heavy metals from polluted soils (Rorrer, 1999).

Soil vitrification

Soil vitrification (conversion to glass) is an aggressive method of soil treatment that is useful for metal contaminated soils. Electrodes are placed vertically in the soil, and the soil is heated electrically to melting (1660–2000 °C) in a zone, which grows downwards and outwards from the electrodes. Inorganic hazardous metals are incorporated into the vitrified mass, while organic species of heavy metals are pyrolized and volatilized. The advantage is that this technology can be applied in field conditions without preliminary handling and transportation of the soil.

Vitrification is particularly useful for remediation of soils contaminated with radioactive heavy metals. The radioactive heavy metals are encapsulated in a highly inert, nonporous matrix, which can be stored permanently in underground secured radioactive waste storage facilities currently under development in some Asian countries. Vitrification can also be used for sediments and sludge polluted by heavy metals.

Soil washing

This process is used to concentrate the contaminants into a fraction, which is treated further. The technology involves subjecting the soil to high pressure jets of water, thus breaking down the soil structure. Surfactants are often used to assist the solubilization of the contaminants into aqueous phase. One of such technologies with application of soluble chelation agents is shown in Figure 17.

Bioleaching processes use microorganisms to solubilize heavy metals or heavy-metal oxides from mineral matrices by biological catalyzed oxidation-reduction processes. In both soil washing or bioleaching processes, the eluted heavy metal ions are then removed from the aqueous stream by the various chemical or biological process technologies (see details in Rorrer, 1999).

Phytoremediation

Pytoremediation is an intriguing process for the biologically promoted removal of heavy metals from both polluted soils and wastewater. Phytoremediation is broadly defined as the use of living green plants to remove, contain, or render harmless environmental contaminants. Phytoremediation processes are capable of removing heavy metals directly from soil or water. Phytoextraction is the use of plants to remove heavy metals from aqueous streams.

The basic mechanism of phytoremediation for removal of heavy metals involves two steps (Galiulin et al, 2001):

- active uptake of heavy metal by the root system, and

- translocation of heavy-metal ions to green shoots.

Plants which can sequester high concentrations of heavy metals are called hyper-accumulator metallophytes. For example, the hyper-accumulator *Thalaspi caerulescens* can accumulate several heavy metals, including Ni, Zn, and Cd. In this plant, over 18,000 mg of zinc and 1000 mg of cadmium per kg of shoot biomass can be accumulated without inhibition of growth (Rorrer, 1999). The mechanisms for active uptake for heavy-metal ions involve the use of biologically expressed phytochelatins that complex with heavy metals and thus protect the plant's biochemical processes from the toxic heavy metal effects. Genetic engineering offers a means to improve growth rates of hyper-accumulators or introduce heavy metal remediation activity into plants. For example, the bacterial gene for mercury reductase, which reduces Hg^{2+} to Hg^o, was recently cloned into Arabidopsis.

Engineered reed-bed and constructed wetland systems for removal of heavy metals from wastewater using phytoremediation are in use in some developed Asian countries. The root system of the hyper-accumulator plant penetrates a permeable rock bed. The wastewater is introduced into one end of the bed and flows through the permeable rock layer. The rock layer should be inert to heavy metals binding so that it does not unwittingly serve as a sink for heavy metals. These metals are sequestered by the root system and translocated to the shoots. Periodically, the metal-containing shoots are harvested. The biomass can be burned off or composted to yield a low volume of metal-rich ash.

For example, this system is used for phytoremediation of arsenic-contaminated cooling wastewater from the Mae Moh power plant burning Thai lignite with high content of As and other heavy metals (see Chapter 12). Other examples are shown in Box 1.

Box 1. Heavy metal phytoextraction from contaminated waters (after Sen and Mondal, 1990 and Shrivastava and Rao,1997)

Sen and Mondal (1990) studied the Cu absorption by floating salvinia (*Salvinia natans*) abundant in ponds and lakes of India. The Cu concentrations ($C_0 = 1-100 \, \mu g/ml$) were created in the culture medium where adult plants (with a biomass of 7.5 g/l) were exposed for five days. The plants absorbed the highest amounts of Cu (92.8–97.0%) from the solutions purified within one day at low metal concentrations (1–20 μg/ml); only 82.5 and 67.4% were absorbed at concentrations of 50 and 100 μg/ml, respectively. When the solutions contained 50 and 100 μg/ml Cu, the plants perished within three days and one day, respectively. At low concentrations, Cu was accumulated mainly in roots; its content in the macrophyte shoots increased with time at the high metal rates. The authors attempted to release the metal accumulated using the ignition of the plant matter and subsequent boiling in a 1–2 N H_2SO_4 solution. Crystalline copper sulfate 94–96% pure was obtained. The recovery of Cu from the plants exposed in different solutions of its salt ($C_0 = 10$–$100\,\mu$g/ml) ran to 96.7–99.3%. The second stage of the study involved the phytoremediation of the wastewater from the Indian Copper Complex in the city of Bihar (Bihar state) with pH 7.1–7.6 and containing 0.75–1.2 μg/ml Cu. Practically total absorption of the metal was reached when 20 g of plants were kept in 1L of waste liquid for a day.

Shrivastava and Rao (1997) estimated the effectiveness of the Hg absorption from water ($C_0 = 1$ mg/l) by macrophytes of five genera: *Eichhornia, Salvinia, Hydrilla, Chara*, and *Vallisneria*. It was found that after the four-day exposure of the plants in the solution to be purified, the maximum decrease in the Hg content (by 80%) was observed when a combination of macrophytes was used. Next in descending order were individual plant species of genera *Eichhornia* (by 60%), *Salvinia* (by 51%), *Chara* (by 40%), *Hydrilla* (by 30%), and *Vallisneria* (by 27%). The authors believe that the combined macrophyte system can effectively absorb Hg from all water layers owing to the peculiarities of habitat of the plants studied. *Eichhornia* and *Salvinia* plants will purify the surface layers; *Chara* and *Hydrilla* will operate in the water bulk, and *Vallisneria* will purify the bottom layers by its root system.

Phytoremediation processes have two major advantages. First, phytoremediation is an environmentally friendly process for *on-site* remediation of heavy metals. The only potential drawback is that the species of hyper-accumulator plant introduced into the contaminant site may not be endogenous to the local ecosystems. Second, phytoremediation processes are simple in configuration, require minimal chemical or energy inputs, and are inexpensive to operate. However, like many biologically based processes, they are difficult to control.

5.2. Remediation of Soils Polluted by Persistent Organic Compounds
Bioremediation

Remediation of soils polluted by persistent organic compounds is based mainly on biological processes and thus called bioremediation. The ultimate goal of bioremediation is to biodegrade toxic compounds into nontoxic forms, eliminating the need for further hazardous waste cleanup. Biological treatment has been used in the Asian region for centuries to degrade various organic wastes. Biological degradation exploits the ability of micro-organisms to produce enzymes that metabolize specific compounds. Some micro-organisms break down a contaminant to carbon dioxide, water, methane, inorganic salts, and/or biomass, from which they derive carbon and energy for growth. Other microorganisms only partially break down a compound, resulting in incomplete degradation. These intermediate compounds accumulate or are further degraded by other micro-organisms. Often, although not always, these intermediate compounds are less toxic than the parent compounds. The exception is DDT metabolites DDD and DDE, which are more toxic than DDT itself (see Chapter 14).

A common method of degradation for highly chlorinated, highly recalcitrant compounds is cometabolism. Cometabolism is the degradation of a compound by a micro-organism that derives no nutritional benefit from that compound. The micro-organism population grows on other substrates present in the environment, and the broadly specificity enzymes produced to degrade these food sources fortuitously degrade the contaminant compound. The contaminant degradation rate is very slow and often incomplete, because the microbial population does not increase in biomass or numbers relative to the contaminant's concentration.

Two approaches have been explored for the selection of micro-organisms for degradation of persistent organic compounds, such as organochlorines or polycyclic aromatic hydrocarbons:

- optimize the growth of the consortia of organisms already present at, and already adapted to the site;

- developing microbial isolates or genetically engineered organisms with a particular ability to degrade the contaminant from remediated soils.

There are many biotechnological companies producing the super-organisms, along with nutrients such as nitrogen and phosphorus to optimize microbial growth. However, experience with super-organisms has been generally disappointing, because the newly introduced micro-organisms fail to compete with the populations of microbes already established in the soil and therefore die out. Alternatively, organisms which have been selected in the laboratory to use, for example PAH or 2,4-D pesticide as their sole carbon source, may rapidly lost that ability when placed into soil, where other, more easily available carbon sources exist.

Thus, instead of introducing foreign organisms to a site, a better option is to optimize the growth conditions for the consortia of microorganisms already present in the soil. Natural soil micro-organisms degrade a wide variety of contaminants. Micro-organisms, principally aerobic and anaerobic bacteria and fungi play an important role in the natural degradation of various POPs. These can be transformed oxidatively under aerobic conditions as the conversion of penthachlorophenol to 2,3,5,6-tetrachlorohydroquinone, chloranil and ring cleavage product, or reductively under anaerobic conditions, as dechlorination of PCB contaminants in the sediments.

Environmental factors, which greatly influence biodegradation rates, include suitable temperature, oxygen, nutrient availability, pH, moisture content, organic matter concentrations, contaminant concentrations, and the presence of other contaminants. Indigenous micro-organisms are usually the most reliable source of contaminant degraders, and only the environmental conditions need to be adjusted for adequate

Table 7. Environmental parameters for aerobic microbial degradation of PCP (Diehl and Borajani, 1999).

Parameter	Value
Temperature	Biodegradation detected as low as 0–10 °C, most applicable is mesophilic range 15–45 °C, thermophilic range is 40–60 °C
Soil moisture	40–70% of field capacity for unsaturated soils
Dissolved oxygen	> 0.2 mg L^{-1}; minimum air-filled pore space of 10%
Nutrients (C : N : P)	100–120 : 10 : 1
pH	6–8
Bacteria numbers	1×10^3–1×10^7 colonies per gram of soil
Redox potential	400 to 800 mV for well-aerated soils; −100 to 100 mV for moderate soils; −300 to −100 mV for reduced anaerobic conditions

degradation to occur. An example of the optimized environmental conditions for biodegradation of PCP is shown in Table 7.

Bioremediation may be carried out *in situ*, with suitable engineering systems for pumping air and nutrients in, and if necessary, leached out. Alternatively, the soil may be removed and treated in an enclosed bioreactor, when the outdoor growing season for the micro-organisms is short, like in the north of Japan or China.

Physical-chemical remediation

Ozonation of polycyclic aromatic hydrocarbons (PAHs) in the unsaturated soils is one of the advanced technologies for soil treatment. The PAHs are resistant to typical *in situ* remediation technologies including soil flushing and bioremediation. Ozonation of PAHs in soils is used for overcoming these limitations. *In situ* ozonation is based on the delivery of gaseous ozone molecules to contaminated media to destroy the contaminants by converting them into innocuous compounds commonly found in nature.

A Korean study was demonstrated that the three ring condensed phenanthracene has the highest removal efficiency followed by the four ring condensed pyrene and then the three ring condensed anthracene (Kim et al, 1999). The molecular structure of each PAH significantly affects the reaction rate with ozone rather than the molecular weight. Increasing of the moisture content results in the ozone demand because of the dissolution of gaseous ozone into soil pore water and subsequent decomposition of ozone. However, the presence of water increases the removal rate of PAHs as compared to that of dry soils.

Surfactant-enhanced soil flushing is also an effective procedure for remediation of PAH-contaminated soils. The removal efficiency of phenanthrene by this treatment was up to 90% after 25 pore volumes. The toxicological test with earthworm *Eigenia foetida* showed that the response of this test organism was uniform after 15 pore volumes.

Generally we can note that polychlorinated organic compounds like PCBs are very resistant to various bioremediation procedures and physical-chemical treatments are the most used for heavily contaminated sites. For remediation of PCB-contaminated sites the following technologies can be recommended.

Thermal desorption involves volatilization of contaminants from soil by heating the soil in various kinds of drying kilns, and is the solid phase analog of air or stream stripping. Temperatures in range 200–500 °C are used. The off-gases must then be treated before release to the atmosphere. This method can be applied to hydrocarbon-contaminated soils and to soils contaminated with organochlorine pesticides (Bunce, 1994).

Soil incineration is another method for treating soils that are contaminated with organic compounds. This method is very expensive but the soil can be completely cleaned of organic pollutants.

FURTHER READING

1. Bunce N., 1994. *Environmental Chemistry*. 2nd edition, Wuerz Publishing Ltd, Winnipeg, Canada, Chapter 8.

2. ESCAP, 2000. *State of the Environment in Asia and Pacific*. United Nations, Chapter 17.

3. Diemer R. B., Ellis T. D., Silcox G. D., Lighty J. D., and Pershing D. W., 1999. Hazardous waste: incineration. In: Meyers R. A. (Ed.), *Encyclopedia of Environmental Pollution and Cleanup*, John Wiley, NY, 754–761.

4. Van den Broek W., Wienke D., and Buydens L., 1999. Plastic among nonplastics, identification in mixed waste. In: Meyers R. A. (Ed.), *Encyclopedia of Environmental Pollution and Cleanup*, John Wiley, NY, 1283–1289.

5. Satoh H., Iwamato Y., Mino T. and Matsuo T., 1998. Activated sludge as a possible source of biodegradable plastics. *Water Science Technology*, 38(2), 103–109.

6. Yang J.-W., 1999. Transport of heavy metal in soils under elektrokinetic field. In: *Proceedings of the 2nd International Symposium on Advanced Environmental Monitoring*. KJIST, Kwangju, South Korea, 57–62.

7. Rorrer G. I., 1999. Heavy-metal ions, removal from wastewater. In: Meyers R. A. (Ed.), *Encyclopedia of Environmental Pollution and Cleanup*, John Wiley, NY, 771–779

8. Galiulin R. F., Bashkin V. N., Galiulina R. R. and Birch P., 2001. A Critical Review: Protection from Pollution by Heavy Metals—Phytoremediation of Industrial Wastewater, *Land Contamination & Reclamation*, 9(4), 349–357

WEBSITES OF INTEREST

1. Environmentally friendly technologies for the Asian region,
   http://www.un.org/escap/apctt

QUESTIONS AND PROBLEMS

1. Characterize the hot problems related to waste generation in the whole Asian region and in your country in particular.

2. Describe the generation of municipal solid waste in your country and compare these values with those typical for the whole Asian region.

3. Characterize the typical chemical composition of municipal wastewater and point out the possible sources of these contaminants.

4. Estimate the sources of hazardous solid wastes in your locality and compare with the total generation of these wastes in your country.

5. Discuss the typical chemical composition of industrial wastewater and point out the possible sources of these contaminants.

6. Discuss the production of agricultural waste in the Asian region and stress the similarities and differences between these and other types of wastes.

7. Describe the transboundary shipping of hazardous wastes in the Asian region with discussion of the relevant laws in your country regulating this shipping.

8. Characterize different methods of MSW disposal and note the most applicable methods in your locality.

9. Describe the construction of modern landfills and environmental problems related to the landfilling.

10. Discuss the application of incineration to the MSW, hospital and industrial waste treatment in your country.

11. Discuss the environmental problems of the most concern related to incineration and the ways to minimize these problems.

12. Characterize the advantages of recycling and estimate the role of waste recycling in the waste disposal system.

13. Discuss the generation of plastic waste and plastic recycling in your country. Focus attention on chemical and biological recycling treatments.

14. Discuss the problem of waste recycling both in your country and in the whole Asian region.

15. Discuss the general approaches to soil remediation. Draw the similarities and differences between remediation of sites polluted by heavy metals and persistent organic pollutants.

16. Characterize the physical and chemical principles of electrokinetic treatment of polluted soils. Give examples.

17. What chemical and physical technologies can be applied for remediation of HM polluted sites?

18. Discuss the phytoremediation technology for polluted soils and wastewater. Draw advantages and drawbacks.

19. Describe the principles of bioremediation for soils contaminated by POPs and present a relevant example.

20. Characterize physical-chemical treatments of POP-contaminated soils and its applicability for different compounds.

References

Abdullah M. H., 1996. Malaysia country report, in Proceedings of the ASEAN Network on Environmental Monitoring (ASNEM) on the *3rd Workshop on Air Quality Monitoring and Analysis with Emphasis on Polycyclic Aromatic Hydrocarbons*, Environmental Research and Training Center (ERTC), Pathumthani, Thailand, pp. 177–188.

ADB (Asian Development Bank), 1992. *Environmental Risk Assessment*. Asian Development Bank, 182 pp.

ADB (Asian Development Bank), 1998. *The Bank Policy on Water*. Working Paper. ADB, Manila, The Philippines.

Akimoto H. and Narita H., 1994. Distribution of SO_2, NO_x and CO_2 emissions from fuel combustion and industrial activities in Asia with $1° \times 1°$ resolution. *Atmospheric Environment*, 28(2), 213–215.

Aminuddin B. Y., Teng C. S. and Deratil B., 1994. The collection and analysis of land degradation data in Malaysia. FAO Bangkok Publ., No 3.

Andrea M. O., 1993. In: Oremland R. S. (Ed.), *Biogeochemistry of Global Change: radioactively active trace gases*. Chapman & Hall, New York, 113–150.

Anh P. T. V., 1996. Vietnam country report, in *Proceedings of the ASEAN Network on Environmental Monitoring (ASNEM) on the 3rd Workshop on Air Quality Monitoring and Analysis with Emphasis on Polycyclic Aromatic Hydrocarbons*, Environmental Research and Training Center (ERTC), Pathumthani, Thailand, pp. 177–188.

Aschner M., 1999. Manganese Homeostasis in the CNS. *Environmental Research Section A*, 80: 105–109.

Ayers G. P., 1991. Atmospheric acidification in the Asian region. *Environmental Monitoring and Assessment*, 19: 225–250.

Ayers G. P, Gillet R. W., Ginting N., Hooper M., Selleck P. W. and Tapper N., 1995. Atmospheric sulfur and nitrogen in West Java. *Water, Air and Soil Pollution*, 85: 2083–2088.

Ayers G. P. and Yeung K. K., 1996. Acid deposition in Hong Kong. *Atmospheric Environment*, 30: 1581–1587.

Bagchi L., 1999. Landfills. In: Meyers R. A. (Ed.), *Encyclopedia of Environmental Pollution and Cleanup*, The Wiley, NY, 881–887.

Bai Naibin, 1997. Estimation of emissions of SO_2 per $1° \times 1°$ grid square in China from 1990 to 1995. In: *Proceedings of International Workshop on Monitoring and Prediction of Acid Rain*, September 29–October 1, 1997, Seoul, 29–38.

Baird C., 1999. *Environmental Chemistry*. second edition, W. H. Freeman and Company, NY.

Ballschmitter K. and Zell D., 1980. *Fresenius Z. Anal. Chem.*, 302: 20–31.

Barnes I., 1999. Hydrocarbons in the atmosphere. In: Meyers R. A. (Ed.), *Encyclopedia of Environmental Pollution and Cleanup*, The Wiley, NY, 794–812.

Bashkin V. N., Kozlov M. Ya., Priputina I. V., Abramychev A. Yu. and Dedkova I. S., 1995. Calculation and Mapping of Critical Loads of S, N and Acidity on Ecosystems of the Northern Asia. *Water, Air, and Soil Pollution*, 85: 2395–2400.

Bashkin V. N., Kozlov M. Ya., Abramychev A. Yu. and Dedkova I. S., 1996a. Regional and Global Consequences of Transboundary Acidification in the Northern and Northern-East Asia. In: *Proceedings of International Conference on Acid Deposition in East Asia*, Taipei, May 28–30, 1996, 225–231.

Bashkin V. N., Kozlov M. Ya. and Abramychev A. Yu., 1996b. The application of EM GIS to Quantitative Assessment and Mapping of Acidification Loading in Ecosystems of the Asian Part of the Russian Federation. *Asian-Pacific Remote Sensing and GIS Journal*, Vol. 8(2): 73–80.

Bashkin V. N. (Ed.), 1997. *Heavy Metals in the Environment*. ONTI Publishing House, Pushchino, 321 pp.

Bashkin V. N., 1997a. The critical load concept for emission abatement strategies in Europe: a review. *Environmental Conservation*, 24(1): 5–13.

Bashkin V. N., 1997b. Acid Deposition and Ecosystem Sensitivity in East Asia. In: *Proceedings of International Workshop on Monitoring and Prediction of Acid Rain*, Seoul, 29.09–1.10.1997, 147–161.

Bashkin V. N. and Park Soon-Ung (Eds.), 1998. *Acid Deposition and Ecosystem Sensitivity in East Asia*, Nova Science Publishers, Ltd, 427 pp.

Bashkin V. N. and Kozlov M. Ya., 1999. Biogeochemical approaches to the assessment of East Asian ecosystem sensitivity to acid deposition. *Biogeochemistry*, 47: 147–165.

Bashkin V. N., Kouprianov V. I., Chullabodhi C., Towprayoon S., Khummongkol P., Wangkiat A., Wongyai K., 2000. Ecological Monitoring of the Thai Lignite-Fired Power Plant. In: *Ecology and Energy-2000.* Moscow, MSEI, 67–68.

Bashkin V. and Radojevic M., 2001. Acid check in Asia. *Chemistry in Britain,* No 6, 38–42.

Bashkin V. and Wongyai K., 2002. Environmental fluxes of arsenic from lignite mining in the north Thailand. *Environmental Geology,* Vol. 24(2): 234–239.

Bashkin V. and Howarth R., 2002. *Modern Biogeochemistry.* Kluwer Academic Publishers, 604 pp.

Bashkin V. N., Park S.-U. Choi M. S., Lee C. B., 2002. Regional budget of nitrogen in North-East Asia. *Biogeochemistry,* 57(1): 387–403.

Bhattacharyya D., Mangum W. C. and Williams W. E., 1999. Reverse osmosis. In: Meyers R. A. (Ed.), *Encyclopedia of Environmental Pollution and Cleanup,* The Wiley, NY, 1444–1450.

Bosch D. J. and Wolfe M. 1., 1999. Soil and water quality in agriculture. Assessing and improving. In: Meyers R. A. (Ed.), *Encyclopedia of Environmental Pollution and Cleanup,* The Wiley, NY, 1549–1565.

Boyer E. W. and Howarth R. W. (Eds), 2002. *The nitrogen cycle at regional to global scale.* Kluwer Academic Publishers, 519 pp.

Brassey A., 1989. The last Voyage (1886–1887). Longman & Co., London.

Brimlecombe P., 1987. *The big smoke: a history of air pollution in London since medieval times.* Methuen, London.

Bruijnzeel L. A., 1989. Nutrient content of bulk precipitation in south-central Java, Indonesia. *J. Tropical Ecology,* 5: 187–202.

Bruijnzeel L. A., Waterloo M. J., Proctor J., Kuiters A. T. and Kotterink B., 1993. Hydrological observations in montane rain forests on Gunung Silam, Sabah, Malaysia, with special reference to the "Massenerhebung" effect. *J. Ecology,* 81: 145–167.

Burghouts T. van Straalen and Bruijnzeel L. A., 1993. Contributions of throughfall, litterfall and litter decomposition to nutrient cycling in dipterocarp forest in the upper Segama area, Sabah, Malaysia, In: *Spatial Heterogeneity of Nutrient Cycling in Bornean Rain Forest,* Vrije Universiteit, Amsterdam, 47–79.

Bunce N., 1994. *Environmental Chemistry.* 2nd edition, Wuerz Publishing Ltd, Winnipeg, Canada.

Bunce N. J. and Petrulis J. R., 1999. Dioxin-like compounds: screening assays. In: Meyers R. A. (Ed.), *Encyclopedia of Environmental Pollution and Cleanup,* The Wiley, NY, 440–449.

Butcher S. S., Charlson R. J., Orians G. H., and Wolfe G. V. (Eds.), 1992. *Global Biogeochemical Cycle.* Academic Press Limited, Californoa.

Cachier H. and Ducret J., 1991. Influence of biomass burning on equatorial rains. *Nature* 352: 228–230.

Carmichael G. R., Hong M. S., Ueda H., Chen L. L., Murano K., Park J. K., Lee H., Kim Y., Kang C. and Shim S., 1997. Aerosol composition at Cheju island, Korea. *Journal of Geophysics Research,* Vol. 102, No D5, 6047–6061.

Crutzen P. and Andreae M. O., 1990. Biomass burning in the tropics: impacts on atmospheric chemistry and biogeochemical cycles. *Science,* 250, 1669–1678.

Cha H. J., Kim J. Y., Koh C. H. and Lee C. B., 1998. Temporal and spatial variation of nutrient elements in surface seawater off the West Coast of Korea. *The Journal of the Korean Society of Oceanography,* 3(1): 25–33.

Charasaiya T., 1996. Air quality monitoring in Pathum Thani, Thailand. In: *Proceedings of the ASEAN Network on Environmental Monitoring (ASNEM) on the 3rd Workshop on Air Quality Monitoring and Analysis with Emphasis on Polycyclic Aromatic Hydrocarbons,* Environmental Research and Training Center (ERTC), Pathumthani, Thailand, 245–254.

Chem B. H., Hong C. J., Pandey M. R. and Smith K. R., 1990. Indoor air pollution in developing countries. *WHO Statistics Quarterly,* 43(3): 127–138.

Chen Z. S., Liu J. C. and Cheng C. Y., 1998. Acid deposition effects on the dynamic of heavy metals in soils and their biological accumulation in the crops and vegetables in Taiwan. In: Bashkin V. N. and Park S.-U. (Eds.), *Acid Deposition and Ecosystem Sensitivity in East Asia,* NovaScience Publishers, USA, 189–228.

Cho H.-M., Choi J.-C., Chun Y.-S., Kim J.-Y. and Park K.-J., 1998. Chemical composition of aerosol in Seoul during the spring of 1998. In: *Proceedings of the 1st K-JIST International Symposium on Advanced Environmental Monitoring,* September 11–13, 1998, Kwangju, Korea, 122–123.

Choi M. S., 1998. Distribution of Trace Metals in the Riverine, Atmospheric and Marine Environments of the Western Coast of Korea. PhD thesis, Department of Oceanography, Seoul National University, 338 pp.

Choi K.-C. and Chun E.-C., 1998. Yellow sand composition analyzed by PIXE method. In: *Proceedings of the 1st K-JIST International Symposium on Advanced Environmental Monitoring*, September 11–13, 1998, Kwangju, Korea, 124–125.

Chung S. W., Bae J. O., Koh K. S., et al, 1992. Effects on the structure on productivity of *Pinus rigida* stands and the responses of *Pinus densiflora* seedlings to simulated acid rain. In: *A study on the assessment of damage by air pollutants and acid rain (II)*, Report of National Institute of Environmental Research Republic of Korea, 137–209.

Chung Y.-S., Kim T.-K. and Kim K.-H., 1996. Temporal variation and cause of acidic precipitation from a monitoring network in Korea. *Atmospheric Environment*, 30: 2429–2435.

Chungpaibulpatana S., Lemmeechokchai B., Aye T. T., Ongsakul W. and Sripadungtham C., 1997. *Establishment of a Country Specific Database for Thailand*. A Final Report on Research Contract No. 9277/RB, Sirindhorn International Institute of Technology, Thammasat University, Thailand.

Cleveland C. C., Townsend A. R., Schimel D. S., Fisher H., Howarth R. W., Hedin L. O., Perakis S. S., Latty E. F., Von Fisher J. C., Elseroad A. and Wasson M. F., 1999. Global patterns of terrestrial biological nitrogen (N_2) fixation in natural ecosystems. *Global Biogeochemical Cycles*, 13(2): 623-645.

Climatological Division, 1994. *Data processing*. Meteorological Department, Chiang Mai University, Thailand.

Daniels E. J., Jody B. J., Brockmeier N. F. and Wolsky A. M., 1999. Carbon dioxide recovery from fossil-fueled power plants. In: Meyers R. A. (Ed.), *Encyclopedia of Environmental Pollution and Cleanup*, The Wiley, NY, 271–285.

Davis D. D. and Chameides W. L., 1982. Chemistry in troposphere. *Chemical and Engineering News*, Oct. 4, 1982, 38–52.

Davis J. M., 1998. Methylcyclopentadienyl manganese Tricarbonyl: Health Risk Uncertainties and Research Directions. *Environmental Health Perspectives*, 106(Supp. 1): 191–201.

Davis J. M., Jarabek A. M., Mage D. T. and Graham J. A., 1999. Inhalation Health Risk Assessment of MMT. *Environmental Research Section A.*, 80: 103–104.

De Jong G. and Hollander K., 1998. *Air quality monitoring and risk assessment carried out during haze episode in Malaysia and Brunei (April 1998)*. Shell International Exploration and Production B.V., the Hague.

De Leeuw F. A. A. M. (Ed.), 1994. Modeling and Assessment of Heavy Metals and Persistent Organic Pollutants. In: *Proceedings of EMEP Workshop on European Monitoring*, RIVM Report 722401014, Bilthoven, The Netherlands.

De Vries W., 1996a. Acquisition of data for critical load calculation. 5.5.3. Net growth uptake of base cations and nitrogen. In: UBA (Umbelt Bundes Amt), *Manual on Methodologies and Criteria for Mapping Critical Levels/Loads and Areas Where They are Exceeded*, UNECE Convention on Long-Range Transboundary Air Pollution, 89–93.

De Vries W., 1996b. Methods for the estimation of base cation weathering rates. In: UBA (Umbelt Bundes Amt), *Manual on Methodologies and Criteria for Mapping Critical Levels/Loads and Areas where they are exceeded*. UNECE Convention on Long-Range Transboundary Air Pollution, Annex IV.

De Vries W. and Sverdrup H., 1996a. Acquisition of data for critical load calculation. 5.5.1. Critical ANC leaching. In: UBA (Umbelt Bundes Amt), *Manual on Methodologies and Criteria for Mapping Critical Levels/Loads and Areas where they are exceeded*, UNECE Convention on Long-Range Transboundary Air Pollution, 83–87.

De Vries W. and Sverdrup H., 1996b. Acquisition of data for critical load calculation. 5.5.1. Critical ANC leaching. In: UBA (Umbelt Bundes Amt), *Manual on Methodologies and Criteria for Mapping Critical Levels/Loads and Areas where they are exceeded*, UNECE Convention on Long-Range Transboundary Air Pollution, 84.

De Vries W., Bakker D. J. and Sverdrup H. U., 1997. Effect-based approaches to assess the risks of heavy metal inputs for terrestrial ecosystems. Overview method and models. In: *Background document for the work-shop on critical limits end effect-based approaches for heavy metals and POP's*, 3–7 November 1997, Bad Hazburg, Germany, Umweltbundesamt, 1225–1277.

452

De Vries W. and Bakker D. J., 1998. *Manual for Calculating Critical Loads of Heavy Metal for Terrestrial Ecosystems*. Guidelines for critical limits, calculation methods and input data. SC report 166, DLO Winand Starring Centre. 144 pp.

De Vries W. and Bakker D. I., 1998. *Manual for Calculating Critical Loads of Heavy Metals for Soils and Surface Waters*. DLO Winand Staring Centre, Wageningen, The Netherlands, Report 165, 91 pp.

DEDP, 1995. *Thailand Energy Situation*. Department of Energy Development and Promotion, Ministry of Science, Technology and Environment, Bangkok, 28 p.

Dent F. J., Rao Y. S. and Takeuchi K., 1992. *Womb of the Earth. Regional strategies for assessing soil degradation*. FAO Bangkok Occasional Paper 2, 26 pp.

Dentener F. J and Crutzen P. J., 1994. A three dimensional model of the global ammonia cycle. *Journal of Atmospheric Chemistry*, 19: 331–369.

Deuel L. E., 1999. Salt waste in oil and gas production, remediation. In: Meyers R. A. (Ed.), *Encyclopedia of Environmental Pollution and Cleanup*, The Wiley, NY, 1475–1477.

Dianwu Z., Chuyin C., Julin X., Xiaoshan Z., Zhaohua D., Jietai M., Seip H. M. and Vost R., 1994. *Acid Reign 2010 in China?*, 41 pp.

Diehl S. V. and Borazjani A., 1999. Pentachlorophenol-contaminated soils, bioremediation. In: Meyers R. A. (Ed.), *Encyclopedia of Environmental Pollution and Cleanup*, The Wiley, NY, 1216–1221.

Diemer R. B., Ellis T. D., Silcox G. D., Lighty J. D. and Pershing D. W., 1999. Hazardous waste: incineration. In: Meyers R. A. (Ed.), *Encyclopedia of Environmental Pollution and Cleanup*, The Wiley, NY, 754–761.

Dobrovolski G. V. (Ed.), 1986. Soils of Far East region. In: *Description of Soil Systematic*, Moscow, Nauka Publishers, 180–205.

Dobrovolsky V. V., 1994. *Biogeochemistry of the World's Land*. Mir Publishers, Moscow/CRC Press, Boca Raton-Ann Arbor-Tokyo-London, 362 pp.

Doolgindachbaporn T., 1995. *CALPUFF Dispersion Model Study of Ambient SO_2 Concentrations Around the Moe Moh Thermal Power Plant in Thailand*. Thesis of Master of Engineering, Asian Institute of Technology, Bangkok, 131 pp.

Ducros M., 1854. Observation d'une pluie acide. *J. Pharm. Chim.*, 3: 273–277.

Dutchak S. V., van Pul W. A. J. and Sliggers C. J., 1998. Emission, modeling and measurements of heavy metals and persistent organic pollutants. In: *Critical Limits and Effects based Approaches for Heavy Metals and Persistent Organic Pollutants, Proceeding of UN ECE Workshop*, UBA, Berlin, 3–24.

EMEP/MSC-E, 1996. *Proceedings of Workshop on the Assessment of EMEP Activities Concerning HMs and POPs and their further Development*, Moscow, Russia, MSC-E Report No. 1/97.

Encyclopedia of Malaysia, 1997. *The Environment*. Volume 1, Edited by Prof. Dr. Dato Sham Sani, Archipelago Press.

Environmental Statistics Yearbook, 1998. Ministry of Environment, Republic of Korea, 581 pp.

Erisman J. W., Brydges T., Bull K., Cowling E., Grennfelt P., Nordberg L., Satake K., Scheider T., Smeulders S., van der Hoek K., Wisniewski J. and Wisniewski J., 1999. Summary statement. In: International Nitrogen conference, Elsevier Science.

ESCAP, 1995. *State of the Environment in Asia and Pacific*. United Nations, 638 pp.

ESCAP, 1998. *Sources and Nature of Water Quality Problems in Asia and the Pacific*. New York, United Nations, 164 pp.

ESCAP, 2000. *State of the Environment in Asia and Pacific*. United Nations, 905 pp.

Fenchel T., King G. M. and Blackburn T. H., 1998. *Bacterial Biogeochemistry*. Academic Press, London et al, 307 pp.

Finkelman R. B., 1993. Trace and Minor Elements in Coal. In: M. H. Engel and S. A. Macko (Eds.), *Organic Geochemistry*, Plenum Press, New York, pp. 593–607.

Forth H. D. and Schafer J. W., 1980. *Soil Geography and Land Use*. John Wiley and Sons, New York-Chechister-Brisbane-Toronto, 484 pp.

Fumoto T. and Iwama H., 1996. Sulfate adsorption and chemical weathering in volcanic ash soils. In: NIES. *Proceedings of the International Symposium on Acid Deposition and Its Impacts*, Tsukuba, Japan, 10–12, December, 1996, 345–348.

Galiulin R. V. and Bashkin V. N., 1996. Organochlorinated compounds (PCBs and insecticides) in irrigated agrolandscapes of Russia and Uzbekistan. *Water, Air and Soil Pollution*, 89: 247–266.

Galiulin R. V., Bashkin V. N and Galiulina R. R., 2001a. Behavior of 2,4-D herbicide in coastal area of Oka River, Russia. *Water, Air and Soil Pollution*, 129: 1–12.

Galiulin R. F, Bashkin V. N., Galiulina R. R. and Birch P., 2001b. A Critical Review: Protection from Pollution by Heavy Metals—Phytoremediation of Industrial Wastewater. *Land Contamination & Reclamation*, 9(4): 349–357.

Galiulin R. V., Bashkin V. N., Galiulina R. A. and Birch P., 2001c. The theoretical basis of microbiological transformation and degradation of pesticides in soil. *Land Contamination & Reclamation*, 9(4): 367–376.

Galloway J. N., Zhao D., Jiling X. and Likens G. E., 1987. Acid rain: China, United States, and a remote area. *Science*, 236: 1559–1562.

Galloway J. N., Schelezinger W. H., Levy H., Michaels A. and Schnoor J. L., 1995. Nitrogen fixation: Anthropogenic enhancement, environmental response. *Global Biogeochemical Cycles*, 9: 235-252.

Gao Y., Arimoto R., Duce A., Lee D. S. and Zhou M. Y., 1992. Input of atmospheric trace elements and mineral matter to the Yellow Sea during the spring of a low-dust year. *Journal of Geophysics Research*, Vol. 97, No D4, 3767–3777.

Glazovsky N. F., 1997. An integrated approach to inter-regional co-operation and major activities within the inter-regular programme of action to control desertification and drought. *Desertification Control Bulletin*, No 31.

Goh K. J. and Voon O. L., 1996. Brunei Darussalam country report. In: Proceedings of the ASEAN Network on Environmental Monitoring (ASNEM) on the *3rd Workshop on Air Quality Monitoring and Analysis with Emphasis on Polycyclic Aromatic Hydrocarbons*, Environmental Research and Training Center (ERTC), Pathumthani, Thailand, 143–152.

Granat L., Suksomsankh K., Simachaya S., Tabucanon M. and Rodhe H., 1996. Regional acidity and chemical composition of precipitation in Thailand. *Atmospheric Environment*, 30: 1589–1596.

Grojean D., Grojean E. and Williams E. L., 1994. Thermal decomposition of PAN, PPN, and vinyl-PAN. *J. Air Waste Management Association*, 44: 391–396.

GSC, 1995. The Significance of Natural Sources of Metals in the Environment. A Report prepared by the Geological Survey of Canada (GSC) for the UN-ECE LRTAP (Heavy Metals) Convention, 29 pp.

Gundersen P. and Bashkin V. N., 1994. Nitrogen. In: Moldan B. and Cherny J. (Eds.), *Biogeochemistry of Small Catchments*. John Wiley and Sons, 255–283.

FAO, 1996. *FAO Production Yearbook*. Vol. 50, Rome.

Freney J. R., 1996. Control of nitrogen emission from agriculture. In: Lin H. C., et al (Eds.), Proceedings of SCOPE/ICSU Nitrogen Workshop: *The Effect of Human Disturbance on the Nitrogen Cycle in Asia*, 85–99.

Hao J, Xie S. and Lei D., 1998. Acid deposition and ecosystem sensitivity in China. In: Bashkin V. N. and Park Soon-Ung (Eds.), *Acid Deposition and Ecosystem Sensitivity in East Asia*, Nova Science Publishers, Ltd, 267–312.

Hara H., 1993. Acid deposition chemistry in Japan. *Bull. Inst. Public Health*, 43: 426–437.

Hara H., 1997. Acid deposition chemistry, paper B8010, presented at the 7th Asian Chemical Congress, 16–20 May 1997, Hiroshima, Japan.

Hart B. R., Powell M. A. and Fyfe W. S., 1995. Geochemistry and mineralogy of fly-ash from the Mae-Moh lignite deposit, Thailand. *Energy Sources*, 17: 23–40.

Hashimoto M., 1989. History of air pollution control in Japan, in *How to Conquer Air Pollution: A Japanese Experience*, Studies in Environmental Science 38, edited by H. Nishimura, Elsevier, Amsterdam, pp. 1–94.

Hedin L. O., Likens G. E., Postek K. M. and Driscoll C. T., 1990. A field experiment to test whether organic acids buffer acid deposition. *Nature*, 345: 798–800.

Hiscock R., 1997. Groundwater pollution and protection. In: T. O'Riordan (Ed.), *Environmental Science for Environmental Management*. Chapter 13, Addison Wesley Longman Ltd, England.

Hong S.-K. and Nakagoshi N., 1996. Biomass changes of a human-influenced pine forest and forest management in agricultural landscape system. *Korean J. Ecology*, 19(4): 305–320.

Hong S.-K., 1998. Changes in landscape patterns and vegetation process in the Far-Eastern cultural landscapes: human activity on pine-dominated secondary vegetation in Korea and Japan. *Phytocoenologia*, 28(1): 45–66.

Hong S.-K., Nakagoshi N. and Kamada M., 1995. Human impacts on pine-dominated vegetation in rural landscapes in Korea and western Japan. *Vegetatio*, 116: 161–172.

Howarth R. W. (Ed.), 1996. Nitrogen Cycling in the North Atlantic Ocean and its Watersheds. Kluwer Academic Publishers, Dordrecht-Boston-London, 304 pp.

Huan T. H. and Boo L. S., 1996. Singapore country report. In: *Proceedings of the ASEAN Network on Environmental Monitoring (ASNEM) on the 3rd Workshop on Air Quality Monitoring and Analysis with Emphasis on Polycyclic Aromatic Hydrocarbons*, Environmental Research and Training Center (ERTC), Pathumthani, Thailand, pp. 206–223.

Huyen D. T., 1996. Trace Elements in Mae Moh Soil Samples. In: *Proceedings of International Symposium 'Ecology and Environment'*, Bangkok, 95–106.

Hyman M. H., 1999. Groundwater and soil remediation. In: Meyers R. A. (Ed.), *Encyclopedia of Environmental Pollution and Cleanup*, The Wiley, NY, 684–714.

Iijima T., 1997. Overview of Acid Deposition in Japan and Acid deposition Monitoring Network in East Asia, paper presented at the *4th ASEAN Workshop on Air Quality Monitoring and Analysis with Emphasis on Acid Deposition*, Technopolis, Pathumthani, Thailand, 25 February–4 March 1997.

IIED, 1994. *Environmental Synopsis of Indonesia*. International Institute for Environment and Development, Overseas Development Administration, London.

IPCC, 1997. Guidelines for National Greenhouse Gas Inventories. OECD/ICDE, Paris.

Iserman K., 1991. Share of agriculture in nitrogen and phosphorus emission into the surface waters of Western Europe against the background of their eutrophication. *Fertilizer Research*, 26: 253–269.

Independent Newspaper, 1999. *Headline: Half of the World's rivers polluted or running dry*. Byline by Mary Dejevsky in Washington. November 30, 1999 issue, London.

Joseph U., 1997. Country report (Myanmar), paper presented at the *4th ASEAN Workshop on Air Quality Monitoring and Analysis with Emphasis on Acid Deposition*, Technopolis, Pathumthani, Thailand, 25 February–4 March 1997.

Ishak A. and Hamzah W. N. W., 1997. Air quality monitoring and acid deposition in Malaysia, paper presented at the *4th ASEAN Workshop on Air Quality Monitoring and Analysis with Emphasis on Acid Deposition*, Technopolis, Pathumthani, Thailand, 25 February–4 March 1997.

ISSS, 1998. The Major Soil Groups: Key and Description. In: *World Reference Base for Soil Resources*, 1–113.

Kabata-Pendias A. and Pendias H., 1992. *Trace elements in soil and plants*. 2nd edition, CRC Press, Boca Raton, Florida, USA, 365 p.

Kaewruenrom A., 1990. *Soil of Thailand: Properties, distribution and use*, Division of Soil Science Faculty of Agriculture, Kasetsart University, Bangkok, 99–475.

Kato N. and Akimoto H., 1992. Anthropogenic emissions of SO_2 and NO_x in Asia: emission inventories. *Atmospheric Environment*, 26a: 2997–3017.

Karu A. E., Lesnik B. and Haas R. A., 1999. Polychlorinated biphenyls: detection by immunoassay. In: Meyers R. A. (Ed.), *Encyclopedia of Environmental Pollution and Cleanup*, The Wiley, NY, 1231–1238.

Kellner O., Sawano J., Yoshii T. and Oku K., 1894. Amounts of ammonium and nitrate in rainwater. *Nokadaigaku Gakujyutsushiken Iho*, 1: 28–42.

Kennish M. J., 1999. Marine pollution. In: Meyers R. A. (Ed.), *Encyclopedia of Environmental Pollution and Cleanup*, The Wiley, NY, 909–943.

Khan M. A., 1993. *Problem and Prospect of Sustainable management of Urban Water Bodies in the Asian and Pacific Region*, Bangkok.

Khanh N. H., 1993. *Investigation. Evaluation of the state of pH value of rainfall*. Research Report, Hydrometeorological Service, Hanoi, Vietnam.

Khummongkol P., 1999. Acid Deposition Problems and Related Activities in Thailand. In: *Proc. East Asian Workshop on Acid Deposition*, 6–8 Oct. 1999, Bangkok, Thailand, 60–73.

KIGAM, 1995. *Geological Map of Korea*, 1:1,000,000.

Kim J.-U., 1995. *Ecophysiplogical Responses of Quercus mongolica Stands to Air Pollution and a Model for its Productivity Affected by Ambient Ozone*, Seoul National University, 135 pp.

Kim K.-D., 1994. *Nutrients Input by Precipitation, throughfall and Stemflow in stands of Pinus densiflora and Quercus mongolica*, Seoul National University, 77 pp.

Kim K.-W., Kim S.-O., Son A.-J., Lee B.-T. and Ko I., 1999a. Chemical composition of Yellow Sand particles during 1999 Spring episode. In: *Proceedings of the 2nd International Symposium on Advanced Environmental Monitoring*. KJIST, Kwangju, South Korea, 118–119.

Kim K. W., Kim Y. J., Oh S. J. and Kim M. O., 1999b. Contribution of light absorption by elemental carbon to visibility degradation in an urban atmosphere, Kwangju, Korea. In: *Proceedings of the 2nd K-JIST International Symposium on Advanced Environmental Monitoring*, November 4–5, 1998, Kwangju, Korea, 85–86.

King H. B. and Yang B. Y., 1984. Precipitation and stream water chemistry in Pi-Lu-Chi watersheds. January 1981–December 1982. *Bulletin of the Taiwan Forestry Research Institute*, Taipei, No. 427.

Kira T., 1977. Production rates. In: Shido T. and Kiro T. (Eds.), *Prymary Productivity of Japanese Forests: Productivity of Terrestrial Communities*, JIBP Synthesis, Vol. 16, 101–162.

Knisel W. G. (Ed.), 1980. *CREAMS: a flied-scale model for chemical runoff and erosion*. USDA-SEA Conservation Research Report, 26, US Governmental Printing Office, Washington, DC.

Koch G. W. and Mooney H. A. (Eds.), 1996. *Carbon Dioxide and Terrestrial Ecosystems*. Academic Press, New York-Boston, 443 pp.

Kohno Y., Matsumura H. and Kobayashi T., 1998. Differential sensitivity of 16 tree species to simulated acid rain or sulfur dioxide in combination with ozone. In: Bashkin V. N. and Park S.-U. (Eds.), *Acid Deposition and Ecosystem Sensitivity in East Asia*, Nova Science Publishers, Ltd, 143–188.

Korean Soil Series, 1995. *Areas of distribution, physical and chemical properties of soils*, 217 pp.

Kouprianov V. I. and Bashkin V. N., 2000. Environmental Impact of Heavy Metals Contained in Fly Ash Emitted from the Thai Lignite-Fired Power Plant, In: Feeley T. J., III (Ed.), *Particulate Matter and Stationary Sources*, Division of Fuel Chemistry, 219-th American Chemical Society Meeting, Vol. 45, No. 1: San Francisco, CA, pp. 83–87.

Kozlov M. Ya. and Towprayoon S., 1998. Sensitivity of Thailand's ecosystems to acidic deposition. In: Bashkin V. N. and Park S.-U. (Eds.), *Acid Deposition and Ecosystem Sensitivity in East Asia*, Nova Science Publishers, Ltd, 335–378.

Kurita H., Hori J., Hamada Y. and Ueda H., 1991. Decrease of pH of river and lake water in mountainous region of central Japan and its relation to acid rain. *J. Jpn. Assoc. Air Pollut.*, 28: 308–315.

Kuylenstierna J. C. I., Cambridge H. M., Cinderby S. and Chadwick M. J., 1995. Terrestrial Ecosystem Sensitivity to Acidic Deposition in Developing Countries. *Water, Air and Soil Pollution*, 85: 2319–2324.

Lacaux J.-P., Cachier H. and Delmas R., 1993. Biomass burning in Africa; an overview of its impacts on atmospheric chemistry. In: Crutzen J. P and Goldammer J. G. (Eds.), *Fire in the Environment: the ecological, atmospheric and climatic importance of vegetation fires*. Wiley, Chichester, 159–191.

Lan N. T., 1996. Vietnam country report, in Proceedings of the ASEAN Network on Environmental Monitoring (ASNEM) on the *3rd Workshop on Air Quality Monitoring and Analysis with Emphasis on Polycyclic Aromatic Hydrocarbons*, Environmental Research and Training Center (ERTC), Pathumthani, Thailand, 258–262.

Lee B. K., Hong S. H. and Lee D. S., 2000. Chemical composition of precipitation and wet deposition of major ions on the Korean peninsula. *Atmospheric Environment*, 34: 563–575.

Lee B. H., Bae J. O., Chung S. W., Kim E. S., Koh K. S., Huh I. A., Kim D. H., Lee J. B. and Cha B. J., 1990. A comparison of the structure and productivity of plant communities around industrial complexes and an unpolluted area. In: *A study on the Evaluation and the Prevention of Damage to Plant Communities by Air Pollution (II)*. Report of National Institute of Environmental Research Republic of Korea, Vol. 12, 181–235

Lee H. J., Lee M. H., Kim D. H. and Kim S. D., 1992. Influence of acidic deposition on forest soils. In: *A study on the assessment of damage by air pollutants and acid rain (II-5)*. Report of National Institute of Environmental Research Republic of Korea, 14: 255–278.

Lee H. J., Lee M. H., Kim D. H., Kim S. D., Yun J. K. and Kim Y. H., 1993. Effects on the forest soils. In: *A study on the assessment of damage by air pollutants and acid rain (II)*. Report of National Institute of Environmental Research Republic of Korea, 191–252.

Lee K. W. and Young J. C. (Eds.), 1998. Advance Environmental Monitoring, *Proceedings of the 1st K-JIST International Symposium*, Kwangju, Korea, 176 pp.

Lee K. W., et al (Eds.), 1999. Advance Environmental Monitoring, *Proceedings of the 2nd K-JIST International Symposium*, Kwangju, Korea, 187 pp.

456

Leong C. P., Lim S. F. and Lim J. T., 1988. *Report on Rain Acidity Analysis Based on Data from the National Acid Rain Monitoring Network*, Malaysian Meteorological Service.

Levine J. S. (Ed.), 1991. *Global Biomass Burning*. Cambridge, MA, MIT Press.

Lin H.-C., Yang S.-S., Hung T.-C. and Chou C.-H. (Eds.), 1996. The Effect of Human Disturbance on the Nitrogen Cycle in Asia, Proceedings of SCOPE/ICSU Nitrogen Workshops.

Lin T. C., 1998. Acid deposition and forest ecosystem sensitivity in Taiwan. In: Bashkin V. N. and Park S.-U. (Eds.), Acid Deposition and Ecosystem Sensitivity in East Asia, NovaScience Publishers, USA, 379–412.

Lubis S. M. and Aprishanty R., 1996. Indonesia country report. In: Proceedings of the ASEAN Network on Environmental Monitoring (ASNEM) on the *3rd Workshop on Air Quality Monitoring and Analysis with Emphasis on Polycyclic Aromatic Hydrocarbons*, Environmental Research and Training Center (ERTC), Pathumthani, Thailand, 153–176.

Mahatnirunkul V., Towprayoon S. and Bashkin V., 2002. Application of EPA hydrocarbon spill screening model to the hydrocarbon contaminated site of Thailand. *Land Contamination & Reclamation*, Vol. 10, No 1, 17–24.

Manahan S. E., 1994. *Environmental Chemistry*. Sixth Edition, Lewis publishers, Boca Raton etc, 868 pp.

Manahan S. E., 2000. *Environmental Chemistry*. Seventh Edition. Lewis publishers, Boca Raton etc, 898 pp.

Manokaran N., 1980. The nutrient contents of precipitation, throughfall and stemflow in a lowland tropical rain forest in peninsular Malaysia. *The Malaysian Forester*, 43: 266–289.

Mantell L., 1998. World Socialist Web Site News and Analysis: Asia-Indian Subcontinent. December 2, 1998. http://wsws.org/news/1998/dec1998/bang-d02.shtml

Matson P. A., McDowell W. H., Townsend A. R. and Vitousek P. M., 1999. The globalization of N deposition: ecosystem consequences in tropical environments. Biogeochemistry, 46: 45–65.

Matsushita H., 1996. Atmospheric pollution by PAHs in Japan. In: Proceedings of the ASEAN Network on Environmental Monitoring (ASNEM) on the *3rd Workshop on Air Quality Monitoring and Analysis with Emphasis on Polycyclic Aromatic Hydrocarbons*, Environmental Research and Training Center (ERTC), Pathumthani, Thailand, 84–104.

Mavalankar D. V., Trivedi C. R. and Gray R. H., 1991. Levels and risk factors for perinatal mortality in Ahmedabad, India, WHO, Bulletin.

McElroy M. B. and Salawith R. J., 1989. Changing composition of the global stratosphere. *Science*, 243: 763–770.

McKenzie L., 1999. Smoldering biomass fuels, measuring and modeling. In: Meyers R. A. (Ed.), *Encyclopedia of Environmental Pollution and Cleanup*, The Wiley, NY, 1536–1548.

Meybeck M., 1982. Carbon, nitrogen, and phosphorus transport by World rivers. *American Journal of Science*, 282: 401–450.

Meyers R. A. (Ed.), 1999. *Encyclopedia of Environmental Pollution and Cleanup*, The Wiley, NY, Vol. 1– Vol. 2, 1890 pp.

Misch A., 1994. Assessing environmental health risk. In: Brown L. R. (Ed.), *State of the World*. W.W. Norton and Company, NY.

Miyake Y., 1939. The chemistry of rain water. *J. Met. Soc. Japan, II*, 17: 20–37.

Mo T., Gu Q. and Zhao K., 1988. Criterion of acid rain formation. *Huanjing Kexue Xuebao*, 8: 32–39.

Moldan B. and Cherny J. (Eds.), 1994. *Biogeochemistry of Small Catchments*. John Wiley and Sons, 419 pp.

Mosier A., et al (Eds.), 1998. *Nutrient Cycling in Agroecosystems*. Kluwer Academic Publishers, 313 pp.

Murray J. L. and Lopea A. D., 1994. Global and regional causes of death patterns in 1990. *Bulletin of the World Health Organization*, 72(3).

Muraleedharan T. R., Radoejeic M., Waugh A. and Caruana A., 2000. Chemical characterization of the haze in Brunei Darussalam during the 1998 episode. *Atmospheric Environment*, 34: 2725–2731.

Muraleedharan T. R., Radoejeic M., Waugh A. and Caruana A., 2000. Emissions from the combustion of peat: an experimental study. *Atmospheric Environment*, 34: 3033–3035.

Muraleedharan T. R. and Radoejvic M., 2000. Personal particle exposure monitoring using nephelometry during haze in Brunei. *Atmospheric Environment*, 34: 2733–2738.

Murano K., 1997a. Some part of draft technical manual for monitoring wet deposition. Acid Deposition Network in East Asia, paper presented at the *4th ASEAN Workshop on Air Quality Monitoring and Analysis with Emphasis on Acid Deposition*, Technopolis, Pathumthani, Thailand, 25 February–4 March 1997.

Murano K., 1997b. Activity of JEA for East Asian Acid Precipitation Monitoring Network, paper presented at the *4th ASEAN Workshop on Air Quality Monitoring and Analysis with Emphasis on Acid Deposition*, Technopolis, Pathumthani, Thailand, 25 February–4 March 1997.

Muttamara S. and Sales C., 2000. A study on disinfection of filtered water using ozone and hydrogen peroxide. In: *Abstract of Sixth Eurasian Congress on Chemical Sciences*, Brunei Darussalam, 17.

National Academy of Sciences, 1978. An Assessment of Mercury in the Environment. National Academy of Sciences, Washington, DC.

Narita Y., Chai H. and Tanaka S., 1997. *A study on acid rain in China on the basis of the literature investigation*, paper B9P09, presented at the 7th Asian Chemical Congress, 16–20 May 1997, Hiroshima, Japan.

NEPA (National Environmental Protection Agency), 1994. *China Environmental News*. 4 June 1994.

Neudachin A. P., 1999. Agrogeochemistry of Ameliorated Peat Soils in the Northwest Part of Middle Amur low plain. PhD thesis, Biology-Soil Institute RAS, Vladivostok, 25 pp.

Nilsson I. and Grennfelt P. (Eds.), 1988. Critical Loads for Sulfur and Nitrogen. Report from a Workshop Held at Stokhoster, Sweden, March 19–24, 1988, Miljo Rapport 1988: 15. Copenhagen, Denmark, Nordic Council of Ministers, 418 pp.

Nikolaev Yu. A. and Fomin P. A., 1997. The nature of nature of noctilucent clouds and the Earth's ozone layer. *Combustion, Explosion, and Shock Waves*, 33(4): 393–402.

Nishimura H. and Sadakata M., 1989. Emission control technology. In: *How to Conquer Air Pollution: A Japanese Experience*, Studies in Environmental Science 38, edited by H. Nishimura, Elsevier, Amsterdam, 115–156.

Nixon S. W., Ammerman J. W., Atkinson L. P., Berounsky V. M., Billen G., Boicourt W. C., Boynton W. R., Church T. M., Ditoro D. M., Elmgren R., Garber J. M., Giblin A. E., Jahnke R. A., Owens N. J. P., Pilson J. H. and Seitzinger S. P., 1996. The fate of nitrogen and phosphorus at the land-sea margin of the North Atlantic Ocean. In: R. W. Howarth (Ed.), *Nitrogen Cycling in the North Atlantic Ocean and its Watersheds*, Kluwer Academic Publishers, 141–180.

Nriagu J. O., 1989. A global assessment of natural sources of atmospheric trace metals. *Nature*, Vol. 338, 47–49.

Nriagu J. O. and Pacina J. M., 1989. Quantitative assessment of worldwide contamination of air, water and soils by trace metals. *Nature*, 333: 134–139.

Oonk H. and Kroeze C., 1999. Nitrous oxide emissions and control. In: R. A. Meyers (Ed.), *Encyclopedia of Environmental Pollution and Cleanup*, The Wiley, 1055–1069.

Oshihima Y., 1977. Primary production. In: Kitazawa Y. (Ed.), *Ecosystem Analysis of the Subalpine Coniferous Forest of the Shigayama IBP Area, Central Japan*, JUBP Synthsis, Vol. 15, 125–134.

Pacina J. M., Voldner E., Keeler G. J. and Evans G. (Eds.), 1993. *Proceedings of the First Workshop on Emissions and Modeling of Atmospheric Transport of Persistent Organic Pollutants and Heavy Metals*, EMEP/CCE Report No. 7/93, Norwegian Institute for Air Research, Norway.

Pacyna J. M., 1995. Emission inventory for heavy metals in the ECE. In: Heavy Metals Emissions. State-of-the-Art Report. Economic Commission for Europe. Convention on Long-Range Transboundary Air Pollution, Prague, June 1995, 33–50.

Pandey M. R., et al, 1989. Domestic smoke pollution and acute respiratory infection in a rural communities of Hill region of Nepal. Environmental Institute, 15, 337–345.

Part H.-S., Kang J.-W., Oh H.-J. and Kim W.-J., 1998. Continuous monitoring of ozone consumption for application of ozonation. In: *Proceedings of the 1st K-JIST international Symposium on Advanced Environmental Monitoring*, Kwangju, Korea, 135–136.

Park S.-U., Lee T.-Y., Lee D.-S., Shim S.-K., Chao S.-Y., Lee C.-B. Sunwoo Y. and Moon S.-E., 1997. *Research and Development on Technology for Monitoring and Prediction of Acid Rain*. Ministry of Environment, Progressive Report 1 on Second Stage, 542 pp.

Park S.-U. (Ed.), 1998. *Research and Development on Basic Technology for Atmospheric Environment in Global Scale: Development of Technology for Monitoring and Prediction of Acid Rain (G-7 project)*. Ministry of Environment, Republic of Korea, 602 pp.

Park S.-U., In H.-J. and Lee Yu-H., 1999. Parameterization of wet deposition of sulfate by precipitation rate. *Atmospheric Environment*, 33: 4469–4475.

458

Park S.-U. and Lee Y-H., 1999. Mapping of base cation deposition in South Korea. In: *Proceedings of 5th International Joint Seminar on Regional Deposition Processes in Atmosphere*, Seoul, October 12–16, 177–188.

Park S.-U., In H.-I., Kim S.-W. and Lee Y.-H. (2000). Sulfur deposition in South Korea. *Atmospheric Environment*, 34: 3259–3269.

Park S.-U. and Bashkin V. N. (2001). Critical loads of sulfur on Korean ecosystems. *Water, Air and Soil Pollution*, 132(1–2): 19–41.

PDC, 1999. *Proceedings of East Asian Workshop on Acid Deposition*. 6–8 October 1999, Bangkok, 276 pp.

Peakall D. B. and Shugart L. R., 1999. Biomarkers. In: Meyers R. A. (Ed.), *Encyclopedia of Environmental Pollution and Cleanup*, The Wiley, NY, 223–229.

Peart M. R., 1995. The occurrence of acid rain in Hong Kong. In: *Proceedings of the International Symposium on Climate and Life in the Asia Pacific*, University of Brunei Darussalam, 10–13 April 1995, edited by Sirinanda K. U., 10–19.

Pepelko W. E. and Valcovic L., 1999. Carcinogens: identification and risk assessment of. In: Meyers R. A. (Ed.), *Encyclopedia of Environmental Pollution and Cleanup*, The Wiley, NY, 285–291.

Phien T. and Siem N., 1994. *Land degradation and its data collection and analysis in Vietnam*. FAO Bangkok Publ., No 3.

Phonboon K. (Ed.), 1998. *Health and Environmental Impacts from the 1997 ACEAN Haze in Southern Thailand*. Health System Research Institute, 198 pp.

Posch M., Hettelingh J.-P., de Smet P. A. M. and Downing R. J. (Eds.), 1997. *Calculation and Mapping of Critical Thresholds in Europe*. RIVM, Bilthoven, The Netherlands, 163 pp.

Posch M., Hettelingh I.-P., Alcamo J. and Krol M., 1996. Integrated Scenario of Acidification and Climate Change in Asia and Europe. *Global Environmental Change*, 6(4): 375–394.

Posch M., de Smet P. A. M., Hettelingh J.-P. and Downing R. J. (Eds.), 1999. *Calculation and Mapping of Critical Thresholds in Europe*. Status Report 1999. Coordination Center for Effects, RIVM Report No. 259101009, Bilthoven, the Netherlands, 165 pp.

Potisuwan U., 1994. *Collection and analysis of land degradation data in Thailand*. FAO Bangkok Publ., No 3.

Rachmawati E., 1996. NO_2 concentration in local cities in Indonesia. Result of the nationwide monitoring by the TEA plate method, EMC Air Quality Laboratory, Indonesia.

Radojevic M., 1995. Taking a rain check, *Analysis Europa*, August 1995, 35–38.

Radojevic M., 1996. Sea changes. *Chemistry in Britain*, 32(11): 47–49.

Radojevic M., 1998. Opportunity NO_x. *Chemistry in Britain*, 34(3): 30–33.

Radojevic M. and Bashkin V. N., 1999. *Practical Environmental Analysis*. RSC, UK, 466 pp.

Radojevic M. and Harrison R. M., 1992. *Atmospheric Acidity: Sources, Consequences and Abatement*. Elsevier Applied Science, London.

Radojevic M. and Lim L. H., 1995a. A rain acidity study in Brunei Darussalam. *Water, Air and Soil Pollution*, 85: 2369–2374.

Radojevic M. and Lim L. H., 1995b. Short-term variation in the concentration of selected ions within individual tropical rainstorms. *Water, Air and Soil Pollution*, 85: 2363–2368.

Radojevic M., Lim L. B. B., Ling V. O. and Lim L. H., 1997. A report on the studies of rain acidity and the concentration of selected ions in Negara Brunei Darussalam, paper presented at the *4th ASEAN Workshop on Air Quality Monitoring and Analysis with Emphasis on Acid Deposition*, Technopolis, Pathumthani, Thailand, 25 February–4 March 1997.

Radojevic M., Tan K. S., Makarimi A. and Medan R., 1998. Unpublished data.

Radoejvic M. and Tan K. S., 2000. Impacts of biomass burning and regional haze on the pH of rain water in Brunei Darussalam. *Atmospheric Environment*, 34: 2739–2744

Radoejvic M. and Hassah H., 2000. Air quality in Brunei Darussalam during the 1998 haze episode. *Atmospheric Environment*, 33: 3651–3658.

Rahman A. and Chin T. G., 1984. A preliminary study on the acidity of rainfall in Singapore. In: *Proceedings of 3rd Symposium on Our Environment*, 27–29 March 1984, edited by Lin K. L. and Sin H. C., Faculty of Science, National University of Singapore, 344–361.

Rasmussen P. E., 1994. Current methods of estimating atmospheric mercury fluxes in remote areas. *Environmental Science Technology*, 28(13): 2233–2241.

Ratanasthien B. and others, 1993. Chemistry Balance of Mae Moh Fly Ash. In: *Proceedings of the Seminar on Potential Utilization of Lignite Fly Ash*, EGAT, Bangkok, Chapter 1, 1–2.

Richardson S. D., 1999. Drinking water disinfection by-products. In: Meyers R. A. (Ed.), *Encyclopedia of Environmental Pollution and Cleanup*, The Wiley, NY, 454–468.

Richter O., 1999. Pesticides: environmental fate. In: Meyers R. A. (Ed.), *Encyclopedia of Environmental Pollution and Cleanup*, The Wiley, NY, 1224–1231.

Rinda N., 1996. SO_2 *concentration in local cities in Indonesia. Result of the nationwide monitoring by the* PbO_2 *method*, EMC Air Quality Laboratory, Indonesia.

Rodhe H., Galloway J. and Dianwu Z., 1992. *Acidification in Southeast Asia-Prospects for the coming decades*, AMBIO, 21: 148–149.

Rorrer G. I., 1999. Heavy metal ions, removal from wastewater. In: Meyers R. A. (Ed.), *Encyclopedia of Environmental Pollution and Cleanup*, The Wiley, NY, 771–779.

Ross D. G., Lewis A. M. and Koutsenko G. D., 1998. Predicting and Forecasting of Sulfur Dioxide Impact Events at Mae Moh, Thailand. In: *Proceedings of the 14th International Clean Air & Environment Conference*, Australia, 468–474.

Ryaboshapko A. G., Sukhenko V. V. and Paramonov S. G., 1994. *Assessment of wet sulfur deposition over the former USSR*, Tellus, 46B: 205–219.

Saefudin A., 1997. Analysis of rain water at EMC station Serpong, Indonesia, paper presented at the *4th ASEAN Workshop on Air Quality Monitoring and Analysis with Emphasis on Acid Deposition*, Technopolis, Pathumthani, Thailand, 25 February–4 March 1997.

Saheed S. M., 1992. *Environmental issues in land development in Bangladesh*. FAO Bangkok Publ., No 8, 105–127.

Sanhueza E., Arias M. C., Donoso L., Graterol N., Hermosos M., Marti I., Romero J., Rondon A. and Santana M., 1992. *Chemical composition of acid rains in the Venezuelan savannah region*. Tellus, 44B: 54–62.

Satake K., Inoue T., Kasasaku K., Nagafuchi O. and Nakano T., 1998. Monitoring of nitrogen compounds on Yakushima island, a world natural heritage site. *Environmental Pollution*, 102, S1, 107–113.

Sato K., Wakamatsu K. and Takahashi A., 1998. Changes in distribution of aluminum species in soil solution due to acidification. In: Bashkin V. N. and Park Soon-Ung (Eds.), *Acid Deposition and Ecosystem Sensitivity in East Asia*, Nova Science Publishers, Ltd, 125–142.

Satoh H., Iwamato Y., Mino T. and Matsuo T., 1998. Activated sludge as a possible source of biodegradable plastics. *Water Science Technology*, 38(2): 103–109.

Scherr S. J., 1999. *Soil degradation*. International Food Policy Research Institute, Washington, D.C.

Schiele R., 1991. Chapter II.19. Manganese. In: Merian E. (ed.), *Metals and Their Compounds in the Environment: Occurrence, Analysis, and Biological relevance*. VCH Publishers, USA, 1035–1044.

Schlesinger W. H., 1991. *Biogeochemistry*. An Analysis of Global Changes, Academic Press, 443 pp.

Schultz Institute, 1991, *Environmental Impact Assessment for Mae Moh Mine and Power Plant Expansion Project*. Final Report by Schultz Inst., November, Bangkok.

Schweithelm J., 1999. The fire this time: an overview of Indonesia's forest fire in 1997/1998. WWF-Indonesia, Jakarta.

Seung Y. H. and Park Y. C., 1990. Physical and environmental character of the Yellow Sea. In: Park C. H., Kim D. H. and Lee S. H. (Eds.), *The Regime of the Yellow Sea*, Institute of East and West Studies, Seoul, 9–38.

Sierra P., Loranger G., Kennedy G. and Zayed J., 1995. Occupational and environmental exposure of automobile mechanics and non-automotive workers to airborne manganese arising from the combustion of methylcyclopentadienyl manganese tricarbonyl (MMT). *American Industrial Hygiene Association Journal*, 56: 713–716.

Sierra P., Chakrabarti S., Tounkara R., Loranger S., Kennedy G., and Zayed J., 1998. Bioaccumulation of manganese and its toxicity in feral pigeons (Columba Livia) exposed to manganese oxide dust (Mn_3O_4). *Environmental Research, Section A*, 79: 94–101.

Shindo J., 1998. Model application for assessing the ecosystem sensitivity to acidic deposition based on soil chemistry changes and nutrient budgets. In: Bashkin V. N. and Park Soon-Ung (Eds.), *Acid Deposition and Ecosystem Sensitivity in East Asia*, Nova Science Publishers, Ltd, 312–334.

Shindo J, Bregt A. K. and Takamata T., 1995. Evaluation of Estimation Methods and Base Data Uncertainties for Critical Loads of Acid Deposition in Japan. *Water, Air, and Soil Pollution*, 85: 2571–2576.

460

Shindo J., 1998. Model application for assessing the ecosystem sensitivity to acid deposition based on soil chemistry changes and nutrient budget. In: Bashkin V. N. and Park Soon-Ung (Eds.), *Acid Deposition and Ecosystem Sensitivity in East Asia*, Nova Science Publishers, Ltd, 313–335.

Semenov M. Yu., Bashkin V. N. and Sverdrup H., 2000. Application of biogeochemical model "PROFILE" for an assessment of North Asian ecosystem sensitivity to acid deposition. *Asian Journal of Energy & Environment*, 1(2): 143–162.

Sen A. K. and Mondal N. G., 1990. Removal and uptake of copper (II) by Salvinia natans from wastewater. *Water, Air and Soil Pollution*, 49(1–2): 1–6.

Sequeira R., Peart M. R. and Lai K. H., 1996. A comparison between filtered and unfiltered atmospheric depositions from a rural area in Hong Kong. *Atmospheric Environment*, 30: 3221–3224.

Shim S.-G., Kim Y.-P. and Kang C.-H., 1997. Acid rain research activities in Korea, paper no. B8011, presented at the 7th Asian Chemical Congress, 16–20 May 1997, Hiroshima, Japan.

Shin E. B., Lee S. K. and Song D. W., 1989. Acidity of rainwater in Seoul area. In: *Proceedings of the 8th World Clean Air Congress*, Vol. 2, edited by Brasser L. J. and Moulder W. C., Elsevier, Amsterdam.

Shrivastava S. and Rao K. S., 1997. Observation on the Utility of Integrated Aquatic Macrophyte Base System for Mercury Toxicity Removal. *Bulletin of Environmental Contamination and Toxicology*, 59(5): 777–782.

Siador C. S., Jr., 1996. Philippines country report, in Proceedings of the ASEAN Network on Environmental Monitoring (ASNEM) on the *3rd Workshop on Air Quality Monitoring and Analysis with Emphasis on Polycyclic Aromatic Hydrocarbons*, Environmental Research and Training Center (ERTC), Pathumthani, Thailand, 189–205.

Siador C. S., Jr. and Calderon I. G., 1997. Environmental monitoring in the Philippines with emphasis on acid deposition, paper presented at the *4th ASEAN Workshop on Air Quality Monitoring and Analysis with Emphasis on Acid Deposition*, Technopolis, Pathumthani, Thailand, 25 February–4 March 1997.

Siriswasdi J., 1997. Acid rain deposition monitoring in Thailand. Monitoring by Pollution Control Department, paper presented at the *4th ASEAN Workshop on Air Quality Monitoring and Analysis with Emphasis on Acid Deposition*, Technopolis, Pathumthani, Thailand, 25 February–4 March 1997.

Sloan J. J., Dowdy R. H., Dolan M. S. and Linden D. R., 1997. Long-term effects of biosolids applications on heavy metals bioavailability in agricultural soils. *J. Environ. Qual.*, 26: 966–974.

Smith D. K., Niemeyer S., Estes J. A. and Flegal A. R., 1990. Stable lead isotope evidence of anthropogenic contamination in Alaskan sea otters. *Environmental Science & Technology*, 24: 1517–1521.

Smith K., Apte M., Ma Y., Wongsekirttirat W. and Kulkarni A., 1994. Air pollution and the energy latter in Asian Cities. *Energy*, 19(5).

Sophy M., 1997. Kingdom of Cambodia report, paper presented at the *4th ASEAN Workshop on Air Quality Monitoring and Analysis with Emphasis on Acid Deposition*, Technopolis, Pathumthani, Thailand, 25 February–4 March 1997.

Sophal P. (1999). Kingdom of Cambodia report. In: *Proceedings of East Asian Workshop on Acid Deposition*, Thailand, 6–8 October 1999, 146–150.

Soviev M., 1998. Numerical modeling of acid deposition on Eurasian continent. In: Bashkin V. N. and Park Soon-Ung (Eds.), *Acid Deposition and Ecosystem Sensitivity in East Asia*, Nova Science Publishers, Ltd, 5–48.

Spain J. D., 1982. *Basic computer models in biology*. Advanced book program. Addison-Wesley Publishing Co., Reading, Mass.

Stern D. I. and Kaufman R. K., 1996. Relative rates of methane emission sources. *Chemosphere*, 33: 159–176.

Subue T., 1990. Association of indoor air pollution and lifestyle with lung cancer in Osaka. *Int. J. of Epidemiology*, 19 (Suppl. 1), 62–66.

Sverdrup H. and de Vries W., 1994. Calculating critical loads for acidity with the simple mass balance method. *Water, Air, and Soil Pollution*, 72: 142–162

Ter Haar G. L., Griffing M. E., Brandt M., Oberding D. G., and Kapron M., 1975. Methylcyclopentadienyl Manganese Tricarbonyl as an Antiknock: Composition and Fate of Manganese Exhaust Products. *Journal of Air Pollution Control Association*, 25: 858–860.

Tha Tun O. and Swe M., 1994. *Collection and analysis of land degradation data in Myanmar*. Paper for expert consultation on collection and analysis of land degradation data. FAO Bangkok Publ., No 3.

Thuc T. and Yen N. H., 1997. Acid rain study in Vietnam, paper presented at the *4th ASEAN Workshop on Air Quality Monitoring and Analysis with Emphasis on Acid Deposition*, Technopolis, Pathumthani, Thailand, 25 February–4 March 1997.

Thomas V. M. and Spiro T. G., 1999. Dioxin emission inventory. In: Meyers R. A. (Ed.), *Encyclopedia of Environmental Pollution and Cleanup*, The Wiley, NY, 436–440.

Tillman D. A., 1994. *Trace Metals in Combustion Systems*. Academic Press, Inc., USA, 97–109.

Tobacova S., 1986. Maternal exposure to environmental chemicals. *Neurotoxicology*, 7(2): 421–440.

Toda E., 1999. Overview of Acid Deposition in Japan. In: *Proceedings of East Asian Workshop on Acid Deposition*, 6–8 Oct. 1999, Bangkok, Thailand, 83–112.

Thornton I., Ramsey M. and Atkinson N., 1995. *Metal in the global environment: facts and misconceptions*. International Council on Metals and Environment, 103 pp.

UBA, 1996. *Manual on Methodologies and Criteria for Mapping Critical Levels/Loads and geographical areas there they are exceeded*. Berlin, 142 pp. + Annex

Um K. T., 1985. *Soils of Korea*. Soil Survey Materials No. 11, Agricultural Science Institute, Rural Administration, Korea, 66 pp.

UNEP/GEMS, 1992. The Impact of Ozone Layer Depletion. UN, Nairobi.

UNEP, 1996. Groundwater: a threatened resource. *UNEP Environmental Library*, No 15, UNEP Nairobi, Kenya.

UNESCO, 1978. *World Water Balance and Water Resources of the Earth*. Studies and Reports in Hydrology, UNESCO

U.S. EPA, 1995. *How to evaluate alternative cleanup technologies for underground storage tank sites: A guide for corrective action plant reviewers*. U.S. Environmental Protection Agency, Washington, D.C., EPA 510-B-95-007.

U.S. EPA Chemical Profile. Online, 1997. Available from http://www.epa.gov/swercepp/ehs/profile/12108133.txt [Accessed 1999 Nov 18].

Van den Broek W., Wienke D. and Buydens L., 1999. Plastic among nonplastics, identification in mixed waste. In: Meyers R. A. (Ed.), *Encyclopedia of Environmental Pollution and Cleanup*, The Wiley, NY, 1283–1289.

Vearasilp T. and Songsawad K., 1991. *Soil Information System for Thailand*. Department of Land Development, Bangkok, 253 pp.

Vialard-Goudou A. and Richard C., 1956. Etude pluviometrique, physiochimique et economique des eaux de pluie a Saigon (1950–1954). *l'Agronomie Tropicale*, 11: 74–92.

Vitousek P. M., Aber J. D., Howarth R. W., Likens G. E., Matson P. A., Schinder W. H., Schlezinger W. H. and Tilman D. G., 1997. Human alteration of nitrogen global cycles: Sources and consequences. *Ecol. Appl.*, 7: 737–750.

Voitkevich G. V., 1986. *Chemical Evolution of the Earth*. Nauka Publishing House, Moscow, 235 pp.

Vongphosy T., 1997. Lao PDR coutry report, paper presented at the *4th ASEAN Workshop on Air Quality Monitoring and Analysis with Emphasis on Acid Deposition*, Technopolis, Pathumthani, Thailand, 25 February–4 March 1997.

Walker J. C. G., 1977. *Evolution of the atmosphere*, Plenum Press, New York.

Wang W. and Wang T., 1996. On acid rain formation in China. *Atmospheric Environment*, 30: 4091–4093.

Wilkins E., 1999. Oil spill contaminant, terrestrial. In: Meyers R. A. (Ed.), *Encyclopedia of Environmental Pollution and Cleanup*, The Wiley, NY, 1148–1152.

Wolff G., 1999. Air Pollution. In: Meyers R. A. (Ed.), *Encyclopedia of Environmental Pollution and Cleanup*, The Wiley, NY, 48–65.

Wongsiri B., Haraguch K., and Yamada K., 1999. Dephosphorization from Aqueous Solution by Fly Ash from Mae Moh Power Plants, Thailand. In: Towards to the year 2000, *Symposium on Mineral, Energy, and Water Resources of Thailand*, Bangkok, 402–407.

World Bank, 1994. RAIN/ASIA. User's Manual, IISAA: Washington, D.C., 138 pp.

World Resources Institute, 1994. World Resources 1994–1995, New York.

World Resources Institute. *World Resources 1998–1999*. The World Bank and the United Nations Environment Program, 1999. www.wri.org

Xing G. X. and Zhu Z. L., 2002. Regional nitrogen budget for China and its major watersheds. In: Boyer E. W. and Howarth R. W. (Eds.), *The nitrogen cycle at regional to global scale*. Kluwer Academic Publishers, 405–427.

462

Yadav J. S. P., 1986. *Management of saline and alkaline soils of South Asia.* Report FAO Bangkok.

Yanagisawa Y., 1989. Monitoring and simulation. In: *How to Conquer Air Pollution: A Japanese Experience,* Studies in Environmental Science 38, edited by H. Nishimura, Elsevier, Amsterdam, 157–196.

Yang J.-W., 1999. Transport of heavy metal in soils under elektrokinetic field. In: Proceedings of the 2nd International Symposium on Advanced Environmental Monitoring. KJIST, Kwangju, South Korea, 57–62.

Yong L. M. and Eng N. B., 1997. Singapore country report, paper presented at the *4th ASEAN Workshop on Air Quality Monitoring and Analysis with Emphasis on Acid Deposition,* Technopolis, Pathumthani, Thailand, 25 February–4 March 1997.

Yoon S.-C. and Won J. G., 1998. Monitoring of atmospheric aerosols in Seoul using a micro pulse lidar. In: *Proceedings of the 1st K-JIST International Symposium on Advanced Environmental Monitoring,* September 11–13, 198. Kwangju, Korea, 66–69.

Zayed J., Bande H. and Gilles L. E., 1999. Characterization of manganese-containing particles collected from the exhaust emissions of automobiles running with MMT additive. *Environmental Science and Technology,* 33: 3341–3346.

Zayed J., Gerin M., Loranger S., Sierra P., Begin D. and Kennedy G., 1994. Occupational and environmental exposure of garage workers and taxi drivers to airborne manganese arising from the use of methylcyclopentadienyl manganese tricarbonyl in unleaded gasoline. *American Industrial Hygiene Association Journal,* 55(1): 53–58.

Zayed J., Mikhail M., Loranger S., Kennedy G. and Esperance G. L., 1996. Exposure of taxi drivers and office workers to total and respirable manganese in an urban environment. *American Industrial Hygiene Association Journal,* 57: 376–380.

Zhang J., Huang W. W., Liu S. M. and Wang J. H., 1992. Transport of particulate heavy metals towards the China Sea: a preliminary study and comparison. *Marine Chemistry,* 40: 161–178.

Zhang J., Liu S. M., Lu X. and Huang W. W., 1993. Characterizing Asian wind-dust transport to the Northwest Pacific Ocean. Direct measurements of the dust flux for two years. *Tellus,* 45B: 335–345.

Zhao D. and Sun B., 1986. Air pollution and acid rain in China. *Ambio,* 15: 2–5.

Zhao D. and Xiong J., 1988. Acidification in Southwestern China. In: *Acidification in Tropical Countries,* Rodhe H. and Herrera R. (Eds.), John Wiley & Sons Ltd, 1988, 317–346.

Zhao D., Xiong J., Xu Y. and Chan W. H., 1988. Acid rain in Southwestern China. *Atmospheric Environment,* 22: 349–358.

Zhu Z. L. (Ed.), 1997. *Nitrogen Balance and Cycling in Agroecosystems of China,* Kluwer Academic Publisher, 355 pp.

Zitong, 1994. *Problem of soils and their evolution in China.* Paper for expert consultation on collection and analysis of land degradation data. FAO Bangkok Publ., No 3.

INDEX